Studies in Logic
Logic and Cognitive Systems
Volume 25

The Analytical Way
Proceedings of the 6th European Congress
of Analytic Philosophy

Volume 16
Foundations of the Formal Sciences VI. Probabilistic Reasoning and Reasoning with Probabilities.
Benedikt Löwe, Eric Pacuit and Jan-Willem Romejin, eds.

Volume 17
Reasoning in Simple Type Theory. Festschrift in Honour of Peter B. Andrews on His 70th Birthday.
Christoph Benzmüller, Chad E. Brown and Jörg Siekmann, eds.

Volume 18
Classification Theory for Abstract Elementary Classes
Saharon Shelah

Volume 19
The Foundations of Mathematics
Kenneth Kunen

Volume 20
Classification Theory for Abstract Elementary Classes, Volume 2
Saharon Shelah

Volume 21
The Many Sides of Logic
Walter Carnielli, Marcelo E. Coniglio, Itala M. Loffredo D'Ottaviano, eds.

Volume 22
The Axiom of Choice
John L. Bell

Volume 23
The Logic of Fiction
John Woods, with a Foreword by Nicholas Griffin

Volume 24
Studies in Diagrammatology and Diagram Praxis
Olga Pombo and Alexander Gerner

Volume 25
The Analytical Way: Proceedings of the 6th European Congress of Analytical Philosophy
Tadeusz Czarnecki, Katarzyna Kijania-Placek, Olga Poller and Jan Woleński , eds.

Studies in Logic Series Editor
Dov Gabbay   dov.gabbay@kcl.ac.uk

The Analytical Way
Proceedings of the 6th European Congress of Analytic Philosophy

Edited by

Tadeusz Czarnecki,

Katarzyna Kijania-Placek,

Olga Poller,

and

Jan Woleński

© Individual author and College Publications 2010.
All rights reserved.

ISBN 978-1-84890-014-1

College Publications
Scientific Director: Dov Gabbay
Managing Director: Jane Spurr
Department of Computer Science
King's College London, Strand, London WC2R 2LS, UK

http://www.collegepublications.co.uk

Typesetting by Jacek Wawer
Original cover design by orchid creative   www.orchidcreative.co.uk
Printed by Lightning Source, Milton Keynes, UK

All rights reserved. No part of this publication may be reproduced, stored in a retrieval system or transmitted in any form, or by any means, electronic, mechanical, photocopying, recording or otherwise without prior permission, in writing, from the publisher.

Preface

This volume documents the 6th European Congress of Analytic Philosophy, held in Krakow, 21–26 August 2008; the Congress was organized by the European Society of Analytic Philosophy. The book contains the plenary talks, invited lectures (unfortunately, not all these speakers submitted their papers), as well as selected contributions delivered during the course of special workshops. Previous European Congresses have not culminated in the publication of proceedings. We therefore hope the work at hand will install a new tradition with regards to the meetings of European analytic philosophers. Here we would like to express our gratitude to Dov Gabbay for taking these proceedings under the wing of College Publications, and to Jane Spurr for her assistance.

Jan Woleński

CONTENTS

Preface   v

PART I
PLENARY SPEAKERS

Timeless Properties of Persistent Objects   3
Andrea C. Bottani

The Ontological Attitude   21
Zoltán Gendler Szabó

PART II
INVITED SPEAKERS

HISTORY OF PHILOSOPHY
**Hundred Years After:
How McTaggart Became a Thing of the Past**   47
Kristóf Nyíri

LOGIC AND COMPUTATION
Independence of Quantifiers in Classical Logic   67
Gabriel Sandu

METAPHYSICS
Carving Nature at Our Joints   85
Achille C. Varzi

PHILOSOPHY OF SCIENCE
On Propensity-Frequentist Models for Stochastic Phenomena; with Applications to Bell's Theorem 105
Tomasz Placek

PHILOSOPHY OF MIND
Independent Intentional Objects 147
Katalin Farkas

SOCIAL PHILOSOPHY, POLITICAL PHILOSOPHY AND PHILOSOPHY OF LAW
Market Efficiency and Contractual Justice 167
Peter Koller

PART III

WORKSHOPS

FORMAL METHODS IN PHILOSOPHY

Shooting Right Without Collateral Damage 191
Pascal Engel

Formal Methods in Philosophy of Science 203
Thomas Müller

If in Doubt, Treat'Em Equally
A Case Study in the Application of Formal Methods to Ethics 219
Wlodek Rabinowicz

Extending the Hegselmann-Krause Model II 245
Alexander Riegler and Igor Douven

STRUCTURED MEANINGS AND CONCEPTS

Concepts as Structured Meanings   261
Marie Duží and Pavel Materna

Hyperintensions and Procedural Isomorphism: Alternative (½)   299
Bjørn Jespersen

Against Propositional Radicals   321
Marián Zouhar

BEYOND MINIMALISM

Truth's Role in Understanding   339
Filip Buekens

Deflationism and Reducibility   357
Martin Fischer

On a Necessary Use of Truth in Epistemology   371
Leon Horsten

METAPHYSICS AND ITS METHODS

Generals and Particulars   379
Andrea Borghini

INDEX OF AUTHORS   395

Part I
Plenary Speakers

Timeless Properties of Persistent Objects

Andrea C. Bottani

Whatever it may be, persistence involves both existing at many times and having properties at those times. Theories of persistence thus fall into two closely related parts. On the one hand they have to say how persistent entities can exist at more than one time (i.e., what exactly existence at many times *is*). On the other hand, they have to say what role time plays in the properties they have (for example, what kind of property something that is flat at a time *t* has in virtue of being flat at that very time). A theory of persistence can fail to include a theory of intrinsic change, provided it is admitted that there can be persistence without intrinsic change. And it can fail to include a theory of change (intrinsic or extrinsic), provided it is admitted that something can persist even in an unchanging universe. But even in those cases, the theory should give an account of the properties that unchanging persistent things have and say what role time plays in those properties.

After the long debate over temporary intrinsic properties of persistent things, originating from Lewis's famous argument against endurantism, recent discussions about persistence have increasingly shifted the focus from the properties of persistent things to their existence in time. Examples of this tendency include discussions about the so-called 'argument from vagueness,' arguments related to the so-called 'paradoxes of coincidence,' Hofweber's and Velleman's claim that it is a conceptual truth that something that exists at more than one time has more than one temporal part, and Sider's argument to the effect that three-dimensionalists are hopelessly confused when they say that persistent objects exist *entirely* at each time of their lifespan.[1]

There is a widely shared, sometimes implicit and often ungrounded confidence that, once a theory of persistence resolves problems related to coincidence, composition and existence in time, ingenious paraphrases of our familiar sentences about ordinary persistent things can be devised that allow one to interpret those familiar sentences as speaking of the less familiar technical entities of the ontological theory. In Sider's words: "Any philosophical theory of persistence, change, property instantiation, and related matters must save the appearances. What is certain is that things persist, somehow, that things change, somehow, and that things have properties at times, somehow. It is not part of the common belief just how this occurs." (Sider, 2001, p. 215)

I agree, but one must be clearer about what exactly "save the appearances" means. For it must be noticed that the sentences we ordinarily use to state that everyday objects persist, change and have properties at times do not come in isolation. They are tied together by inferential relations that cannot get lost in the ontological theory. To "save the appearances"—which is what matters for a theory of how ordinary objects persists—is thus not *merely* to find a systematic method able to yield, for any familiar sentence concerning some 'ordinary' object, a paraphrase of that sentence that has the same truth-value but that speaks uniquely of the technical entities of the ontological theory (stages, worms, continuants, states of affairs or the like). For, if the devised paraphrases do not additionally bear to each other the same inferential relations that the corresponding paraphrased sentences bear to each other, there is no reason to believe that the paraphrases tell anything about the familiar objects the paraphrased sentences speak of. Indeed, one cannot assume that p and q have the same *truth conditions* as r and s if one can infer r from s but not p from q (or *vice versa*).

There are the inferential ties among everyday sentences about ordinary objects, manifested by our current *use* of ordinary language. And there are our pre-analytic intuitions about changing things and their properties. Once all this is taken into due account, the confidence that the technical sentences of an ontological theory can be taken as stating how ordinary objects like cats, trees and tables persist, change and have properties at times becomes much more difficult to sustain than is usually believed. This is why the issue of what properties a theory of persistence ascribes to ordinary persistent objects remains crucial for evaluating the theory.

---

[1] See Lewis (1986, pp. 202–205 and 211ff.), Sider (2001, pp. 63–68 and 120ff.), Hofweber and Velleman (2007), Hawley (2001, pp. 140ff.).

In what follows, I shall try to argue that a number of well known theories of how ordinary objects persist ascribe to those objects properties that they do not have—they are wrong regarding what properties cats, trees and tables have. In doing this, I shall mainly focus on timeless properties rather than on temporally qualified properties. According to a number of theories of persistence, persistent entities can have no timeless properties like *being flat*; all they can have are temporally qualified properties—either tensed like *having been flat yesterday*, or untensed, like *being flat at 8.00 a.m.* I shall argue that ordinary persistent objects *do* have timeless properties. But I shall also argue that a number of theories of persistence that maintain that ordinary persistent objects have timeless properties are wrong regarding what timeless properties ordinary objects have. As I shall maintain, in particular, there is no way of construing 'being timelessly flat' as 'being flat at one or more particular times' (for example as 'being flat now' or as 'being permanently flat during its lifespan').

Endurantists are often accused of reducing non relational, timeless properties of persistent objects to relations they bear to times. The standard charge is that endurantists can only eliminate timeless properties of persistent objects, but can give no account of them. I will show that perdurantists have to face difficulties at least as big concerning timeless properties of ordinary objects. Surprisingly enough, endurantists are probably better placed than perdurantists as regards giving an account of timeless properties of cats, chairs and the like.

Before approaching this topic, let me make just one short preliminary remark. I do not think that whether or not something has a property (timeless or otherwise) can depend on what having a property is—exemplifying a universal, including a trope, belonging to a set and so on. Different general views of properties can in principle completely agree on the truth-value of all subject-predicate sentences. In order to make what I shall say as independent as possible of one or another particular theory of what having a property is, I shall use 'having a property' in a very minimal sense. In this sense, nothing else is required in order to have a property than belonging to a set. Having a property might well consist in something other than belonging to a set—exemplifying a universal, embracing a trope or something else in virtue of which an entity belongs to the set. And it might be that one entity can belong to the same set in virtue of having two distinct properties (it might even be that one entity can belong *to different* sets in virtue of having at different times or possible worlds *the same* property). Nonetheless, whenever something has, had or will have a property, there is a set such that something has, had or will have the property just in case it belongs to that set. So, whatever having a property may be, one can wonder whether something belongs to a certain set rather than wondering whether it has, had or will have a property.

Mentioning sets instead of properties, the problem of what timeless properties ordinary objects have—if any—can be formulated as follows. Some sets—for example, the set of flat things—can be described by making no mention of times and no use of tenses (in general, making no reference to particular individuals). Under what conditions does a persistent entity belong to one or another of those sets?

Commitment to sets is not ontologically neutral. Let me invite those who are sceptical about sets to reconstrue talk of sets as innocent "talk of pluralities that are nothing 'over and above' the several entities comprising them," and talk of an entity "being a member of" a set as innocent "talk of that entity 'being one of' the entities in question" (Lowe, 2006, p. 35). Once this is done, any commitment to sets completely vanishes.

1. The Set of (Timelessly) Flat Things

Talk of sets is particularly suitable to introduce the so-called 'problem of change.' Suppose someone crumples up a sheet of paper, and asks what just happens in virtue of this. One might be tempted to answer that something (namely, the sheet) ceases to belong to a set of which it was previously a member (namely, the extension of the property of being flat). Something must be wrong with this answer (for no set can have different members at different times), but it is not easy to say exactly what. Here, in a nutshell, is the so-called 'problem of change.' According to *perdurantists*, what is wrong with the answer is the presumption that the entity which was within the extension of the property of being flat is numerically the same as the entity which is outside of it. According to *endurantists*, what is wrong is rather the presumption that the set to which the sheet belonged when it was flat is the same to which it fails to belong when it is crumpled—the former being the set of all things that *were* flat at some past time while the latter is the set of all things that *are* flat now. The endurantist approach has two versions depending on whether or not its proponents think that a property always has the same extension. If so, the two above sets are extensions of different properties; if not, they are extensions of the same property at different times, which allows the same entity both to belong and to fail to belong to the extension of the same property at different times (this is the so-called 'adverbialist' approach).[2] One purported drawback of the endurantist

---

[2] See Lowe (1998), Haslanger (1989, 2003), Lewis (2002).

approach in both versions is that it says nothing about the set *F* of flat things (those that are not merely flat at one time or another, but flat timelessly, flat "full stop," flat *simpliciter*). Does the changing sheet belong to *F* or not? And what entities belong to *F*, if any? The endurantist approach, at least if taken in conjunction with an eternalist conception of time, seems to say nothing about this. This looks most strange. Having a flat shape does not seem to be a matter of bearing some relation with anything else but, according to the second approach, ordinary things like a sheet seem unable to have a shape *simpliciter*, but only to bear shape-relations with other things called 'times.'

This is Lewis's intuition that some properties are timeless (in Lewis's words, had *simpliciter* by objects) because they are intrinsic. As is well known, this intuition is not universally shared and has been the subject-matter of a well-known discussion about which I shall say nothing here.[3]

Regardless of whether Lewis is right or not in thinking that a treatment of temporary intrinsics should be given in terms of timeless properties had *simpliciter* or not had *simpliciter* by objects, the problem I am interested in remains. When a sheet is crumpled, something seems to 'go out' of a set (that of flat things), but this cannot be the case, for nothing can belong to one set at some times but not at others. Thus, the changing sheet can only belong permanently to *F*, or belong permanently to its complement, or (perhaps) neither. Which of these options holds? Maybe there is a fourth option: embracing the idea that the question of whether the changing sheet belongs to *F* or not is hopelessly confused, or at least grounded in false presuppositions. Perhaps, one might suggest, *F* is misconceived. Perhaps, the right move is simply to dissolve the problem instead of trying stubbornly to resolve it.

Some theories of persistence reduce ordinary persistent things to things that do not persist in a literal sense. They treat things that are supposed to exist at a number of times (cats, chairs, etc.) as nothing more than a number of things, each of which exists at just one time. Let me call theories of persistence of this variety 'reductionist.' According to reductionist theories of persistence, there are plenty of sheets that belong to *F*. Both friends of the so-called 'stage view' and presentists, for example, are generally happy to grant that, if a sheet is now flat, a sheet that exists (only) now is timelessly flat, and so belongs to *F*, even if it was non-flat a little while ago (Sider, 2001; Merricks, 1994). Anti-reductionist theorists of persistence, according to which ordinary things exist at many times (partly or wholly, whatever

---

[3] See Lewis (1986, 1988), Hawley (2001), Haslanger (2003), Wasserman (2003).

that may mean), are less at ease with timelessly flat sheets belonging to *F*. Sometimes, they are inclined to deny that anything in space and time can be timelessly flat, or even to claim that sets like *F* are completely misconceived (Lowe, 1998). Sometimes, they reduce *timelessly* flat sheets to *permanently* flat sheets along supervaluational lines, construing membership of *F* in terms of membership of other less troublesome sets—a suggestion of this sort can be found in van Inwagen (1990). Sometimes, they seem to avoid the issue.

A short scheme of some proposed treatments of timeless properties of persistent objects can be given as follows.

1. Things in time (among which, persistent entities) can tenselessly have no timeless properties. Only entities out of time—for example, numbers—can (Lowe, 1998). Therefore, no persistent entity can belong to sets like *F*, the set of (timelessly) flat entities.

2. A persistent entity belongs to *F* just in case it is flat at the only moment at which it exists by itself (not through the medium of a temporal counterpart) (Sider, 2001; Hawley, 2001).

3. A persistent entity belongs to *F* just in case it is flat now (Zimmerman, 1998; Merricks, 1994).

4. A persistent entity belongs to *F* just in case it is flat at each moment at which it exists—in other words, just in case it belongs to the set of things that are flat at *t* for any *t* at which they exist (van Inwagen, 1990).

Let me consider in detail 1–4.

2. Persistent Entities with No Timeless Properties

In many places, among which his book *The Possibility of Metaphysics*, J. Lowe argues that things in time (among which, persistent entities) can have tenselessly no properties. Only entities out of time—for example, numbers—can. Lowe says: "I do not believe that properties can be tenselessly predicated of concrete historical events but only of abstract entities which do not exist in time (or space) at all." (Lowe, 1998, p. 89).[4]

---

[4] Lowe's claim concerns entities in time in general and not only events.

Lowe says 'tenselessly' instead of 'timelessly,' so his thesis is stronger than it might appear at first glance. It is the A-theoretical idea that entities existing in time either have properties in the present or (inclusive) had properties in the past or (inclusive) will have properties in the future, but there is no other way they can have a property. Since having a property in the present (or in the past, or in the future) requires having a property at a time, the A-theoretical idea implies a weaker idea that even a B-theorist can share, namely, the 'adverbialist' thesis that everything existing in time can have a property only at a time. If the adverbialist thesis is true, nothing in time can belong to F—the set of (timelessly) flat things. For the set of things that have at a time t the property of being flat is just the set of things that have the property of being flat-at-t—which is not F. If the property of being flat can be tenselessly predicated of no entity existing in time, no sheet can be said to belong to F.

What might suggest that entities existing in time have tenselessly no properties? One can easily refuse to take the verb "barks" in "Bobby barks at midnight of July 24th, 2002" and even the verb "is" in "Bobby is a dog" as semantically tenseless. However, the sentences

1) The set of dogs contains Bobby

2) The set of beings barking at midnight of July 22th, 2002 contains Bobby

do not seem to be semantically tensed,[5] but certainly they qualify a particular dog called "Bobby" in various meaningful ways. Sure, they do not (strictly speaking) predicate any properties of Bobby alone. Rather, they predicate tenselessly a certain two-place set-theoretical relation of a particular set and Bobby. But the set-theoretical relation is one that the set can tenselessly bear to Bobby just in case Bobby has tenselessly a certain property (for example, being a dog). Therefore, neither 1) nor 2) could be true if entities existing in time had *tenselessly* no properties (timeless or otherwise). And 1) couldn't be true (even though 2) could), if entities existing in

---

[5] I assume what seems to me to be evident, i.e. that in the context of 1) and 2) 'contains' cannot be more tensed, semantically speaking, than "=" in the context of "Napoleon = Napoleon." Moreover, I assume that no tense is implicitly nested in the definite descriptions 'the set of dogs' or 'the set of beings barking at midnight of July, 2002.' In particular, I assume that the definite description 'The set of Fs' cannot denote the set of things that were, are or will be F, for in that case something that has become F (or has ceased to be F) would belong both to the set of Fs and to the set of things that are non-F; and I assume that the description cannot denote the set of things that are F *now*, for in that case 1) and 2) would be false in case Bobby no longer exists.

time had tenselessly no timeless properties. The above sentences, however, seem to be just as true as they might be, provided that Bobby, a certain dog, barked at midnight of July 22th, 2002.

Even if one believes that timeless properties can sometimes be tenselessly predicated of persistent things, one might naturally feel some deep unease when the property is *being flat* and the persistent thing is a sheet that is sometimes flat and sometimes crumpled. One might say: "If you mean that a sheet has the property of being flat-at-some-times, I understand what you say; if you mean that a sheet has-at-some-times the property of being flat, still I understand what you say; but if what you are trying to say in some mysterious way is that a sheet has *simpliciter* (that is, in a temporally unqualified mode) the property of being timelessly flat, I no longer understand. You just cannot tenselessly predicate of a sheet the property of being tenselessly flat!"

It is far from easy to understand what exactly one might mean by saying that a timeless property cannot be tenselessly predicated of a persistent entity. One cannot mean that, whenever a timeless property is tenselessly predicated of something existing in time, the resulting sentence is non-true. For, in that case, both the timeless property of being flat and the timeless property of failing to be flat are non-true of the sheet when they are predicated tenselessly of it, which means that the sheet falls neither in the set of flat things nor in the set of things that do not fall in the set of flat things— a very unhappy result.

In other words, let us assume what seems to be most natural, namely, that a thing has the property of failing to be flat just in case it fails to have the property of being flat. In that case, the extension of the latter property is the complement of the extension of the former, which implies that nothing can fail to belong to the extension of one of them. But this is exactly what should happen if every timeless property were non-true of the sheet when it is predicated tenselessly of it. For surely both the property of being flat and the property of failing to be flat are timeless properties.

So, it can hardly be the case that timeless properties are invariably non-true of any persistent entities. Perhaps one might suggest that, if a timeless property is tenselessly predicated of a thing in time, the resulting sentence is meaningless. Still, this is hard to believe. For the resulting sentence is true if and only if the sheet falls into the set of flat things. And why on Earth should it be meaningless to say that a certain entity belongs to a certain set (leaving aside semantic paradoxes)? One answer might go like this. "The expression 'flat' is a two-place predicate, the property of being flat is a two-place relation between individuals and times and the set of flat things is a set of individual-time couples; hence, you cannot tenselessly predicate the property of being flat of the sheet just as you cannot predicate the property of being a brother of."

The answer does not seem to be very convincing, even leaving aside Lewis' intuitions about intrinsicness. No single thing has or lacks the property of being a brother of, but some entities fall timelessly into the set of flat things (like the spatial region the sheet occupied at some time before it was crumpled, or better its two-dimensional surface at that time) and many entities (concrete entities like mountains and abstract entities like numbers) fall into the complement of that set (leaving aside instantaneous slices of four-dimensional worms, if there are any). There is one set (the set of flat things) to which the number 4 does not belong. Why should it be meaningless to ask whether a sheet belongs to that set or not? The set of flat things is not a set of pairs, it is just that we have (metaphysical) difficulties in establishing whether a sheet sometimes flat and sometimes crumpled belongs to it or not. This is the problem I am interested in: when is it the case that a changing thing has a timeless property? On what conditions does it belong, for example, to the set of flat things?

3. Timeless Properties of Worms

Four-dimensionalists have an easy answer to the above question. For, according to them, sheets are four-dimensional worms and nothing that is thick in four dimensions can be flat. Therefore, the sheet is timelessly non-flat (it is timelessly non-flat inasmuch as it is timelessly four-dimensional). This is a weird answer, however: had the sheet not been crumpled, it would have been *permanently* flat, that is, flat at every moment included in its lifespan (disregarding its negligible thickness), but nonetheless timelessly non-flat—in other words, at no time non-flat but non-flat. This utterly contrasts with some fundamental intuitions on what having a property timelessly must be.

Here is why. If 'timelessly' means anything, it means 'regardless of time.' If anything is timelessly non-flat, then its being flat cannot be affected by the sequence of moments—at least no more than the Japanese constitution can be affected by the sequence of Swedish kings. Therefore, everything timelessly non-flat must be such that at every time t it is, was or will be the case that it *is* non-flat.[6] But everything such that it is always the case that it is non-flat must also be non-flat-at-t for any time t at which it exists (existed, will exist). For certainly, for any such t, a will belong to the set of things that are, were or will be non-flat when t is, was or will be present.

---

[6] I have put 'is' in italics to show that here the verb is semantically untensed.

In a nutshell, if an *x* is timelessly *P*, then for any time *t* such that *x* exists (existed, will exist) at *t*, *x* is *P-at-t*. These intuitions ground the validity of the following argument:

ARG
(1) The sheet is timelessly non-flat
 (that is, it belongs to the set of non-flat things)
(2) The sheet existed (will exist) at time *t*

(3) The sheet is non flat-at-*t*
 (that is, *x* belongs to the set of things that were, or will be, non-flat at time *t*)

According to the above intuitions, the argument is formally correct, which means that, if its premises are true, its conclusion must also be true. According to the perdurantist, on the contrary, if the second premise of the argument is true, the argument is formally incorrect (hence, by contraposition, if the argument is formally correct, the second premise is false).[7] Here is why. If (2) means that a temporal part of the sheet existed (or will exist) at *t*, (2) comes out true according to the perdurantist, but in this interpretation of (2), (3) does not follow from (1) and (2). On the contrary, if (2) means that the sheet itself—that is the whole worm—existed (or will exist) at *t*, the argument is correct but in this reading (2) cannot be true according to the perdurantist, for no worm can exist at just one moment of its long lifespan. (This explains why the conclusion is also false, according to the perdurantist). For, in his opinion, the sheet is permanently flat, so it is not the case that it is non-flat-at-*t*). According to the perdurantist, then, the timeless properties of the sheet can have no bearing on its temporally qualified properties, which utterly contrasts with our intuitions of what a timeless property of a persistent entity is.

Only very thin temporal parts of four-dimensional worms can be *flat*, four-dimensional worms themselves can't. Other properties, however, can be enjoyed both by worms and by their temporal parts. Let me say just a couple of words to show that endurantists are no more at ease with timeless properties of this sort than they are with timeless properties like 'flat.' It seems reasonable that familiar properties like *straight* or *bent*, for ex-

---

[7] This implies that, according to four-dimensionalists, the following scheme of inference:

(1) *x* is timelessly *P*;
(2) *x* existed (will exist) at *t*;

(3) *x* was (will be) *P-at-t*

have no sound instance.

ample, can be exemplified by something independently of whether it extends in two or more dimensions, and so both by a four-dimensional worm and by its momentary three-dimensional parts. Now, just as a bent three-dimensional entity can be a sum of straight two-dimensional entities, a bent four-dimensional worm can be a sum of straight three-dimensional worm-stages, which means that a permanently straight worm can be timelessly non-straight. If ordinary objects are worms, a permanently straight object—say a brush-handle—can nonetheless be timelessly bent. This exactly replicates the above difficulties with *being flat*. The point is that, according to perdurantists, 'x is timelessly P and x exists at t' does not generally imply 'x is P-at-t,' which annihilates any inferential tie between timeless properties and temporally qualified properties of ordinary objects.

An important related point, which does not directly concern timeless properties, is that a worm that is bent in a temporal interval can nonetheless be straight at each moment included in that interval. Consider a broom handle moving during a temporal interval (say, today) along a zigzag path. If it is a worm, the worm it is must be bent today, but straight at any moment included in today. But no broom-handle can be bent today and straight at any moment included in today. Moreover, suppose that a chair regularly changes colour in the course of its lifespan. If chairs are four-dimensional worms, surely that chair will be a *striped* four-dimensional worm. But at no moment of its lifespan is the chair striped. So, it is striped in an interval (that long interval that is its lifespan) but at no moment included in the interval. This cannot happen to any chair. If a chair was non-striped at every moment included in the last year, surely it was non-striped the last year.

Let me call 'temporal distributivity of predication' the principle according to which predication of ordinary objects distributes relative to time just as multiplication distributes relative to addition. The distributivity of multiplication relative to addition is the well-known principle according to which $n \times (k + j) = (n \times k) + (n \times j)$. A similar distributivity holds for predication relative to time:

(DPT) Px at $(i^1 + i^2)$ iff $[(Px$ at $i^1)$ and $(Px$ at $i^2)]$

where i^1 and i^2 are (non-scattered) temporal intervals, and the variable ranges over ordinary objects. DPT entails that if an object—for example, a broom-handle—was straight at any moment included in yesterday, then it was straight yesterday, and vice versa. The principle seems to be intuitively non-negotiable whenever the property of being P can be had by a persistent object both during a long interval and at a single unextended moment (which is the case with respect to 'flat,' but not, for example, with respect to a lingering property like 'eating') and it seems central in our current way of making inferences about the properties that persistent objects have at times. Perdurantism, however, seems to imply a radical negation of this principle, even in case of properties like 'flat.'

Let me say a couple of words about the intended bearing of this argument. I do not think this is an argument against perdurantist theories of persistence as such. Rather, it is an argument against perdurantists theories of how *ordinary objects* persist. It might well be that everything is a spatiotemporal worm, persisting just as spatiotemporal worms do. The argument suggests nothing to prevent this possibility. What the argument shows is just that, in such a case, ordinary objects do not exist. For timeless properties of ordinary persistent objects do not behave inferentially as timeless properties of worms. And the inferential ties between properties had over long times and properties had during shorter sub-times of those longer times are different according to whether those properties are had by ordinary objects or by worms. This is an interesting conclusion because perdurantist, stageist, presentist and endurantist theories of persistence normally aim to say how ordinary objects—cats, trees and tables—persist.[8]

In order to avoid problems of this kind, perdurantists have often replicated a typical endurantist move and denied that a persisting entity like a sheet can be said to be timelessly flat or to be timelessly crumpled, but only to be flat at moments, or to be crumpled at moments. As I have just argued in the preceding section, it is hard to understand what this might mean. For it cannot mean that the sheet falls outside the extensions of all timeless properties (which requires that the sheet falls both outside the set of flat things and outside the set of things that fall out of it); and it cannot mean that it is meaningless to ask whether the sheet falls into a certain set (provided that it is true that some entities belong to this set and others do not).

4. Timeless Properties of Stages

Imagine now that one adopts a reductionist stance on persistence. This is the idea that things that are supposed to exist at a number of times are nothing over and above a number of things, each of which exists at just one time. Suppose, for example, that sheets of paper are stages. Everything that is a stage exists by itself just at one moment, say t, and it is timelessly P just in case it is P at that very moment. Friends of the so-called 'stage view' are thus generally happy to grant that, if a sheet is now flat, a sheet that exists

---

[8] There are exceptions. See for example Heller (1990).

(only) now is timelessly flat, and so belongs to the set of flat things, even if it was non-flat a little while ago.

On the other hand, suppose presentists are right in believing that all that exists exists now, and at no other time. Presentists are generally happy to grant that, if a sheet is now flat, a sheet that exists (only) now is timelessly flat, and so belongs to F (the set of flat things), even if it was non-flat a little while ago. This is the core of Prior's well-known 'redundancy theory of the present tense,' namely, the idea that no grounded distinction can be drawn between present tensed and tenseless sentences (Prior, 1957, 1967). Presentists and 'stageists' diverge dramatically concerning the nature of time but converge in believing that, for any ordinary object a, there is one time t such that a is timelessly P just in case a is P at time t (according to stageists, t varies from stage to stage, while according to presentists, t is the same for all entities). In a nutshell, they converge in embracing a *reductionist* theory of persistence (a theory according to which what persists exists at just one time). Unfortunately, approaches of this variety are not in a position to grant the validity of ARG any more than their perdurantist rival.

According to the 'stageist,' indeed, the second premise of ARG is true just in case it means that a *temporal counterpart of the sheet* existed (or will exist) at t, but, under this interpretation of (2), (3) does not follow from (1) and (2). (3) follows from (1) and (2) only if (2) means that the sheet itself (that is, the stage that exists *now*) existed or will exist at t, but (2) is utterly false in this reading. According to the presentist, (3) does not follow from (1) and (2) because things change properties, and the sheet might well have now properties it didn't have in the past (or will not have in the future). Once again, these treatments of timeless properties are unable to recognize obvious inferential connections between the timeless properties of the sheet and its temporally qualified properties. This utterly conflicts with our intuitions of what a timeless property *of a persistent entity* is. Let me repeat that this does not show that nothing persists in the 'stageist' or presentist sense, but only that ordinary objects are not among the entities that persist in that way.

5. A Supervaluational Treatment of Timeless Properties

The provisional moral to be drawn from all this seems to be that there is not much hope of construing 'being timelessly flat' as 'being flat at one particular time' (for example as 'being flat now' or as 'being flat at the only moment at which something exists'). Non-reductionist theories of persistence, however, can perhaps try to give a completely different kind of approach. For, if a sheet exists at a number of times in the same sense of 'exists' (no matter whether partly or wholly, whatever that might mean), it can make no sense to suppose that it belongs to the set of timelessly flat things just in case it is flat at some *specific* moment. The only available solution seems to be to assume that it belongs to the set of timelessly flat things if and only if it is *permanently* flat, that is, flat at every moment of its lifespan. And this holds for 'flat' just as for every timeless property, regardless of whether it is simple or complex. A version of this idea is sketched, though not developed, by Peter van Inwagen. He says: "To say that Descartes had the property of being human is to say that he had that property at any time at which he existed" (van Inwagen, 1990, p. 250).

So one might say: a persistent entity a falls within the extension of a timeless property P—regardless of whether P is simple or complex—when it is permanently P, it falls outside the extension of P when it permanently fails to be P, and it falls neither inside nor outside the extension of P when it is sometimes P and sometimes not. Accordingly, the sentence 'Pa' is timelessly true just in case it is true at each time at which a exists, timelessly false just in case it is false at each time at which a exists and neither timelessly true nor timelessly false when it is true at some time at which a exists and false at some other.

Now we have a perfect analogue of standard supervaluational treatments of vagueness applied to time and predication. In place of supertruth we have timeless truth, in place of precisifications we have times, in place of the range of admissible precisifications, we have the limits of a lifespan. The sheet belongs to the extension of 'large' if and only if it is large on all admissible precisifications of 'large,' just as it belongs to the extension of '(timelessly) flat' if and only if it is flat at all the moments that are included in its lifespan. And the same kind of deviation from classical logic and bivalence is embraced that is typical of supervaluationism. Being flat at some times and crumpled at others, the sheet falls neither within the set of (timelessly) flat things nor outside of it. And the timeless sentence "the sheet falls in the set of flat things" is neither true nor false, it just has no truth value.

This treatment of timeless properties may be more precisely described as follows. For any moment t^n at which x exists (existed or will exist), there is a certain set s^n, which is the set of things that are (were or will be) flat at t^n. Let us call 'I' the (possibly empty) set of things that belong to each s^n (the intersection of all the s^n). The idea is that, for any persistent entity x, x is timelessly flat if and only if $x \in I$. In other words, I is the set timelessly flat things. The same holds, *mutatis mutandis*, for every timeless or temporally unqualified predicate like 'flat,' *regardless of whether it is simple or complex*.

No problem arises with ARG and DPT. Is there anything else wrong with this approach? Here is an argument. Suppose someone introduces a predicate 'Q' as follows:

$$(\forall x) x \text{ is } Q \text{ if and only if } x \text{ fails to belong to } I.$$

The extension of the timeless property of being Q is the complement of the extension of the timeless property of being flat, which means that everything that does not belong to the set of flat things belongs to the set of Qs (and vice versa). But it is not the case that everything that is Q (in other words, everything that does not belong to I) is Q-at-t^n for any moment t^n at which it exists. If something is sometimes flat and sometimes non-flat, it does not belong to I and so it is Q, but it is not Q-at-t^n for any moment t^n at which it exists (for some t^n it is flat-at-t^n).[9] Therefore, a predicate like Q could *not* receive the same supervaluational treatment as 'flat.' But certainly, the property of being Q is just as timeless as the property of being flat, which entails that the described supervaluational account of timeless properties is inadequate for every language in which a predicate like 'Q' exists or is definable. In a nutshell, timeless properties cannot be reduced to permanent properties along the above lines. This does not mean that there is no connection between being timelessly P and being permanently P. If something is timelessly P, then it is permanently P, but the converse, as I have argued, does not hold.

6. Timeless Properties as Time-Transcendent

I conclude that there is no way of construing 'being timelessly flat' as 'being flat at one or more particular times' (for example as 'being flat now' or as 'being permanently flat during its lifespan'). An obvious consequence seems to be that being timelessly P does not depend on what happens at one or another time. This is the basis of a distinction that Kit Fine has drawn in recent years between *sempiternal* and *eternal* truths. "An eternal truth," he says, "will be true *regardless* of the time, i.e. regardless of how things are at the time, while a sempiternal truth will be true whatever the time, i.e. however things are at the time" (Fine, 2005, p. 203). Here is why

---

[9] This is *not* an argument against supervaluational treatments of vague terms like "bald." The main difference is that in such cases no analogue of I can be defined in the object language.

eternal, but not sempiternal, truths can be said to be 'transcendent' (i.e., to transcend the temporal framework). The property of being *P*, it might thus be said, is a timeless property of some *a* just in case it is an *eternal* truth that *a* belongs to the set of *P*s. In this sense, timeless truth is not truth at each time, possible world or index, but just truth *simpliciter*, truth period, truth at no particular index. So, a dog has the timeless property of being a dog just because it is a dog *simpliciter*, a dog *period*, a dog in virtue of what it is to be the sort of thing it is, not because it is a dog at some or each particular moment of its lifespan. In papers such as "Essence and Modality" and "Necessity and Non-Existence" Fine has elaborated in depth this notion of an eternal, time-transcendent property in the framework of a distinction between necessary and essential properties that is now under wide discussion. I shall refrain here from entering in this modal dimension, let me just say that the impossibility of construing timeless properties in terms of temporally qualified properties, for which I have argued here, lends a strong support to this idea of timelessness as time-transcendence.

One might wonder what exactly this idea can suggest about the case discussed above of the crumpled sheet of paper and the timeless property of being flat. This largely depends on how exactly the idea is spelt out. What the idea suggests, one might think, is that an entity falls into the extension of the timeless property of being *P just in case* it is essentially *P* (that is, just in case it is *P* only in virtue of what it is to be the sort of thing it is, regardless of how things are arranged at one time or another). So, flat space-regions can be said to be timelessly flat, for certainly they are flat just in virtue of what they are. And mountains can be said to be timelessly non-flat, for certainly they are non-flat just in virtue of what they are. But sheets of paper are neither flat in virtue of what they are nor non-flat in virtue of what they are (they are sometimes crumpled but very often flat, according to how things are arranged at the time). Therefore, one might conclude, sheets of paper are neither timelessly flat nor timelessly non flat—they fall neither into F (the set of timelessly flat things) nor into F^* (the set of timelessly non-flat things).

There must be something wrong with this way of spelling out the idea of timelessness as time-transcendence. Let me explain why. If sheets of paper fall neither into F nor into F^*, F^* cannot be the complement of F. Call 'F^C' the complement of F and introduce a predicate 'Q' in the following way:

$$(\forall x) x \text{ is } Q \text{ if and only if } x \text{ belongs to } F^C.$$

If sheets do not fall into F, they must fall into F^C. If they fall into F^C, they must be timelessly Q. If they are timelessly Q, they must be essentially Q.

But how could they be essentially Q, given that they are non-Q whenever they are flat? Mountains can be said to be essentially non-flat just because they can *never* be flat (they are non-flat regardless of the time). But certainly sheets of paper are very often non-Q (they are non-Q whenever they are flat, according to how things turn out to be at the time).

The idea of timelessness as time-transcendence must therefore be spelt out in some other way. As far as I can see, a good way of articulating the idea goes as follows. Something belongs to F, the set of timelessly flat things, in case[10] it is essentially flat (that is, flat regardless of how things are arranged at one time or another) and it belongs to F^*, the set of timelessly non-flat things, in case it is essentially non-flat (that is, non-flat regardless of how things are arranged at one time or another). If an entity is neither essentially flat nor essentially non-flat, however, then it is *radically indeterminate* whether that entity is timelessly flat or not. In my intended use of 'radically indeterminate,' being radically indeterminate is *not* being neither true nor false: a radically indeterminate sentence is just a sentence such that it is indeterminate whether it is true and indeterminate whether it is false. In this sense, 'p is radically indeterminate' entails neither 'p is not true' nor 'p is not false' (so, *a fortiori*, it does not entail 'p is neither true nor false'). And the existence of radically indeterminate sentences is compatible with the validity of all principles of classical logic, including bivalence.[11]

There would be a lot to say about radical indeterminacy, but it is not my aim here to elaborate this notion in detail. The important point seems to be that the essence or nature of a persistent thing determines only *partially* its timeless properties, rather as a story determines only partially the properties of a character. This seems to be a necessary condition for change. For, as I have argued above, if a persistent object is timelessly P, it must be permanently P. And if it is timelessly non-P, it must be permanently non-P. Therefore, something can be, for example, flat at one time and non-flat at another only if it is radically indeterminate whether it is timelessly flat or not: change necessarily requires timeless indeterminacy. There is ontic indeterminacy, therefore, but there is no ontic vagueness. For the kind of timeless indeterminacy that I have argued for does not provide ground for any sorites paradox.

---

[10] Both in this sentence and in the following, I mean precisely 'in case,' and NOT 'just in case.' This is precisely the difference between the good and the bad way of spelling out the idea of timelessness as time-transcendence.

[11] Many instances of fundamental classical principles can well come out indeterminate, but this does not mean 'non-true.'

A related important point is that the same timeless property can be true of one object, false of a second object and indeterminate of a third object according to the sorts of objects they are. There are persistent things that belong to the set of flat things (they are flat *simpliciter*, like surfaces and spatial regions), persistent things that belong to the complement of F (they are timelessly non-flat like Mount Everest) and persistent things whose nature requires membership neither of F nor of its complement (like a sheet, that can be flat at one time and crumpled at another). Things of the last sort are such that it is radically indeterminate whether they are timelessly flat or not, they can be determinately flat only relative to times.

References

Fine, K. (1994). Essence and modality. In *Logic and Language*, volume 8 of *Philosophical Perspectives*. Ridgeview Publishing Company.
Fine, K. (2005). Necessity and non-existence. In *Modality and Tense. Philosophical Papers*. Clarendon Press, Oxford.
Haslanger, S. (1989). Endurance and temporary intrinsics. *Analysis*, 49, 119–125.
Haslanger, S. (2003). Persistence through time. In M. Loux and D. Zimmerman (Eds.), *The Oxford Handbook of Metaphysics*. Oxford University Press, Oxford.
Hawley, K. (2001). *How Things Persist*. Clarendon Press, Oxford.
Heller, M. (1990). *The Ontology of Physical Objects: Four Dimensional Hunks of Matter*. Cambridge University Press, Cambridge.
Hofweber, T. and Velleman, D. (2007). How to endure. online paper.
Lewis, D. (1986). *On the Plurality of Worlds*. Blackwell, Oxford.
Lewis, D. (1988). Rearrangement of particles: Reply to Lowe. *Analysis*, 48, 65–72.
Lewis, D. (2002). Tensing the copula. *Mind*, 111, 1–14.
Lowe, E. J. (1988). The problems of intrinsic change: Rejoinder to Lewis. *Analysis*, 48, 72–77.
Lowe, E. J. (1998). *The Possibility of Metaphysics*. Clarendon Press, Oxford.
Lowe, E. J. (2006). *The Four-Category Ontology. A Metaphysical Foundation for Natural Science*. Clarendon Press, Oxford.
Merricks, T. (1994). Endurance and indiscernibility. *Journal of Philosophy*, 91, 165–184.
Prior, A. (1957). *Time and Modality*. Oxford University Press, Oxford.
Prior, A. (1967). *Past, Present and Future*. Clarendon Press, Oxford.
Sider, T. (2001). *Four-Dimensionalism. An Ontology of Persistence and Time*. Clarendon Press, Oxford.
van Inwagen, P. (1990). Four-dimensional objects. *Noûs*, 24, 245–255.
Wasserman, R. (2003). The argument from temporary intrinsics. *Australasian Journal of Philosophy*, 81, 413–419.
Zimmerman, D. (1998). Temporary intrinsics and presentism. In P. van Inwagen and D. Zimmerman (Eds.), *Metaphysics: the Big Questions*. Blackwell, Cambridge, MA.

The Ontological Attitude

Zoltán Gendler Szabó

1. The Received Ontology

Philosophy is more modest than it used to be. It no longer strives to be the queen of the sciences, or the arbiter of sound judgment. Most of us have learned to value conformity with the bulk of received opinion outside of philosophy while safeguarding the boundaries of our discipline. We defer to mathematicians when they claim that interesting theories are incomplete but not when they infer from that that reasoning is non-algorithmic, to biologists when they uncover our genetic makeup but not when they draw far-reaching consequences about morality, to literary theorists when they trace the role of political power in the establishment of artistic canon but not when they argue that all categories are socially constructed. Likewise, we respect the views of "ordinary folk" most of the time but we preserve the right to disregard them when they tell us which numbers will win the lottery, or which houses are haunted by ghosts.

Many have thought and some still suspect that modesty is bound to undermine philosophy as we know it. The concerns are based on the view that philosophy either lacks a subject matter or is concerned with absolutely everything, and hence, that its contributions differ from those of the other disciplines only in emphasis or methodology. But the facts don't support this view: philosophers rarely make noted contributions to mature disciples other than their own and it is also unusual for distinguished physicists, economists, or architects to advance the state of a live philosophical debate. It appears that there are questions which are uniquely suited for philosophical answers if they are answerable at all—received opinion cannot crowd out philosophy unless it is philosophically embellished. Of course, modesty does not mean slavishness. There probably are countless areas where the received view is wrong and philosophers can do great service in sniffing them out. But short of a stunning insight or a momentous discovery, we should not expect to overcome established opinion.

Unfortunately, there is an area where straightforward accommodation between philosophy and received opinion seems impossible. This area is ontology. Just about every important ontological question has an answer trivially derivable from truisms. There are infinitely many primes, which entails that there are numbers and also that there are infinitely many of them. In other words, nominalism and finitism seem refuted without any help from philosophers. There is one hand and here is another, which means that there are at least two objects external to the mind. This quick argument seems to refute both idealism and monism. There are ways the world could be that are different from the way it actually is, and so merely possible worlds exist. Thus it seems that skepticism about possible worlds is absurd. Similar one-liners can be used to establish the existence of shapes, directions, unobservables, causes, mental states, rights, values, properties, propositions, events, concepts, thoughts, and much else philosophers have argued about throughout the ages. Respecting the received view requires respecting the fundamental categories we employ, which in turn requires regarding them as instantiated. While there is arguably no such thing as the received epistemology, ethics, or aesthetics there really is a received ontology.[1] And it is surprisingly bloated.[2]

The stock response to this observation is that philosophers should heed Berkeley's advice. Addressing the objection that his rejection of matter flies in the face of received opinion (even though conformity with common sense is something he valued highly), in the *Principles* Berkeley famously makes the following remark:

---

[1] There are, of course, received opinions about what is known about various things, what should be done in various circumstances, and what is worth our attention among various artifacts. But these don't make up a theory of knowledge, goodness, or beauty. The trouble with ontology is that once we settle what there is, there seems to be nothing left to do.

[2] Arguments for the non-existence of philosophical exotica—the mereological sum of Costa Rica and my left toe, grue grass in the backyard, in-cars in the garage, etc.—are not too hard to come by either. These pruning one-liners, however, pose significantly less threat to philosophical ontologies than their bloating cousins. Philosophical exotica have no conventional names—they all have to be introduced via definition, which makes them more or less palatable. While the man on the street hardly believes in the mereological sum of Costa Rica and his left toe, he is likely to make peace with this putative entity once he sees that in a certain (not particularly clear) sense it is "nothing over and above" Costa Rica and his left toe. The same can obviously not be said about non-philosophical exotica—Santa Claus, Atlantis, the fountain of youth, ghosts, unicorns, etc. The received ontology rules these out while it remains ultimately non-committal about the exotica.

> ... in such things we ought think with the learned, and speak with the vulgar. They who to demonstration are convinced of the truth of the Copernican system, do nevertheless say the sun rises, the sun sets, or comes to the meridian: and if they affected a contrary style in common talk, it would without doubt appear ridiculous. (Berkeley, 1711/1948–1957, §51)

We speak as if we were really committed to a host of entities and when we take this speech at face value we end up constructing a fantastic ontology. The advice is that we should not be taken in by language.

This sounds reasonable at first, but perplexing on second thought. Consider the example of the motion of the sun across the sky. Does Berkeley think the sun does not rise in the morning, come to the meridian at noon, and set in the evening? If so, his view *is* in conflict with received opinion. We don't just speak as if we thought the sun moves across the sky—we really think it does. And the "we" doesn't just include the vulgar: the learned can explain that movement is relative to a frame of reference and that given our usual one, the earth, the sun *strictly and literally* moves across the sky.³ Berkeley's analogy fails: when it comes to the movement of the sun there is no conflict between the vulgar view and the learned one.

When it comes to a *genuine* conflict, Berkeley's advice to the advocate of revisionary ontologies is harsh. If she has decent arguments against holes, values, or propositions, she should stick with them and she should say that although we think there are holes in our cheese, values in our lives, and propositions in our theories, in fact there are none. (Of course, we are permitted to speak as if there were as long as we don't mean what we say.) But what is the basis for paying less respect to received views about ontology than to received views about other matters? And, since ontological claims are inferentially connected to countless others, how are we to stop the spillover of immodesty to other areas? Philosophers who deny the authority of received ontology must arbitrarily restrict the scope of their denial or end up rejecting deference to received opinion altogether.

The alternative to Berkeley's attitude is to stick to modesty and declare that any form of revisionism is bunk. This stance was made famous by G. E. Moore. It is characterized by an incredulous look and a bemused tone that are hard to reproduce in print. But repetition, italics, and heavy punctuation do help:

---

³ For a helpful discussion of Berkeley's example, see Jackson (2007). It is interesting that the view that ordinary talk somehow carries commitment to a geocentric theory is remarkably widespread. Resistance to context-sensitivity may be the culprit. But the view that seeks to minimize context-sensitive items in natural language is by no means a received view.

> ... if I can prove that there exists now both a sheet of paper and a human hand, I shall have proved that there are now "things outside of us"; if I can prove that there exists now both a shoe and a sock, I shall have proved that there are "things outside of us": etc.; and similarly, I shall have proved it if I can prove that there exists now two sheets of paper, or two human hands, or two shoes, or two socks, etc. ... I certainly did at the moment *know* that which I expressed by a combination of certain gestures with saying the words 'There is one hand and here is another.' ... How absurd it would be to suggest that I did not know it, and only believed it, and that was perhaps not the case! (Moore, 1939, pp. 449–450)

I don't think this gives us a stable intellectual outlook. If you find yourself doubting the existence of the external world you are bound to reject either the inference from "There is a hand and here is another" to "There are things outside us" (as Berkeley did) or that you know the premise "There is a hand and here is another" without proof (as Descartes did).

Non-standard ontologies are backed by genuine arguments. Some of these are epistemological, like Descartes's claim that unless we have reason to believe that we are not dreaming we should not believe anything incompatible with the assumption that we are. Some are metaphysical, like Berkeley's contention that matter, being inherently passive cannot be the cause of anything, and hence, would be unobservable if it existed. There are also broad methodological considerations. If you have a piece of clay occupying a spatial region you can't have in addition a statue there; if you have a physical state upon which a particular mental state supervenes there is no causal work for the mental state to do; if you have absolute space you are committed to a fundamental component of physical reality that is undetectable, etc. To dismiss such arguments via a direct appeal to the received ontology is quite unhelpful.

This is an impasse. On the one hand, there seems to be good reason to want to hold on to modesty, and hence, to accept all the truisms about what there is. On the other hand, there seems to be equally good reason not to expect a satisfactory resolution of the problems of ontology by insisting on the truth of those truisms. How accommodation between received views and philosophy is possible remains elusive. The project of this paper is to propose a remedy—one that avoids both Berkeley's and Moore's stance.

In the next section, I discuss the notion of ontological commitment. I argue that we can abstract away from controversial features of Quine's proposal and come up with an attractive core conception of what it is for someone to be ontologically committed to certain entities. I use this conception to identify three types of approaches towards reconciling the received ontology with its philosophical rivals, and I argue that the least popular of the three—the *attitude-based approach*—has the best chance of delivering a satisfactory reconciliation. In section 3, I argue that the standard version of the attitude-based reconciliation—*fictionalism*—faces serious difficulties. In section 4, I propose and defend my own version of the attitude-based reconciliation, and argue that it delivers all the goods the fictionalist approach can without its shortcomings. I close in section 5 with a brief discussion of a possible historical precedent for the type of view I advocate.

2. Ontological Commitment

'Ontological commitment' is a technical term introduced by Quine to describe an attitude one bears towards putative entities when they are included in one's complete view of the world. Quine used a number of terms in writing about this attitude: it is the mental state we are in when we 'acknowledge,' 'admit,' 'assume,' 'believe in,' 'countenance,' 'hypostatize,' 'posit,' 'presuppose,' 'reify,' or 'reckon' certain entities. In the end it is the somber 'ontological commitment' that stuck, probably because unlike its more colloquial rivals it is openly normative. A commitment is something one should but may not live up to. One may be ontologically committed to certain entities while sincerely denying their existence. What a theory of ontological commitment is supposed to do is to help us see how this can happen and help us catch the ontological free-riders.

The fact that ontological commitment attaches first and foremost to thinkers is rarely emphasized because Quine has offered a reductive account of the attitude: a person's commitments are inherited from a theory she accepts, which in turn are inherited from certain sentences the theory entails. What takes center-stage is the sentence, and it is the ontological commitment of sentences for which Quine has offered his famous criterion. The complete Quinean theory comprises three theses:

(i) One is ontologically committed to an F just in case one *believes* a theory that is ontologically committed to an F.
(ii) A theory is ontologically committed to an F just in case the theory *logically entails* a sentence that is ontologically committed to an F.
(iii) A sentence is ontologically committed to an F just in case *if the sentence is true there is an F among the values of its variables.*

Although elements of this theory have been widely criticized, it remains fairly popular. I will go through some of its problems in order to state a less ambitious but more plausible view—one that I think has a strong claim to be common ground in metaphysics today. Then I will use this modified theory to classify attempts to resolve the conflict between philosophy and the received ontology.

(iii) is problematic when applied beyond first-order languages. If we have higher-order variables, we can formulate "$\exists X. X\,(\text{Bill})$"—a sentence whose ontological commitments are opaque. Does it carry commitment to a *set* of which Bill is a member, to a *property* Bill instantiates, or perhaps merely to *individuals* of which Bill is one? Each of these theses is validated by a semantic theory for the language of monadic second-order logic. There is a related problem when it comes to languages containing temporal or modal operators whose semantic clauses quantify over times or worlds. Do variables employed in the meta-language in interpreting a sentence count as belonging to the sentence? Or can we employ quantification over times or worlds in interpreting these sentences without thereby ascribing ontological commitment? It is not clear how to address these questions.

Even more importantly, (iii) starts delivering perplexing results as soon as we abandon standard formal languages. The English sentence "Someone is happy" clearly carries commitment to a person, even though it contains no variable whatsoever. Many insist that it does at the level of logical form, but the basis for this claim is obscure. Is this an empirical truth? If so, what if it turns out—as some actual linguists believe (cf. Jacobson, 1999)—that our best semantic theory for English does not postulate any variable in interpreting this sentence? Or is the claim based on a priori insight? I certainly don't have either the intuition or the proof that any sentence in any natural language contains a variable. One often hears that postulating a variable in logical form is *unavoidable* if we want to capture the sentence's truth-conditions. But this is surely false. Suppose the semantic value of 'is happy' is the property of being happy, the semantic value of 'someone' is the second-order property of being instantiated by someone, and suppose concatenation is interpreted as property-instantiation. Then "Someone is happy" says that the property of being happy instantiates the property of being instantiated by someone. This may not be the right semantics, but it is hard to see why it would be obvious that it isn't.

(iii) is irremediably parochial. The reason it was taken seriously for a long time is two-fold: philosophers like first-order languages (because they have a nice clear syntax and semantics and because they have attractive meta-logical properties) and when applied to such languages the criterion delivers intuitively acceptable verdicts. Quine himself was fully aware of the parochialism and he did not mind it at all: his recommendation for resolving ontological disagreements was to start by restating our theories in a canonical first-order form. He calls this restatement *regimentation*, a term that has misled many into thinking that he regarded the new theories as logically equivalent to the old ones. Given that he believed in the indeterminacy of translation, it is quite clear that he did not.

Those of us who remain convinced of the legitimacy of semantic notions could state a non-parochial version of (iii):

(iii') A sentence is ontologically committed to an F just in case it *says* that there is an F (i.e. if its *content* is that there is an F).

The verdicts that "$\exists x. \text{dog}(x)$" is ontologically committed to a dog but "$\exists x. \text{cat}(x)$" is not are plausible because the former says that there is a dog while the latter says no such thing. Whenever we are uncertain whether a sentence says that there is an entity of a certain type—for example, whether "$\exists X. X (\text{Bill})$" says that there is a property Bill has or whether "It rained" says that there is a past time—we are equally uncertain whether it is the case that if the sentence is true an entity of that type is among the values of the bound variables in an *appropriate* translation of the sentence to a first-order language.

(iii') cuts down the number of first-order sentences that are ontologically committed to a dog. Take, for example, the sentence "$\forall x. \text{dog}(x)$." If this sentence is true then there is a dog among the values of 'x.'[4] On the other hand, "$\forall x. \text{dog}(x)$" does not say that there is a dog, even though it does follow from what it says that there is. So, according to (iii) the sentence is ontologically committed to dogs but according to (iii') it is not. But this is not a problem. I don't think intuition favors the claim that the *sentence* "$\forall x. \text{dog}(x)$" is ontologically committed to a dog—even though the *theory* that comprises the single sentence "$\forall x. \text{dog}(x)$" surely is. The latter is guaranteed, whether we adopt (iii) or (iii') as long as we conceive of theories as sets of sentences closed under logical entailment.

This brings us to (ii)—Quine's theory about the ontological commitment of theories. Logical consequence, as it is usually understood, is bound up with the expressive power of a language, and this fact makes (ii) parochial in much the same way (iii) is. Consider Peano Arithmetic, formulated in a language whose non-logical constants are '0', '′', '+', and '·'. Is this theory ontologically committed to a natural number? The intuitive answer must surely be "yes"—this is the standard theory of natural numbers and it contains sentences such as "$\exists x. x = 0$," "$\exists x. x = 0'$," "$\exists x. x = 0''$," etc. Intuitively, this theory is committed the every natural number. Still, the theory is not ontologically committed to any number according to (ii), because the

---

[4] The reasoning here assumes that this is a sentence in a *classical* first-order language, i.e. that the domain of quantification is non-empty.

language in which it is formulated does not have a number predicate, and consequently, the theory does not contain a sentence like "$\exists x.\,\text{number}(x)$."[5] The fact that first-order Peano Arithmetic is ontologically committed to 0 and the fact that 0 is a natural number is no help because the theory does not logically entail that 0 is a natural number.

The right response to this problem might be to replace the requirement of logical entailment in (ii) with a notion of entailment that permits the extension of the language of the theory. First-order Peano Arithmetic is ontologically committed to natural numbers, one might say, because there is a way of introducing a number predicate 'N' and stating its relation to the non-logical expressions of the old theory such that the new theory logically entails a sentence that is ontologically committed to natural numbers. Since "$N(0)$" and "$\exists x.\,x=0$" are theorems of the extended theory, it logically entails "$\exists x.\,N(x)$."

But the response is problematic. Take for example a theory (let's say it is Thales's) that contains just one sentence: "$\forall x.\,\text{water}(x)$." By Quine's own account, this theory is ontologically committed to water (for it logically entails "$\exists x.\,\text{water}(x)$") and to nothing else. This sounds right. Suppose we now extend the theory by introducing the one-place predicate "hydrogen" and the two-place predicate "contains" and by specifying the relation of the new expressions to "water" through the following metaphysically necessary truth: "$\forall x\,(\text{water}(x) \to \exists y(\text{hydrogen}(y) \wedge \text{contain}(x,y)))$." The new theory logically entails "$\exists x.\,\text{hydrogen}(x)$," and is thus ontologically committed to hydrogen. But—I take it—we don't want to say that Thales's Theory carries such a commitment. Ontological commitment is a non-trivial matter, but presumably we don't have to make substantive empirical discoveries like that water contains hydrogen to figure out what our own theories are committed to. So, this sort of theory extension does not preserve ontological commitment. And it is rather unclear what exactly the relevant difference between this case and the case of extending first-order Peano Arithmetic with a number predicate might be.

As in the case of (iii), what drives our judgments is an intuitive notion. We think Peano Arithmetic is ontologically committed to natural numbers because we think the theory entails a sentence that says that there are numbers and we think Thales's Theory is not ontologically committed to hy-

---

[5] Of course, the language of first-order Peano Arithmetic does contain a predicate whose extension comprises all and only the natural numbers. Under the standard interpretation, "$x=x$" is satisfied by all and only the natural numbers. But the fact that the theory entails "$\exists x.\,x=x$" can hardly be regarded as a proof that it is ontologically committed to the natural numbers.

drogen because we think it does not entail a sentence that says that there is hydrogen. Whether this sentence can actually be formulated in the language of the theory is neither here nor there. I suggest that we drop the notion of logical entailment and use the intuitive notion entailment instead:

(ii') A theory is ontologically committed to an F just in case the theory *entails* a sentence that is ontologically committed to an F.

This removes the parochialism of (ii) in the way in which helping ourselves to the notion of content did in the case of (iii). Like content, entailment is not language-bound: "Snow is white" has the same content as "Schnee ist weiss" and "A dog is barking" entails "$\exists x (\text{dog}(x) \wedge \text{barking}(x))$" even though these sentences belong to different languages.

Let's consider now the first plank of the Quinean theory, the claim that one's ontological commitments are the ontological commitments of the theories one believes. The first thing to note is that if the theory is to be reductive then (i) has to be strengthened to the claim that we are ontologically committed to an F *in virtue* of believing a theory that is so-committed. The analogous strengthenings of (ii') and (iii') are unproblematic: it seems plausible that a theory carries ontological commitment in virtue of its entailments, and that a sentence carries ontological commitment in virtue of its content. It is not equally obvious that a person is ontologically committed in virtue of her beliefs. The attitude we bear to a theory in virtue of which we assume the ontological commitments of the theory must be *belief-like*—it must be a form of acceptance. But acceptance comes in many forms and varies along different scales. I can accept a theory tentatively or firmly, I can accept it temporarily (e.g. for the sake of discussion or evaluation) or for good, I can accept it on the basis of evidence I posses or on the authority of others, etc. Couldn't ontological commitment be a form of acceptance distinct from belief?

There are cases that suggest even the mere equivalence explicitly stated in (i) might be false. Consider, for example, Max who first hears from his parents about Einstein's theory of special relativity. Suppose what he hears is rather sketchy, imprecise, and in crucial parts simply inaccurate. Nonetheless, Max can surely accept the theory in some sense on the basis of such testimony. Sometimes this sort of acceptance comes short of being a belief—Max may even lack the prerequisite concepts to entertain the theory. Is it clear that in these cases Max is not ontologically committed to whatever Einstein's theory is ontologically committed to? He *subscribes* to the theory and this fact should be taken seriously. *If* we decide that Max is ontologically committed to whatever the theory of special relativity is in virtue of his acceptance of the theory we reject (i). We would say that acceptance weaker than belief is sufficient for ontological commitment.

Here is a different example. Anna is a smart kid but she cruises the web too much. Somehow she picked up the belief that archeologists have determined that dragons were a kind of dinosaur, and so, she also came to believe that there once were dragons on earth. Is she ontologically committed to dragons? Does the world, as she takes it to be, contain dragons? This is unclear. The dragons of her world were dinosaurs, so perhaps the right thing to say is that Anna is committed to some extra dinosaurs, but not to dragons. *If* this is right, (i) is false: a form of acceptance that is stronger than belief is necessary for ontological commitment.

There is nothing in Quine that supports (i) against doubts of this kind and the subsequent literature hasn't much to offer either. In light of this, I think it is reasonable to retreat from (i) to a weaker claim:

(i') One is ontologically committed to an *F* just in case one *accepts* (in the appropriate sense) a theory that is ontologically committed to an *F*.

What exactly the relevant notion of acceptance amounts to is an open question. Putting the three planks together the theory of ontological commitment goes as follows: one is ontologically committed to an *F* just in case one *accepts* (in the appropriate sense) a theory that *entails* (in the ordinary sense) a sentence whose *content* (in the ordinary sense) is that there is an *F*.

Reconciliation between the received ontology and its philosophical rivals can take one of three paths, following one of the three components of the theory of ontological commitment. Traditionally, reconciliatory attempts focused on entailment or content. For example, to reconcile the received view about numbers with nominalism one might take the position that "There are infinitely many primes" does not entail "$\exists x.\,\text{number}(x)$" or that "$\exists x.\,\text{number}(x)$" does not say that there is a number. To reconcile the existence of hands with monism one might argue that "There is a hand and here is another" does not entail "$\exists x \exists y.\, x \neq y$" or that "$\exists x \exists y.\, x \neq y$" does not say that there is more than one thing.

The trouble with traditional approaches is that they save the received ontology by sacrificing the received logic or the received semantics. The fact that received theories can be *rephrased* in a certain way that avoids ontological commitment is in itself irrelevant as long as one does not insist that the rephrased theories have the same entailments with the same contents as the original ones. And once one insists on a tight relation between our familiar theories and the paraphrases one takes a revisionary stance either in logic or in semantics. To do so, will elicit the same reaction from consistent defenders of philosophical modesty as a direct assault on the received ontology:

> ... if the theory seems to say, for example, that every person has a guardian angel in heaven, then the theory is true only if the angels in heaven really exist. (Rosen, 2005, p. 14)

If we want a principled way to address the conflict between received views and philosophical ontology it is best to focus on acceptance. The common core to all attitude-based approaches to reconciliation is that we are not committed to the plethora of entities received theories are because we do not accept those theories *in the sense in which acceptance carries ontological commitment*. This is a more promising line simply because it uncontroversial that there are many forms of acceptance and because there is no received opinion about exactly which of them are ontologically committing. By contrast, the received view is that there is a unitary notion of entailment and content.

I have argued that the key to a successful reconciliation between received views and philosophical ontologies is the recognition that sometimes our attitude towards established theories is non-committal. The question is *how* this is possible, and *when* does this happen.

3. Acceptance Weaker than Belief

The standard way to bring about an acceptance-based reconciliation between a received theory and a conflicting philosophical view is to claim that our usual form of acceptance towards the received theory is something weaker than belief. This is the *fictionalist* position.

What fictionalists about a certain theory T argue is that the virtues of T are independent of its truth. So, according to Hartry Field (1980), the principal virtue of mathematics is that it facilitates deduction within nominalistic theories. It can do so effectively even if it is false. Bas van Fraassen's (1980) opinion is that the aim of science is mere empirical adequacy. False theories may provide empirically adequate accounts of observable phenomena that are simple and elegant. Richard Joyce (2006) thinks many of the virtues of an ethical theory are practical. False theories may be at least as practically useful as true ones.

These three examples illustrate three different versions of fictionalism. Field thinks his target theory is *subsidiary* to another one: the point of mathematics is to serve the inferential needs of physics. Van Fraassen sees physics as a *pumped-up* theory which aims at a smaller target than it seems to: all it is supposed to do is get the entailments about observables right. And Joyce considers ethics a mere *decoy*: a theory whose real function has nothing to do with discovering the truth about anything. What they have in common is the claim that the theories in question are in no way at fault if they happen to be false.

If all the virtues of a theory are independent of its truth it would be a mistake to believe it. If you are clearheaded, you will suspend belief, although you may well continue to act as if you believed in ordinary circumstances. (Let's call your attitude *belief minus*.)[6] When in those circumstances you say that things are as the theory says they are you are not making an assertion, only acting as if you are. (Let's call your speech act *assertion minus*.) *Hermeneutic fictionalists* believe that we are in fact clearheaded: we accept some received theories without believing them and express our attitudes without making assertions.[7] They offer an attitude-based reconciliation between a received ontology and its philosophical rivals.

Hermeneutic fictionalism does not involve commitment to any particular ontology—it is simply a strategy to level the playing field in which ontological debates are conducted. The strategy tends to be deployed by proponents of revisionary ontologies, but this is only incidental.[8] If you were committed to numbers, electrons, or values, you *could* still be a hermeneutic fictionalist with regard to received theories about these entities. But you would forfeit the standard argument in favor of your ontology: you would concede to your opponent that the reason the theory about the entities you believe in is generally accepted is *not* that it is true.

Fictionalists tend to exert the bulk of their efforts in arguing that the virtues of their target theories have nothing to do with truth. But even if one were to grant that, there would remain two further issues to address before they can declare victory.[9] We need an explanation of what belief minus is

---

[6] Fictionalists often think belief minus is a relation we bear to the contents of fictions when we are immersed in them. I regard this as a mere analogy. As I understand it, fictionalists need not insist that the analogy is perfect, or even particularly useful. So, to point out disanalogies between our attitude to Sherlock Holmes and the number 2 is not an effective way to argue against fictionalism.

[7] Fictionalists who think we should be clearheaded (but perhaps are not) are called *revolutionary* fictionalists. The hermeneutic/revolutionary distinction is from Burgess (1983). Revolutionary fictionalists who think we *do* believe the target theory advocate a change of attitude, and hence, abandon modesty.

[8] For example, Yablo (1998) advocates a fictionalist attitude towards mathematics but remains agnostic about the existence of mathematical entities.

[9] I discussed these problems for fictionalism in Szabó (forthcoming).

and a defense of the suggestion that it is at least possible for us to bear such an attitude.

Let's start with the first problem. It is superficially tempting to say that belief minus is the attitude we bear to theories that have been discarded but that continue to be widely employed for certain limited purposes, such as Newton's theory of gravity, Dalton's theory of partial pressures, or Wegener's theory on continental drift, to name just a few. But this gives us a false analogy: the virtues of these theories are *not* independent of their truth. They are false but approximately true, and the fact that they come near enough to truth seems to be the reason why they work in certain settings. We accept these theories tentatively and we regard their falsehood as a strike against them. By contrast, the fictionalist thinks the truth or falsity of his target theory should be of *no concern* to us.

Hermeneutic fictionalists face a dilemma. If they think belief minus is a rational attitude, then it looks like belief minus collapses into a mere *assumption*. For if we really employ a theory for some limited purpose—to aid our complex reasoning, to predict our future observations, to strengthen our wavering resolve, etc.—then we have good reason to adopt it for those exact purposes and for nothing else. But even those who doubt that we are actually pursuing the truth when we are engaged with these theories must acknowledge that we think our acceptance of these theories is open-ended, and thus, intuitively quite different from an assumption. If, on the other hand, they think belief minus is not a rational attitude, they completely abandon modesty. For the view that our attitude towards a received theory, such as mathematics, physics or ethics is not adequately backed by reasons flies in the face of what our ordinary views about these attitudes are.

But let's assume that there is a way to identify belief minus as a rational attitude, distinct from mere assumption. The second problem the hermeneutic fictionalist faces is to explain how belief minus is attainable. The difficulty is that belief minus is supposed to be compatible with philosophical reflection. Someone who bears this attitude towards a theory should be able to endorse it in ordinary contexts and reject it in philosophical ones. In other words, the attitude should be *sensitive* to the difference between those contexts.

It might seem that this sort of sensitivity is a fact of life. There is, after all, an obvious difference between the correctness conditions of ordinary and philosophical existence claims. It would be odd to object to a proposed proof in arithmetic on the grounds of the alleged non-existence of numbers, while such an objection during a philosophical debate would be entirely appropriate. Isn't the fact that we all recognize this enough to show that it is possible for an attitude to be sensitive to the difference between ordinary and philosophical contexts?[10] I don't think so. For a difference in correctness conditions has an alternative explanation, one that does not require there to be a difference *in kind* between the context of a mathematical discussion and the context of a philosophical one.

The alternative explanation is that it is *always* inappropriate to challenge the prevailing assumptions, unless the challenge is signaled explicitly. When we are discussing a mathematical proof we take the existence of numbers for granted. To object that this presupposition is unwarranted would derail the conversation from its current direction—hence it is inappropriate. There is nothing special about philosophical existence claims in this regard. To object to the proof on the grounds that the person advancing it has a bad track record, or on the grounds that the proof is too complicated to merit serious attention, or on the grounds that anyone who disagreed with it would be fired would be equally inappropriate, if the objection came out of the blue.[11]

If philosophy has a proper subject-matter it is hard to believe that there is a difference in kind between ordinary and philosophical contexts. After all, we don't think we transport ourselves into a special context when we start discussing astronomy, philology, or stamp-collecting. Why would philosophy be different? There are more or less relaxed contexts—contexts where we are more or less patient towards far-flung criticism. The fact that raising specifically philosophical objections is in many contexts inappropriate can be adequately explained by assuming that those contexts are not sufficiently relaxed. But belief minus is supposed to be sensitive not to the relaxed/non-relaxed distinction but to the ordinary/philosophical one. If the latter distinction is a myth, so is the non-committal attitude hermeneutic fictionalists think we bear towards various theories.

4. Acceptance Stronger than Belief

Hermeneutic fictionalism rests on the idea that our sincere utterances in ordinary circumstances are sometimes not genuine assertions. Rather, they express a form of commitment that is weaker than belief, and hence does not reveal ontological commitment. We display our genuine commitments only in contexts where matters of ontology are at stake. If so, we should

---

[10] Cf. Chalmers (2009).

[11] It might seem that philosophical contexts are special because *nothing* is shielded from criticism. I think this is an illusion. It is inappropriate to object to the claim that water is necessarily H_2O in a philosophical debate on the grounds that water is a mixture of different substances. Such an objection is not relevant to the topic and accomplishes nothing besides derailing the conversation. The fact that the objection is true is no excuse.

never be reluctant to ascribe ontological commitment to *philosophers* when they make unequivocal pronouncements. But sometimes we are.

Anyone who has taught Berkeley knows that students often read him as denying the existence of tables and chairs, mountains and trees. Berkeley says over and over again that he does no such thing but his words have a hard time getting across. This is because he also says that the familiar objects he thinks are real are nothing more than collections of ideas, and because he denies the existence of matter. Since we don't think tables and chairs, mountains and trees are collections ideas and we do think that they are material, we are naturally reluctant to grant that Berkeley is ontologically committed to them.

Leibniz's insistence that monads cannot interact with one another and change due to an inner principle poses a similar problem. Once the metaphysical picture is clear, it is hard to accept his insistence that he does not deny that a billiard ball can cause the motion of another. Leibniz points out that he has room for the right kind of correlation between aggregates and that he thinks there is nothing more to causation than that, but this does not change the fact that causation is a form of interaction and that he categorically denies that interaction is possible. Again, despite forceful pronouncements to the contrary, it is hard to resist the judgment that Leibniz's ontology does not contain causes and effects.

Lewis famously said in *The Plurality of Worlds* that he is a theist. Those who read him tend to disagree: the fact that he is a modal realist and that he believes that many non-actual worlds contain deities seems insufficient to show that he is ontologically commited to gods. But not according to Lewis's own theory—he thinks that non-actual worlds are as real as this one, even if they are causally and spatio-temporally isolated from us. Quantification appropriate in discussions of ontology cannot be restricted to what is local, so a modal realist of Lewis's stripe really should say with full sincerity that there are gods. Be that as it may, few of us would concede that Lewis was a theist.

It seems that if we are not willing to concede that Berkeley's ontology contains mountains, that Leibniz's had causes, and that Lewis's was full of gods we must either say that these philosophers were mistaken about what they believed or that they were wrong about what they said. Mistaken belief would presumably have come from a shortcoming in conceptual mastery: perhaps Berkley did not have the concept of a mountain, Leibniz failed when it came to the concept of a cause, and Lewis was incompetent with the concept of existence. A mistake about what they said would have to come from a semantic confusion: perhaps these philosophers were wrong about what the words 'mountain,' 'cause,' or 'exist' mean in the public language.

But when we step back for a moment we should find these accusations quite implausible.

Why not take both the philosophers' pronouncements and our own reactions at face value? I think Berkley really did believe that there are mountains, but his ontology does not contain mountains (only ideas of mountains). Similarly, Leibniz was entirely correct when he ascribed to himself the belief that there are causes, but there are no causes in the world as he takes it to be (only correlations). And Lewis did indeed believe that there are countless gods, even though he was not ontologically committed to gods (only to possible gods). To have my cake and eat it too, I propose to decouple ontological commitment from belief.

My suggestion is the mirror image of hermeneutic fictionalism. The hermeneutic fictionalist agrees with Quine that the form of acceptance that carries ontological commitment is belief, but parts way with him in insisting that our attitude towards certain received views is a weaker form of acceptance. I am on board with Quine when it comes to our attitude towards received views: I think we do believe them. What I reject is the contention that the form of acceptance that carries ontological commitment is belief. I think it is a stronger form of acceptance.

The fictionalist thinks the virtues of certain theories are *independent of* their truth. I disagree: I think truth is part of what makes any good theory good. But the virtues of certain theories *go beyond* their truth. When it comes to theories that do not aim at the truth, fictionalists recommend that we should not believe them—our attitude should be something less: an attitude I called *belief minus*. When it comes to theories that aim at more than mere truth, I recommend that we should not merely believe them—our attitude should be something more: an attitude I call *belief plus*.

Like the fictionalist, I will have to say something about the form of acceptance I am postulating. Before I turn to that task, I'd like to highlight what I take to be the chief advantages of the view I am proposing. Like hermeneutic fictionalism, I offer an attitude-based reconciliation between the received ontology and its philosophical rivals. This means that I take the logic and the semantics that governs our discourse at face value. No hidden ambiguities, no unexpected entailments. Unlike hermeneutic fictionalism, I also go along with our ordinary assessment of what our attitude is towards the received ontology. I acknowledge the soundness of the one-liners that show that shapes, directions, unobservables, causes, mental states, rights, values, properties, propositions, events, concepts, thoughts, etc. exist. What I reject is the move from this admission to the claim that these entities are elements of the world as I take it to be. I resist the move from belief to belief plus.

So, how does belief plus differ from mere belief? Consider Berkeley and his belief that there are mountains. It seems natural to say that Berkley does not fully understand the content of his belief. This is not to say he does not understand it at all—he surely passes every reasonable test for possessing the relevant concepts. The problem is that he does not know *what mountains are*: he thinks they are collections of ideas, not material objects. This sort of mistake does not undermine his ability to think about or refer to mountains. What it does is undermine his ability to provide an adequate *explanation* for the truth of his beliefs about mountains. This is, I think, what guarantees that his belief falls short of being belief plus, and this is why he is not ontologically committed to mountains. Belief plus is belief plus full understanding.

Some questions ask for more than true, relevant, and informative answers. When I ask you whether you know what time it is a simple "yes" will not suffice. What I really wanted is for you to tell me what time it is and thereby show that you know it. A more intriguing example of a question that asks for more than a simple answer was given by Sylvain Bromberger (1992, p. 22). Imagine the following situation. Holmes and Watson are investigating a case. A man lies dead in a room with a bullet in his head. He is not wearing a glove and the fingerprints on the gun do not match his. The doors and the windows are locked from the inside by the victim and have not been touched before the investigators entered the room. The chimney is too narrow to allow passage for a grown man. There are no secret doors and hidden outlets and there is nobody hiding in the room.

Watson asks the obvious question: "How did the murderer escape?" and Holmes answers: "Through the chimney." Now suppose Holmes has already figured out that the murderer was a dwarf. Then his answer was true, relevant, and informative. It was nonetheless inappropriate. Watson's question asked not just for an answer—he wanted an explanation of the answer.

Ontological questions, I contend, also ask for explanations. They are simple by form and content—they are questions like "Are there symmetrical relations?," "Do complex numbers exist?," "Are colors real?" But their simplicity is misleading—what they ask is not a just "yes" or "no" but also an account of how the "yes" or "no" could be the right answer. Such an explanation cannot be given without knowing what the entity whose existence is at stake is. One resolves an ontological problem only when one is in a position to give what an ontological question asks for, i.e. when one gives an explanation for the truth of an existential proposition or its negation. So, the conflict between received ontology and philosophical ones is illusory: there are received views about what there is but there is no received ontology.

My view makes genuine ontological disagreements relatively rare. Knowing what an F is turns out to be prerequisite for having an *ontological view* about Fs. By contrast, you can think and talk about Fs—have a *view* about Fs—without knowing that. When you know what an F is and you believe there is an F you bear belief plus to the proposition that there is an F. But you bear belief too, and in the course of an ordinary conversation when ontological matters are not at stake you express nothing more than the weaker attitude. When you think you know what an F is but you are mistaken then you might think that you are revealing your ontological commitments in expressing your beliefs but you are doing no such thing. I suspect this is what happens to a number of philosophers—I think it happened to Berkley, Leibniz and Lewis.

Are the philosophical debates about ontology genuine ontological disagreements? I am not sure. Take the debate about the existence of numbers. Do we know what they are? I think this is open to serious doubt. I believe that natural numbers satisfy the Peano Axioms—this much is surely received view. I also believe that the number 2 is not identical to Julius Caesar. This is another well-established opinion independent of the Peano Axioms. I believe that the number two cannot be identical to both $\{\{\emptyset\}\}$ and $\{\emptyset, \{\emptyset\}\}$ for these sets are different. I am inclined to think that the number two is distinct from both, in fact that it is not a set at all. But I am not too sure about this (cf. Benacerraf, 1965). Some think the number 2 is a property. I like the idea that it is a *sui generis* entity: it is what it is and no other thing. But I worry that this only disguises my ignorance of what it really is. If my worry is well-founded I have no ontological view about natural numbers. Some philosophers may not be better off in this regard than I am. My comfort is that my belief that there are natural numbers is unshaken by these doubts.[12]

5. Clear and Distinct Ideas

While the meta-ontological outlook defended here is unusual, I think it has a historical precedent.[13] Some of his comments suggest that Descartes held

---

[12] What can I offer to a nominalist? I think she should rephrase her view in a slightly more cautious fashion: if numbers are what we normally take them to be then they do not exist. When revisionary ontologies are stated in this way, I think they must be taken seriously.

[13] I have defended a similar view in Szabó (2003). There I have suggested that belief plus is not a propositional attitude—rather it is a mental state whose content is a mere propositional constituent. Ontological commitment to unicorns is expressed by saying that one believes in unicorns, and this, I have argued, is not equivalent to saying that one believes

a similar view. In the *Second Meditation*, immediately after he runs the *cogito* argument, Descartes raises a question that casts doubt on what exactly the argument has achieved. He writes:

> But I do not yet have a sufficient understanding of what this 'I' is, that now necessarily exists. So I must be on guard against carelessly taking something else to be this 'I,' and so making a mistake in every item of knowledge that I maintain is the most certain and evident of all. (Descartes, 1641/1984, AT VII 25)

The passage suggests that while Descartes believes he has shown that he exists, he thinks his knowledge may contain a mistake. What he has in mind seems to be misidentification—since he does not know *what* he is, he may be mistaking himself for something else. And if that other thing fails to exist then he made a mistake in drawing the conclusion of the *cogito*. This is a doubly curious idea. First, it is unclear in what sense one might be able to misidentify oneself in thought. If I am delusional and believe myself to be Julius Caesar, the thought that I exist is still about me, not about the Roman general. Second, even if I can misidentify myself in thought, it is not clear how I could be mistaken that this thing exists, if I have indeed proved that it does. Isn't it obvious that if one knows a proposition then the proposition is true, and hence, it cannot be mistaken?

As it becomes clear from the discussion following this passage, Descartes wants to make sure that nothing is included in his idea of himself that does not properly belong to it. Once he shows that the idea contains only what is required for entertaining the possibility that an evil demon is deceiving him, the possibility of the mistake goes away, and the conclusion of the *cogito* is restated:

> I am, then, in the strict sense only a thing that thinks; that is I am a mind, or intelligence, or intellect, or reason,—words whose meaning I have been ignorant of until now. But for all that I am a thing which is real and which truly exists. (Descartes, 1641/1984, AT VII 27)

This might solve the first puzzle. Despite the slightly misleading phrase "carelessly taking something else to be this 'I' " Descartes is not concerned about misidentification, but about misdescription, which of course is not

---

the proposition that unicorns exist, or any proposition about unicorns, for that matter. While I stand by my conclusions, I have grown dissatisfied with the fact that they are so tightly connected to semantic considerations. The core of my position—a core that withstands a change in argumentative strategy—is that our attitude that carries ontological commitment is stronger than belief. Whether it is a propositional attitude is of secondary importance.

only possible but commonplace. He grants that the proposition that he exists is *about him*—what he is worried about is that the idea of himself within that proposition *misrepresents him.* But the second puzzle is as alive as before. How can we understand the possible mistake in Descartes knowledge he is worried about? The proposition is true, so how is it mistaken?

Here is an idea. We need some way to distinguish between two distinct contents associated with the sentence "I exist." One is the proposition the sentence expresses in a given context—this is what has shown to be true by the *cogito.* The other is a proposition that results from unpacking all the information encoded in the ideas that make up the first proposition. This one can be false if the idea expressed by 'I' misrepresents the person it refers to. Explanation of the truth of the first proposition would need to go though the second, and if the explanation is correct, the second proposition must be true as well. I conjecture that the words 'real' and 'truly existent' in the restatement of the conclusion of the *cogito* after Descartes has answered the question what he is marks the admittance of the self into the Cartesian ontology. Now that I know what I am I don't just know that I exist; I also fully *understand* this proposition.[14]

The *cogito* is not an outlier, when it comes to settling ontological questions in Descartes's philosophy. He has strikingly similar things to say in discussing pain.

> ...when someone feels an intense pain, the perception he has of it is indeed very clear, but it is not always distinct. For people commonly confuse this perception with an obscure judgment they make concerning the nature of something which they think exists in the painful spot and which they suppose to resemble the sensation of pain; but in fact it is the sensation alone which they perceive clearly. (Descartes, 1644/1984, AT VIII 22)

Someone who feels intense pain knows that she is in pain. Nonetheless, there is a possibility of a mistake in this very item of knowledge. Although as clear as any idea can be, this person's idea of pain may not be distinct. This happens if she is mistaken about *what* pain is, for example, if she thinks it is a sensation that is located in the body or resembles something located in the body. As long as her idea of pain remains confused she does not fully understand the proposition that pain exists. Consequently, she is not in

---

[14] Descartes says he was hitherto ignorant of the meaning of words like 'mind,' 'intelligence,' 'intellect' or 'reason,' and perhaps 'I.' This is an overstatement—he surely would not insist that he lacks what we would normally call linguistic competence with these words. I suggest that he means only that he lacked full understanding of these words, which boils down to nothing more or less than that he did not know what these things are.

a position to give an account of the truth of the proposition and, I contend, she cannot address the ontological question about pain.

Descartes thinks philosophical questions in general, and ontological ones in particular can be adequately discussed only after we made our ideas clear and distinct. Clarity for an idea is nothing more than being "present and accessible to the attentive mind," but distinctness is something much more demanding. Descartes says that a distinct idea must be "so sharply separated from all other perceptions that it contains within itself only what is clear." (Descartes, 1644/1984, AT VIII 22). What exactly this amounts to is somewhat hazy, but in practice Descartes always insists that someone who has a clear and distinct idea of something must know what that thing is. As I read him, Descartes thinks we are not ontologically committed to a thing until we have a clear and distinct idea of it. Before one embarks on the journey of philosophical reflection one has no ontology. In this, I fully agree with him.

Let me close the paper by summarizing its main claims. I have argued that the ontological attitude—our attitude that carries ontological commitment—is not mere belief. The fact that I believe that there are Fs does not settle the question whether my ontology contains Fs. Unlike belief, the ontological attitude requires full understanding of the proposition believed. Such a full understanding is by no means guaranteed by semantic and conceptual competence—we can and often do think thoughts whose content we fail to understand fully. Full understanding of the proposition that there are Fs requires knowledge of what Fs are, and such knowledge is a precondition for explaining the truth of that proposition. What ontology aims at are explanations for the truth of true existential propositions. Consequently, ordinary existential beliefs can be reconciled with philosophical ontologies, as long as the nature of the relevant entities is not fully understood. Since I think this is indeed the case with most controversial entities, I conclude that the received ontology poses no threat to continued modesty in matters philosophical.[15]

---

[15] Earlier versions of the paper have been presented at Workshop on Ontological Commitment, University of Paris, IHPST and at the Eidos Conference on 'Because' in Geneva. The final version was given at the 6th European Conference on Analytic Philosophy in Kraków. I thank the participants of these events for discussion and comments. I also thank Tamar Gendler for numerous conversations on the topic.

References

Benacerraf, P. (1965). What numbers could not be. *Philosophical Review*, 74, 47–73.
Berkeley, G. (1948–1957). A treatise concerning the principles of human knowledge. In A. A. Luce and T. E. Jessop (Eds.), *The Works of George Berkeley, Bishop of Cloyne*, volume 1–9. Thomas Nelson and Sons, Ltd., London. (Original work published 1711).
Bromberger, S. (1992). An approach to explanation. In *On What We Know We Don't Know: Explanation, Theory, Linguistics, and How Questions Shape Them*. University of Chicago Press, Chicago.
Burgess, J. (1983). Why I am not a nominalist. *Notre Dame Journal of Formal Logic*, 24, 93–105.
Chalmers, D. (2009). Ontological anti-realism. In D. Chalmers, D. Manley, and R. Wasserman (Eds.), *Meta-Metaphysics*. Oxford University Press, Oxford.
Descartes, R. (1984). Meditations on first philosophy. In J. Cottingham, R. Stoothoff, and D. Murdoch (Eds.), *The Philosophical Writings of Descartes*, volume 2. Cambridge University Press, Cambridge. (Original work published 1641).
Descartes, R. (1984). Principles of philosophy. In J. Cottingham, R. Stoothoff, and D. Murdoch (Eds.), *The Philosophical Writings of Descartes*, volume 2. Cambridge University Press, Cambridge. (Original work published 1644).
Field, H. (1980). *Science without Numbers: A Defense of Nominalism*. Basil Blackwell, Oxford.
Jackson, B. (2007). Truth vs. pretense in discourse about motion (or, why the sun really does rise). *Noûs*, 42, 298–317.
Jacobson, P. (1999). Towards a variable-free semantics. *Linguistics and Philosophy*, 22, 117–184.
Joyce, R. (2006). *The Evolution of Morality*. MIT Press, Cambridge, MA.
Lewis, D. (1986). *On the Plurality of Worlds*. Blackwell, Oxford.
Moore, G. E. (1939). Proof of an external world. In P. K. Moser and A. van der Nat (Eds.), *Human Knowledge: Classical and Contemporary Approaches* (439–452). Oxford University Press, Oxford.
Rosen, G. (2005). Problems in the history of fictionalism. In M. E. Kalderon (Ed.), *Fictionalism in Metaphysics* (14–64). Oxford University Press, Oxford.
Szabó, Z. (2003). Believing in things. *Philosophy and Phenomenological Research*, 66, 584–611.
Szabó, Z. (forthcoming, Noûs). Critical notice on fictionalism in metaphysics.
van Fraassen, B. (1980). *The Scientific Image*. Clarendon Press, Oxford.
Yablo, S. (1998). Does ontology rest on a mistake? *Proceedings of the Aristotelian Society*, 72, 229–261. (Supplement volume).

Part II
Invited Speakers

History of Philosophy

Hundred Years After: How McTaggart Became a Thing of the Past

Kristóf Nyíri

It is a great honour for me to be the invited speaker in the History of Philosophy section of this major Analytic Philosophy conference. It is also, as I increasingly came to realize as the date of the conference approached, a great responsibility. What I particularly came to feel uneasy about was the choice of my topic. Was it important enough? Was it broad enough? Seeking reassurance, I turned, as so often before, to the writings of the man who had been my first mentor in philosophy, one who played a significant role in analytic philosophy from the late 1940s to the 1960s, and one who was renowned for his skill in exploiting the history of philosophy as the background against which to act out philosophical analysis: Wilfrid Sellars. And Sellars did reassure me. I hit on the passage in his "Autobiographical Reflections" where he describes his first serious encounter with philosophy. It happened at Ann Arbor, in 1931/1932, when in a seminar in metaphysics he was introduced, as he reports, "to McTaggart's classic paper on the unreality of Time," and chose to write his term paper on the topic. He was soon "deep in the literature" and found himself "genuinely involved." As he puts it: "Philosophy was no longer a storehouse of alternatives to be explored and evaluated but, from that moment on, an unfinished dialogue in which I might have something to say. I soon became convinced that the problem of time was so intimately connected with other classical problems that it, like the mind-body problem, is one of the major proving grounds for philosophical systems."[1] Sellars continued to work on the topic of time, returning to it again and again; and defending, from the very beginning, "a substantialist ontology of change," that is, a position diametrically opposed to that of McTaggart. I will come back to Sellars on two occasions later in my talk; just now, let me give a summary outline of the same.

McTaggart's paper on "The Unreality of Time" was published in 1908, in the journal *Mind*. The argument of the paper is sufficiently elusive to stand in need of scrutiny before being subjected to criticism. Such scrutiny is what I will attempt to provide in the first section of my talk, under the heading "The McTaggart Motley." In the second section, under the heading "Refuted and Ridiculed," I shall summarize the devastating criticisms that, since the 1920s, C. D. Broad, and others in his wake, have been directing against McTaggart's position, asking, in the third section, how, in the face of such a series of convincing refutations, his argument could still gain, and does still gain, adherents. The answer is, as I will briefly show, that McTaggart's position has become mixed up with, and won undeserved respectability from, the Einstein-Minkowski conception of spacetime, proclaimed in the very same year that McTaggart's paper was published. In the final section of my talk I shall sketch, under the heading "A Future for Time?," the rudiments of an alternative—admittedly adventurous—philosophical strategy, designed to overcome the position represented by McTaggart, that is, to vindicate the common-sense view of the reality of time.

1. The McTaggart Motley

McTaggart's paper exists in two versions—or in two-plus-a-bit versions, if you like. The first one is the *Mind* version.[2] The second, bearing the title "Time," is the text making up chapter XXXIII in the second volume of McTaggart's *The Nature of Existence*, published in 1927. This was a posthumous publication. McTaggart died in 1925, leaving behind a semi-finished draft of the volume, half typescript, half manuscript, bequeathing to C. D. Broad, his successor at Trinity College, Cambridge, the task of preparing it for press. Bringing it into line with the first volume that had been published

---

[1] W. Sellars, "Autobiographical Reflections," in H.-N. Castaneda (ed.), *Action, Knowledge, and Reality: Critical Studies in Honor of Wilfrid Sellars*, Indianapolis: Bobbs-Merrill, 1975, p. 281.

[2] J. McT. E. McTaggart, "The Unreality of Time," *Mind: A Quarterly Review of Psychology and Philosophy*, N. S., no. 68 (October 1908), pp. 457–474.

in 1921, Broad divided the text into numbered sections, constructed an analytical table of contents, but otherwise reports to have made only very minor editorial changes.[3] Perhaps he should have been more thorough. Chapter XXXIII was printed from the typescript part of the draft, but my impression is that the typescript had not been without flaws, with some resulting wordings even more confused than McTaggart's formulations usually were. Also, it is generally unrecognized that the textual differences between the 1908 paper and the *Nature of Existence* version are quite significant. Certainly the latter is not just a re-written text of the former. Rochelle's formula, according to which the "Unreality of Time" paper "[f]orms a substantial part" of *The Nature of Existence* chapter, is closer to the facts.[4] For instance, the so-called "C series," the discussion of which McTaggart clearly saw as playing an important role in the overall argument of the 1908 paper, is introduced only in the last paragraphs of the 1927 "Time" chapter, the topic then recurring, with embellishments, in later chapters of the volume. In the 1927 chapter, there is an extended analysis directed against Russell's treatment of time in his 1903 book *The Principles of Mathematics*, entirely missing in the 1908 paper. More importantly, the 1927 chapter contains a five-page discussion of the criticism C. D. Broad levelled, in his 1923 book *Scientific Thought*, at McTaggart's 1908 position. To mention one more example, while in the 1908 paper the hypothesis that "there might be several independent time-series in reality" is introduced as a possibility raised by Bradley, and the implication that under such conditions "no time would be *the* time—it would only be the time of a certain aspect of the universe" is rejected with reference to the fact that "the theory of a plurality of time-series is a mere hypothesis" and "no reason has ever been given why we should believe in their existence," in the 1927 chapter the name of Bradley is missing, and the observation that under the conditions discussed "no time would be the time—it would only be the time of a certain aspect of the universe" is not followed by the remark that no reason has ever been given for the hypothesis in question. Why the change? Might it not be Einstein, after all, who haunts McTaggart here? Might not, by the 1920s, the news about the special theory of relativity, against all the odds, have reached him? But I am getting ahead of myself. I said McTaggart's paper exists in two-plus-a-bit versions; I managed to list the first two; I am now coming to the plus-a-bit one. This is the reprint of "The Unreality of Time" in the

---

[3] Cf. the "Editor's Preface," p. v, in J. McT. E. McTaggart, *The Nature of Existence*, vol. II, ed. by C. D. Broad, Cambridge: Cambridge University Press, 1927.

[4] G. Rochelle, *The Life and Philosophy of J. McT. E. McTaggart, 1866–1925*, Lewiston, NY: Edwin Mellen, 1991, p. 234.

volume *Philosophical Studies*, a 1934 collection of McTaggart's essays.[5] I am calling it a plus-a-bit version, because although it is indeed a reprint, it is supplemented by a number of notes by the editor S. V. Keeling, indicating the places where the *Nature of Existence* text contains significant additions to the 1908 one. Even if not conveying the full extent of the differences between the first two versions, these notes are interesting. Interesting, or rather, telling, is also the chapter "The Relation of Time and Eternity" in *Philosophical Studies*, following upon the "Unreality of Time" chapter. This is the text of a talk delivered by McTaggart before the Philosophical Union of the University of California on August 23, 1907. I am tempted to call it version zero of the 1908 *Mind* paper, giving a feel, as it were, of the *weltanschauung* behind the latter. As McTaggart here put it: "All existence which presents itself as part of our ordinary world of experience presents itself as temporal. But . . . we have reason to believe that some reality which exists, exists timelessly—not merely in the sense that its existence endures through unending time, but in the deeper sense that it is not in time at all. . . . I do see a possibility of showing that the timeless reality would be, I do not say unmixedly good, but very good, better than anything which we can now experience or even imagine. I do see a possibility of showing that all that hides this goodness from us—in so far as it is hidden—is the illusion of time."[6] This passage, glaringly mystical and devoid of analytic rigour, might give us a foretaste of McTaggart's arguments in "The Unreality of Time." It is an inventory of these arguments I now turn to.

I am speaking of "arguments" in the plural, since I believe that McTaggart's essay cannot be seen—contrary to what standard summaries take for granted—as proceeding along a single train of thought. It consists, rather, of a number of sometimes overlapping, sometimes frayed and only loosely connected, threads—stipulations, arguments, half-arguments, and asides. Attempting to take stock of them here, I cannot avoid repeatedly quoting McTaggart's text directly. Commenting on McTaggart's favourite formula that if an historical event is ever earlier than another, then it always was and will be earlier, Miss Cleugh in her 1937 book *Time and Its Importance in Modern Thought* says that this is "an unsatisfactory way of expressing" whatever McTaggart wishes to convey, "and one which is perilously near nonsense."[7] My impression is that McTaggart's wordings are

---

[5] J. McT. E. McTaggart, *Philosophical Studies*, ed., with an introduction, by S. V. Keeling, London: Edward Arnold, 1934.

[6] *Ibid.*, p. 135.

[7] M. F. Cleugh, *Time and Its Importance in Modern Thought*, London: Methuen, 1937, p. 153.

almost always perilously near nonsense, not yielding to meaningful and yet faithful paraphrase; hence my preference for direct citations. Let me first quote the string of stipulations McTaggart begins his essay with. "Positions in time," writes McTaggart, "as time appears to us *prima facie*, are distinguished in two ways. Each position is Earlier than some, and Later than some, of the other positions. And each position is either Past, Present, or Future. The distinctions of the former class are permanent, while those of the latter are not. If M is ever earlier than N, it is always earlier. But an event, which is now present, was future and will be past." McTaggart then goes on to refer to "the series of positions running from the far past through the near past to the present, and then from the present to the near future and the far future, as the A series"; the "series of positions which runs from earlier to later" he calls "the B series"; and he concludes the passage with the stipulations "[t]he contents of a position in time are called events," and "[a] position in time is called a moment."[8] With this passage—let me list it as THE A AND B SERIES STIPULATION—the stage is set; by accepting it as a point of departure, the reader accepts an idiosyncratic—namely *timeless*—way of speaking about temporal phenomena. McTaggart now continues by pressing the point that "the A series is essential to the nature of time." As he puts it, "a B series without an A series" will not suffice to "constitute time," and, consequently, if "the distinction of past, present and future" is an illusion, then *time* must be an illusion, too. He puts forward here what might be taken as his first attempted proof of the unreality of time—I am listing it as the EVENTS NEVER CHANGE argument. This is how it runs: "It would, I suppose, be universally admitted," writes McTaggart, "that time involves change. ... A universe in which nothing whatever changed ... would be a timeless universe. —If, then, a B series without an A series can constitute time, change must be possible without an A series. Let us suppose that the distinction of past, present and future does not apply to reality. Can change apply to reality? What is it that changes?" McTaggart insists that what *cannot* change are events. "An event," as he puts it, "can never cease to be an event. ... it will always be, and has always been, an event, and cannot begin or cease to be an event." On the other hand, indicates McTaggart, events change in the sense that future events become present events, and present events become past events. I am citing an oft-quoted passage: "Take any event—the death of Queen Anne, for example—and consider what change can take place in its characteristics. That it is a death, that it is the death of Anne Stuart, that it has such causes, that it has such effects—every char-

---

[8] J. McT. E. McTaggart, "The Unreality of Time," p. 458.

acteristic of this sort never changes. ... in every respect but one it is ... devoid of change. But in one respect it does change. It began by being a future event. It became every moment an event in the nearer future. At last it was present. Then it became past, and will always remain so, though every moment it becomes further and further past." Now this kind of change, McTaggart tells us, can only be posited if we assume there to be an "A series." No time without change, and no change without the "A series."[9]

The next step to follow is the introduction of the "C series," a series that is "not temporal, for it involves no change, but only an order."[10] McTaggart puts forward an argument that purports to show that "the A series, together with the C series, is sufficient to give us time. ... It is," he writes, "when the A series, which gives change and direction, is combined with the C series, which gives permanence, that the B series can arise."[11] I do not wish to spend time on this argument here—let me call it the A PLUS C MAKE B argument—but let me just remark, however, that it is quite usual for commentaries not to take note of it, nor even of the "C series" as such. Alexander Gunn in his classic *The Problem of Time*[12] does not; Gregory Currie in his 1992 essay "McTaggart at the Movies"[13] does not; Runggaldier in his 2005 paper "Are There 'Tensed' Facts (A-Series)?"[14] does not; Kanzian in his 2005 paper "Warum McTaggarts Beweis für die Unwirklichkeit der Zeit fehlschlägt"[15] does not; Katalin Farkas in her recent *Synthese* paper "Time, Tense, Truth"[16] does not; Richard Gale in *The Blackwell Guide to Metaphysics*[17] does not. Indeed Gale in his reader *The Philosophy of Time*[18] prints McTaggart's 1927 "Time" chapter with the last pages—the pages where the "C series" are introduced—left out. McTaggart might have believed that his arguments add up to a cohesive whole, but many of his commentators clearly thought otherwise. They were right. Upon the A PLUS

---

[9] *Ibid.*, pp. 458–461.

[10] *Ibid.*, p. 462.

[11] *Ibid.*, pp. 463 f.

[12] J. A. Gunn, *The Problem of Time: An Historical and Critical Study*, New York: Richard R. Smith, 1930, pp. 345–349.

[13] *Philosophy*, vol. 67, no. 261 (July 1992), pp. 343–355.

[14] In F. Stadler and M. Stöltzner (eds.), *Time and History*, Frankfurt/M.: Ontos Verlag, 2006, pp. 77–84.

[15] In F. Stadler and M. Stöltzner (eds.), *Time and History: Papers of the 28th International Wittgenstein Symposium*, Kirchberg am Wechsel: ALWS, 2005, pp. 131–133.

[16] *Synthese*, vol. 160, no. 2 (January 2008), pp. 269–284.

[17] Cf. R. M. Gale, "Time, Temporality, and Paradox," in R. M. Gale (ed.), *The Blackwell Guide to Metaphysics*, Oxford: Blackwell Publishers, 2002, pp. 66–86.

[18] R. M. Gale (ed.), *The Philosophy of Time: A Collection of Essays*, London: Macmillan, 1967.

C MAKE B argument there follows, in the 1908 text, the digression on the possible plurality of time-series[19] I have referred to above—let me list it as THE MULTIPLE TIMES ASIDE; then comes an entirely obscure passage which I shall christen THE A SERIES ARE RELATIONS OF EVENTS half-argument, and which McTaggart concludes with the words, "[t]he relations which form the A series ... must be relations of events and moments to something not itself in the time-series. What this something is might be difficult to say";[20] and upon this half-argument then follows what might be regarded as the main argument of the essay "The Unreality of Time"—I will call it the IMPOSSIBILITY OF THE A SERIES argument.

Presenting this argument I must, again, quote McTaggart at some length. "Past, present, and future," he writes, "are incompatible determinations. Every event must be one or the other, but no event can be more than one. . . . If M is past, it has been present and future. If it is future, it will be present and past. If it is present, it has been future and will be past. Thus all the three incompatible terms are predicable of each event, which is obviously inconsistent with their being incompatible. . ." Now it might be objected, McTaggart says, that this is only a seeming incompatibility. An adversary might point out that "our language has verb-forms for the past, present, and future, but no form that is common to all three. It is never true, the answer will run, that M *is* present, past and future. It *is* present, *will* be past, and *has been*, future. Or it *is* past, and *has been* future and present, or again *is* future and *will be* present and past. The characteristics are only incompatible when they are simultaneous, and there is no contradiction to this in the fact that each term has all of them successively."[21] McTaggart retorts, and purports to prove in some detail, that this objection involves a vicious circle—let me, then, list the passages involved as the VICIOUS CIRCLE argument. I must admit that I am unable to follow him here; that I am glad every time I encounter a commentary refuting the VICIOUS CIRCLE argument; but that, generally speaking, I am not able to follow those refutations either. However, I think I am able to follow, and I take pleasure in, the remaining two arguments, or semi-arguments, that the "Unreality of Time" essay offers. These are, first, the SPATIAL MOVEMENT METAPHOR FOOTNOTE, and, secondly, the SPECIOUS PRESENT argument.

In the SPATIAL MOVEMENT METAPHOR FOOTNOTE, there are unmistakable echoes of Bradley. One is reminded of the *Principles of Logic* pas-

---

[19] J. McT. E. McTaggart, "The Unreality of Time," p. 466.
[20] *Ibid.*, p. 468.
[21] *Ibid.*

sage, "the present is no time[;] ... it is a point we take within the flow of change";[22] or of the *Appearance and Reality* passages, "[i]t is usual to consider time under a spatial form. It is taken as a stream, and past and future are regarded as parts of it... It is natural to set up a point in the future towards which all events run, or from which they arrive, or which may seem to serve in some other way to give direction to the stream. ... We think forward, one may say, on the same principle on which fish feed with their heads pointing up the stream."[23] This is how the SPATIAL MOVEMENT METAPHOR FOOTNOTE runs, and I am not quoting the passage in full: "It is very usual to present Time under the metaphor of a spatial movement. But is it to be a movement from past to future, or from future to past? ... If the events are taken as moving by a fixed point of presentness, the movement is from future to past, since the future events are those which have not yet passed the point, and the past are those which have. If presentness is taken as a moving point successively related to each of a series of events, the movement is from past to future. Thus we say that events come out of the future, but we say that we ourselves move towards the future. For each man identifies himself especially with his present state, as against his future or his past, since the present is the only one of which he has direct experience. And thus the self, if it is pictured as moving at all, is pictured as moving with the point of presentness along the stream of events from past to future."[24] I take the SPATIAL MOVEMENT METAPHOR FOOTNOTE to be understood by McTaggart as a third proof of the unreality of time, further supporting, as it were, the IMPOSSIBILITY OF THE A SERIES argument and the VICIOUS CIRCLE argument. If the passage of time were real, McTaggart must have thought, the direction of time's flow would be unambiguously given. The fact that time appears to us as a movement both "from past to future" and "from future to past" proves that that movement is, indeed, mere appearance. However, I might think of a second, rather more interesting, reading of the SPATIAL MOVEMENT METAPHOR FOOTNOTE. On this reading, Bradley, and subsequently McTaggart, have discovered what later, in the 1980s, became one of the important findings of conceptual metaphor theory, namely that there are two related, but apparently different, ways to conceptualize time: the "time-moving" and the "ego-moving" metaphors. As I will attempt to show in the last section of my talk, this finding could

---

[22] F. H. Bradley, *The Principles of Logic*, London: Oxford University Press, 1883, Bk. I, p. 53.

[23] F. H. Bradley, *Appearance and Reality*, London: Swan Sonnenschein, 1893, pp. 39 and 214.

[24] J. McT. E. McTaggart, "The Unreality of Time," p. 470.

play a significant role in a philosophical strategy designed to demonstrate the *reality* of time. Just now, however, by way of concluding the present section, let me discuss, very briefly, McTaggart's SPECIOUS PRESENT argument.

The term "specious present" was coined by E. R. Clay in 1882, and made more precise by William James in his *The Principles of Psychology*, published in 1890. As James in an oft-cited passage puts it, "the practically cognized present is no knife-edge, but a saddle-back, with a certain breadth of its own on which we sit perched, and from which we look in two directions into time. The unit of composition of our perception of time is a *duration*. . ."[25] To express it in a nutshell, the notion of the specious present is the empirically supported alternative to the age-old speculative notion of the present as a fleeting, momentary boundary between the future and the past. McTaggart of course cannot accept this latter notion, since he does not believe either in the future or in the past; while he does accept the experience of the specious present as an empirical fact. However, as he points out, "the 'specious present' varies in length according to circumstances, and may be different for two people at the same period. The event M may be simultaneous both with X's perception Q and Y's perception R. At a certain moment Q may have ceased to be part of X's specious present. M, therefore, will at that moment be past. But at the same moment R may still be part of Y's specious present. And, therefore, M will be present, at the same moment at which it is past. This," McTaggart says, "is impossible."[26] What the phenomenon of the specious present according to McTaggart demonstrates is, precisely, that time is illusory; accepting the reality of time, he tells us again by way of conclusion, leads to paradoxical results.

2. Refuted and Ridiculed

At the very beginning of his 1908 paper, McTaggart has some lines explaining that the doctrine of the unreality of time is not at all an unheard-of one; in fact "in all ages" it has been "singularly attractive"—or "singularly persistent," as he puts it in the 1927 version, in which these lines are repeated with some slight changes only. McTaggart refers to the philosophy, religion, theology and the mysticism of the East and West; mentioning, in particular, the

---

[25] W. James, *The Principles of Psychology*, vol. I, New York: Henry Holt, 1890, p. 609.
[26] J. McT. E. McTaggart, "The Unreality of Time," p. 472.

philosophers Spinoza, Kant, Hegel, Schopenhauer, and Bradley. He could also have referred to, say, Parmenides, Zeno of Elea, Augustine, or, among the moderns, Leibniz. In fact, the view that time is somehow real has always been a minority position in philosophy,[27] defended, with reservations, by Aristotle, and postulated, rather than demonstrated, by Newton. Time was real, indeed it was the ultimate reality, for Henri Bergson, writing at the turn of the nineteenth and twentieth centuries; but Bergson had, for understandable reasons, almost no impact on analytically minded philosophers. Russell even wrote a pamphlet against him in 1914. But he did influence C. D. Broad; and William James of course adored him. Be that as it may, McTaggart might well have been unaware of Bergson in 1908, and even in later years. And he was entirely right when depicting the doctrine of the unreality of time as a mainstream one. Also, he was right in maintaining that his own arguments—or his own "reasons," as he puts it[28]—for the denial of the reality of time were different from those employed by other philosophers. But he was mistaken in believing that his arguments were sound. I am now coming to the criticism that C. D. Broad, in the 1920s and 1930s, has levelled at McTaggart.

In his "Intellectual Autobiography," Broad recalls his student days at Cambridge, roughly at the time McTaggart published his *Mind* essay. McTaggart was one of the teachers "from whose lectures and personal instruction [he] gained most." However, apparently it was easier to venerate McTaggart than to build on his work. As Broad writes: "No one could fail to be impressed by his extraordinary dialectical power, his wit, and his amazing quickness in discussion; but, though he had many admirers, he had hardly any disciples. For all practical purposes Moore and Russell held the philosophical field and continued to do so for many years."[29] After teaching at St. Andrews, Dundee, and Bristol, Broad became McTaggart's successor at Trinity College in 1923. The same year, he published his book *Scientific Thought*. In this book, he takes up "the alleged difficulty that every event is past, present, and future; that these characteristics are incompatible; and that there is no way of reconciling them which does not either involve an infinite regress, in which the same difficulty recurs at every stage, or a vicious circle. This argument," Broad writes, "has been used by Dr M'Taggart as

---

[27] Cf. the section "A Nutshell History of the Philosophy of Time," in my paper "Time and the Mobile Order," in K. Nyíri (ed.), *Mobile Studies: Paradigms and Perspectives*, Vienna: Passagen Verlag, 2007, pp. 103–105.

[28] J. McT. E. McTaggart, "The Unreality of Time," p. 457.

[29] In P. A. Schilpp (ed.), *The Philosophy of C. D. Broad*, New York: Tudor Publishing, 1959, p. 50.

a ground for denying the reality of Time. It is certainly the best of the arguments which have been used for this purpose; since it really does turn on features which are peculiar to Time, and not, like most of the others, on difficulties about continuity and infinity which vanish with a knowledge of the relevant mathematical work on the subject."[30] May I just interject, though the issue has no direct bearing on our present topic, that Broad is here victim to a widespread error; as Whitrow in his magnificent book *The Natural Philosophy of Time* explains, Cantor did not solve Zeno's problem.[31] But back to McTaggart. Broad goes on by referring to the EVENTS NEVER CHANGE argument, citing the "example of the death of Queen Anne, as an event which is supposed to combine the incompatible characteristics of pastness, presentness, and futurity." Broad's comment is momentous. "[F]uturity," he says, "is not and never has been literally a characteristic of the event which is characterised as the death of Queen Anne. Before Anne died, there was no such event as Anne's death, and 'nothing' can have no characteristics."[32] The criticism levelled at McTaggart, as Broad here advances it, must be seen against the background of the latter's own philosophy of time and change. According to this philosophy, it of course makes sense to speak of the changes of *things*, but not of the changes of events.[33] "When an event, which was present, becomes past," writes Broad, "it does not change or lose any of the relations which it had before; in simply acquires in addition new relations which it *could* not have before, because the terms to which it now has these relations were then simply non-entities.—It will be observed," Broad continues, "that such a theory as this accepts the reality of the present and the past, but holds that the future is simply nothing at all. Nothing has happened to the present by becoming past except that fresh slices of existence have been added to the total history of the world." This increase in "the sum total of existence" is what Broad calls *becoming*.[34] "[T]he laws of logic," Broad maintains, "apply to a fixed universe of discourse... But the universe of actual fact is continually increasing through the becoming of fresh events; and changes in truth, which are mere increases in the *number of truths* through this cause, are logically unobjectionable." Contrary to

---

[30] C. D. Broad, *Scientific Thought*, London: Kegan Paul, Trench, Trubner & Co., 1923, p. 79.

[31] G. J. Whitrow, *The Natural Philosophy of Time*, London: Thomas Nelson, 1961, pp. 135 and 145–148.

[32] C. D. Broad, *Scientific Thought*, pp. 79 f.

[33] *Ibid.*, pp. 62 f.

[34] *Ibid.*, pp. 66 f. Any "complete analysis of the qualitative changes of things," Broad here points out, "is found to involve the coming into existence of events" (*ibid.*, p. 67).

what McTaggart believed, Broad says, "no event ever does have the characteristic of futurity," and it is because of this that the law of the excluded middle does not apply to future events.[35]

Broad repeats these same critical observations in greater detail, and in rather harsher terms, in the second volume of his book *Examination of McTaggart's Philosophy*, published in 1938.[36] The text he there analyzes, in the chapter "Ostensible Temporality," is the 1927 version of McTaggart's paper; but his remarks fully apply to the 1908 version, too. He dwells at some length on McTaggart's attempt to replace all temporal copulas by a single non-temporal one. Referring to the EVENTS NEVER CHANGE argument, and to the McTaggarian formula that if an historical event *ever* precedes another historical event by a given interval, than it *always* precedes the latter by exactly that interval, Broad says that "[n]o one but a philosopher doing philosophy" would use the verb "precedes" in this seemingly non-temporal sense. "Such phraseology," points out Broad, "would suggest that the two events are particulars which (*a*) somehow *coexist* either timelessly or simultaneously, and yet (*b*) stand timelessly or sempiternally in a certain *temporal* relation of precedence. This must be nonsense, and it is most undesirable to use phrases which inevitably suggest such nonsense. I cannot help suspecting," writes Broad, "that there is some muddle of this kind at the back of McTaggart's mind when he says that events cannot be annihilated or generated because this would be incompatible with the fact that they *always* stand in the determinate temporal relation in which they do stand to each other."[37] Coming to the end of the chapter "Ostensible Temporality," Broad sums up McTaggart's main argument against the reality of time as nothing but "a philosophical 'howler' "—a logical blunder "of the same kind as the Ontological Argument for the Existence of God."[38]

Broad's criticism of McTaggart has been very influential. It is exploited in Alexander Gunn's 1930 monograph, with its references to "the reality of changing objects," and to that "fundamental becoming" of the universe which "brings new events into being";[39] and its impact is still, or again, fully there in John Perry's paper "How Real Are Future Events?," given at the 2005 *Time and History* Kirchberg symposium.[40] Also, I would like

---

[35] *Ibid.*, pp. 81 and 83.

[36] Cambridge: Cambridge University Press, 1938. The first volume appeared in 1933.

[37] C. D. Broad, "Ostensible Temporality," in C. D. Broad, *Examination of McTaggart's Philosophy*, I am here quoting from R. M. Gale (ed.), *The Philosophy of Time*, p. 131.

[38] *Ibid.*, p. 142.

[39] J. A. Gunn, *The Problem of Time*, pp. 346 f.

[40] J. Perry, "How Real Are Future Events?," in F. Stadler and M. Stöltzner (eds.), *Time*

to single out specifically the influence Broad had on Sellars. Recalling his time in Oxford in the mid-thirties, Sellars comes to compare G. E. Moore with Broad. "I had long felt," he tells us, "that, although C. D. Broad might not be clearer than Moore, nevertheless he had a more adequate grasp of the problems they shared. I now think," Sellars says, "that this can be traced to Broad's awareness of, and technical competence in, the scientific background of these problems."[41] My impression is that, to some measure at least, it was under Broad's influence that Sellars developed his substantialist ontology of change, opposing the view that "when S changes from being φ to being ψ, S must *really* consist of an event which is φ and an event which is ψ to be the terms for the relation *earlier than*." As Sellars saw the matter, "[t]hings couldn't *consist* of events, because events were the changes of things."[42]

Let me conclude this section by briefly referring to an overlapping, but somewhat different, variety of anti-McTaggart argumentation—the ordinary-language variety—rather well represented by David Pears' 1956 essay "Time, Truth, and Inference." As Pears sees the matter, the paradoxes to which McTaggart's way of thinking about time leads are "the revenge which time takes on philosophers who deprive it of its proper means of expression, temporal verbs."[43] Focussing on the death of Queen Anne example, Pears discusses the EVENTS NEVER CHANGE argument, finding that what McTaggart actually does is to turn, as it were, "the timeless shadows of the future (and the past) into contemporary things." McTaggart achieves this by making the timeless present tense, as Pears puts it, "refer to any time when really it refers to no time."[44] McTaggart's move relies on the misconception of *the eternity of truth*, a bizarre misconception which, Pears believes, might perhaps be psychologically explained by "a strong desire to know the future,"[45] but is, nonetheless, logically untenable. There are no eternal truths, and there are no non-temporal facts. McTaggart was unable, or unwilling, to realize that "temporal predicates are unlike nontemporal predicates and that events are unlike things";[46] he was unwilling to yield to "the natural tendency of ordinary people to use temporal verbs." Had he

---

and History, Frankfurt/M.: Ontos Verlag, 2006, pp. 13–30.

[41] Wilfrid Sellars, "Autobiographical Reflections," p. 284.

[42] *Ibid.*, pp. 281 f.

[43] D. F. Pears, "Time, Truth, and Inference," in A. Flew (ed.), *Essays in Conceptual Analysis*, London: Macmillan, 1956, p. 228.

[44] *Ibid.*, p. 232.

[45] *Ibid.*

[46] *Ibid.*, p. 230.

done so, writes Pears, "his conclusion would have been not the unreality of time, but the unreality of timelessness."[47]

3. Spurious Respectability

As Broad wrote, and indeed as Wittgenstein again and again lamented, philosophers, when doing philosophy, tend to be attracted to phoney language. Even so, the magic of McTaggart's systematically skewed syntax by itself can hardly explain the continuing influence his position exerts. As I suggested by way of introduction, the explanation is, rather, that this position has become systematically conflated with the Einstein-Minkowski conception of space-time, winning, thereby, undeserved esteem. There are innumerable places where McTaggart on the one hand, and relativity theory on the other, are mentioned in one breath; let me single out just a few. In the Einstein volume in the series *The Library of Living Philosophers*, published in 1949, the chapter by Kurt Gödel begins with a note referring to McTaggart's *Mind* paper. Peter Geach in his 1965 essay "Some Problems about Time" feels it his task to indicate that there is no *real* parallel between, on the one hand, the metaphysical genius McTaggart's conviction that time is an illusion, and on the other, the "view of time that is now widely held in one form or another. In its crudest form, this view makes time out to be simply one of the dimensions in which bodies are extended; bodies have not three dimensions but four. . . . Since Einstein," Geach adds, "this sort of view has been very popular with philosophers who try to understand physics and physicists who try to do philosophy."[48] Again, Hugh Mellor in his 1998 book *Real Time II* finds it necessary to argue against, as he puts it, the often-voiced falsehood that McTaggart's so-called "B-theory explains, and may even be entailed by, a key implication of Einstein's special theory of relativity, namely that the four dimensions of spacetime are in reality all alike."[49] Physicist Julian Barbour in his book *The End of Time*, published in 2000, aimed at demonstrating that time is but an illusion, notes that some ideas in McTaggart match his own thinking, although of course the latter's arguments "are purely logical and make no appeal to physics."[50] Very telling is the way Sider begins his 2001 book, bearing the subtitle *An*

---

[47] *Ibid.*, p. 235.

[48] P. T. Geach, "Some Problems about Time," in P. F. Strawson (ed.), *Studies in the Philosophy of Thought and Action*, London: Oxford University Press, 1968, pp. 175 f.

[49] D. H. Mellor, *Real Time II* (1998), London: Routledge, 2006, p. 47.

[50] J. Barbour, *The End of Time: The Next Revolution in Our Understanding of the Universe* (1999), London: Phoenix, 2000, p. 343.

Ontology of Persistence and Time, by announcing that it "articulates and defends four-dimensionalism: an ontology of the material world according to which objects have temporal as well as spatial parts. . . . The philosophy of time defended is the B-theory, the so-called 'tenseless theory of time.' . . . The advent of Minkowski spacetime," writes Sider, "seems to have inspired much interest in [four-dimensionalism], although some versions of the doctrine predate Minkowski spacetime."[51] And to name a very recent publication: Sattig in his book *The Language and Reality of Time* opens by introducing in immediate succession first the McTaggartian notions of "A series" and "B series," and secondly the Minkowski-Einstein idea of spacetime.[52]

It is an historical coincidence that McTaggart's paper on "The Unreality of Time," published in the October 1908 issue of *Mind*, followed so closely upon Minkowski's famous *Raum und Zeit* talk, given at Cologne on September 21, 1908. But it is no more than a coincidence, having neither symbolic, nor indeed factual import. In his book *The Life and Philosophy of McTaggart*, Gerald Rochelle suggests that Einstein was aware of McTaggart's work.[53] This might easily be true, since Einstein probably had a look at Gödel's chapter in the volume I mentioned some two-three minutes ago. But Rochelle also suggests that McTaggart kept himself "in touch with major scientific thinking," and "was most interested in Einstein's work on relativity."[54] Rochelle offers no evidence for this, and I find it hard to believe. Rather, it is Broad who convinces me. This is what he writes in the 1933 "Preface" of his *Examination of McTaggart's Philosophy*: "I am inclined to think that McTaggart's complete lack of acquaintance with contemporary natural science was in certain respects a great advantage to him as a philosopher. The recent advances in physical theory have been so important and spectacular that they have only too obviously 'gone to the heads' of some eminent physicists, and have encouraged them and the public to believe that their pronouncements on technical philosophical problems, for which they have no special training or aptitude, are deserving of serious attention."

So the alleged McTaggart–Einstein connection is spurious. McTaggart's own logic is spurious. I think it is time for us to realize that McTaggart has,

---

[51] T. Sider, *Four-Dimensionalism: An Ontology of Persistence and Time*, Oxford: Clarendon Press, 2001, pp. xiii and 3.

[52] T. Sattig, *The Language and Reality of Time*, Oxford: Clarendon Press, 2006, pp. 19–22.

[53] G. Rochelle, *The Life and Philosophy of J. McT. E. McTaggart*, p. xi.

[54] *Ibid.*, p. 186.

indeed, become a thing of the past. When did he become that? If I had the courage of my convictions, I would say that this happened as early as 1908, when he formulated, in the first passages of his *Mind* paper, THE A AND B SERIES STIPULATION. But certainly it happened by 1923 at the latest, when Broad's *Scientific Thought* saw the light of day. Or if you think that is still too harsh, then let us say it happened in 2005, when several papers at the Kirchberg *Time and History* symposium, most notably the neo-Broadian one given by John Perry, offered some decisive criticisms of McTaggart's position. And if you think I am too partisan, then let us look again, but this time from a different angle, at our much-discussed parallel, between McTaggart on the one hand, and Einstein-Minkowski on the other.

Wilfrid Sellars, in his 1962 paper "Time and the World Order," made the following remark: "The non-perspectival structure which, as realists, we conceive to underlie and support perspectival temporal discourse is, as yet, a partially covered promissory note the cash for which is to be provided not by metaphysics (McTaggart's C-series), but by the advance of science (physical theory of time)."[55] May I here make three comments. First, I do not think physics by itself can give us a theory of time; metaphysics, or more broadly, philosophy, will always play a role in synthesizing the concepts with which science grasps reality. Secondly, major discoveries in science evidently influence the way philosophers think: should the notion of time become really superfluous in science, the philosophy of time would clearly not remain unaffected. Thirdly, the "partially covered promissory note" Sellars refers to, today looks increasingly unlikely to be cashed; the scientific proof of a non-temporal universe does not seem to be forthcoming. The subject of physics, forgive me the pun, is indeed a dark matter today. *Time* may yet have a future.

4. A Future for Time?

Leaving physics aside, but not losing sight of the metaphysical issue, let me now, by way of conclusion, enter the field of psychology, or, rather, of cognitive science.[56] Doubt as to the reality of time can arise because, in contrast to our sense of vision, hearing, touch, and so on, we do not seem to

---

[55] W. Sellars, "Time and the World Order," in H. Feigl and G. Maxwell (eds.), Minnesota Studies in the Philosophy of Science, vol. III: *Scientific Explanation, Space, and Time*, Minneapolis: University of Minnesota Press, 1962, p. 593.

[56] For a more detailed presentation of the argument of the present section see my paper "Film, Metaphor, and the Reality of Time," *New Review of Film and Television Studies*, vol. 7, no. 2 (June 2009), pp. 109–118.

have a *sense of time*. A magisterial presentation of the issue was provided by William James in his *The Principles of Psychology*. "Let one sit with closed eyes," he wrote, "and, abstracting entirely from the outer world, attend exclusively to the passage of time." What do we perceive? Not, as it were, a "pure series of durations," but "[o]ur heart-beats, our breathing, the pulses of our attention, fragments of words and sentences that pass through our imagination."[57] Now heartbeats, breathing, attention, etc. all involve, as James learnt from Hugo Münsterberg in 1889, the play of muscular tension and relaxation. According to Münsterberg, it is feelings in the muscles of the eye, the ear, and also muscles in the head, neck, etc., by which we estimate lengths of time. These perceptions of tension, "triggered off by real muscular contractions or by memories of the same," amount to *a direct sense of time*[58]—a physical encounter with time, we might say. As James puts it, "muscular feelings can give us the object 'time' as well as its measure."[59]

There exists a substantial research tradition which has demonstrated that to muscular sensations there correspond images of one's posture—schematic bodily images. And since the 1980s conceptual metaphor theory invites ever more detailed descriptions of how kinesthetic experiences give rise to so-called *image schemas*. An image schema, as Mark Johnson defines it, is "a recurring, dynamic pattern of our perceptual interactions and motor programs."[60] Now it is image schemata that give rise to a great number of fundamental metaphors. Recall that according to conceptual metaphor theory, metaphor is only incidentally "a device of poetic imagination and the rhetorical flourish," its essence consists in *"understanding and experiencing one kind of thing in terms of another."*[61] Time is a much-discussed topic in conceptual metaphor theory. The essential finding is that "[m]ost of our understanding of time is a metaphorical version of our understanding of motion in space."[62] Earlier in my talk I have referred to the "time-moving" and "ego-moving" metaphors. As Lakoff and Johnson point out, these metaphors are *"figure-ground reversals* of one another."[63] *Figure-ground*

---

[57] W. James, *The Principles of Psychology*, vol. I, pp. 619 f.

[58] H. Münsterberg, *Beiträge zur experimentellen Psychologie*, Heft 2: *Zeitsinn – Schwankungen der Aufmerksamkeit – Augenmass – Raumsinn des Ohres*, 1989, p. 20.

[59] W. James, *op. cit.*, p. 637.

[60] M. Johnson, *The Body in the Mind: The Bodily Basis of Meaning, Imagination, and Reason*, Chicago: The University of Chicago Press, 1987, p. xiv.

[61] G. Lakoff and M. Johnson, *Metaphors We Live By*, Chicago: University of Chicago Press, 1980, pp. 3 and 5.

[62] G. Lakoff and M. Johnson, *Philosophy in the Flesh: The Embodied Mind and Its Challenge to Western Thought*, New York: Basic Books, 1999, p. 139.

[63] *Ibid.*, p. 149.

reversal: this brings us to gestalt psychology—and to film theory. In the 1930s, German-born psychologist Karl Duncker made the following discovery with respect to "figure" and "ground" in moving visual gestalts: the "figure" tends to move, the "ground" to stand still. When observers, say, stand on a bridge and look at the moving water, their perceptions will be veridical; but when they fixate the bridge, they and the bridge may be seen as moving along the river. Duncker explained the phenomenon by pointing out that "the object fixated assumes the character of the 'figure', whereas the nonfixated part of the field tends to become ground."[64] Film theorist Rudolf Arnheim exploits this explanation to come to terms with a trivially well-known phenomenon in film. "[T]he setting photographed by the traveling camera," Arnheim points out, "is seen as moving across the screen, mostly because the viewer receives the kinesthetic information that his body is at rest. Only in extreme cases, e.g., when enough of the entire environment is seen as moving, will the visual input overrule the kinesthetic." Normally however, when our "muscular experiences" tell us that we are at rest, it is "the street [that] is seen as moving. It appears to be actively encountering the spectator as well as the characters in the film, and assumes the role of an actor among actors."[65]

There is a very clear analogy here between, on the one hand, the time-moving metaphor and film's moving road, and, on the other, the ego-moving metaphor and the spectator's perception of moving along in the film's environment. Thinking of time as passing, and seeing the road pass by on the screen, appear to have the same motor background. And the perception of time passing is no more of an illusion than the perception of the road moving towards us, or receding behind us, on film. Our everyday metaphors of the flow of time are grounded in kinesthetic image schemata depicting reality. Contrary to what McTaggart believed, the common-sense view of the reality of time can be vindicated.

---

[64] I am here quoting Duncker from R. Arnheim, *Art and Visual Perception: A Psychology of the Creative Eye* (1954), exp. and rev. ed., Berkeley: University of California Press, 1974, p. 380.

[65] *Ibid.*, pp. 379 and 381.

LOGIC AND COMPUTATION

Independence of Quantifiers in Classical Logic

Gabriel Sandu

ABSTRACT

We offer a survey of several notions of independence of quantifiers proposed in the literature. The focus will be on notions of independence which rely on game-theoretical concepts (uniform strategy, Nash equilibrium).

1. Dependence of Quantifiers

According to a common assumption which goes back to Frege, quantifiers receive their meaning from their "ranging over" a class of entities. An existential quantifier $\exists x$ prefixed to a formula $\varphi(x)$ says that the class of entities satisfying φ is not empty, while $\forall x \varphi$ says that that class coincides with the universe. It is Frege who proposed that quantifiers be interpreted by higher-order predicates.

Although the "ranging over" is an important part of the semantical job of quantifiers, it is not the only one. One of the revolutionary aspects of modern logic consists in its considering valid inferences which involve multiple quantification. When at least two quantifiers, say $\forall x$ and $\exists y$ occur in a formula, we can express that one of them is in the scope of the other: $\forall x \exists y$. The intriguing question is, obviously, how does the "ranging over" paradigm fit with the dependence of quantifiers on one another in a sequence of quantifiers: the former requires to consider one quantifier at a time, while the latter seems to demand a mechanism that "glues" these successive steps together.

The linear ordering of the quantifiers in the Frege–Tarski tradition was felt to be the natural one, because it allowed to give a relatively straightforward answer to this question in terms of a particular way of representing relations. In such an ordering one can consider one individual quantified formula at a time and still account for the dependence of a quantifier on the *previous* one in the sequence. Consider a simple sequence of quantifiers as in the formula

$$\forall x_1 \forall x_2 \exists x_3 R(x_1, x_2, x_3) \tag{1}$$

In the Frege–Tarski tradition one defines the truth-conditions of such a formula in terms of a series of one-place relations that can be represented by a multiple tree. At the top we have $|M|$ distinct children, each labelled with an individual in the domain $|M|$. For each child a_i we have again $|M|$ distinct children b_j. And for each child b_j there is again a child c_k. The first row in the multiple tree represents the first domain of R, that is, the extension of the first of its arguments; the second row represents its second domain, etc. In this way we obtain a sequence of three domains. The first domain is a separate, independent subdomain of the given universe of discourse. The second domain is relativized to individual elements of the first domain; and the third domain is relativized to individuals members of individuals in the second domain. To summarize:

> Different quantifier-prefixes allow different multiple-tree views of relations, but Frege's linear quantification limits the expressive power of quantifier prefixes to properties of relations which are discernible in a multiple-tree representation. (Sher, 1990, p. 401)

2. Independence of Quantifiers: Sher

Against the background for the dependence between quantifiers sketched above, there are several ways to conceive of independence. One such way has been for the first time considered by Sher (1990): it is simply the non-nesting, non-relativization of a quantifier to a separate domain. We consider a very simple example

$$\begin{pmatrix} \forall x_1 \\ \forall x_2 \\ \exists x_3 \end{pmatrix} R(x_1, x_2, x_3) \tag{2}$$

which represents syntactically a formula where the three (monadic) quantifiers $\forall x_1$, $\forall x_2$, and $\exists x_2$ are independent. On the present account, the truth-conditions of (2) are given by

$$\begin{pmatrix} \forall x_1 \\ \forall x_2 \\ \exists x_3 \end{pmatrix} R(x_1,x_2,x_3) \Leftrightarrow$$
$$\forall x_1 \exists x_2 \exists x_3 R(x_1,x_2,x_3) \land \forall x_2 \exists x_1 \exists x_3 R(x_1,x_2,x_3) \land \exists x_3 \exists x_1 \exists x_2 R(x_1,x_2,x_3) \tag{3}$$

The generalization should be obvious: every conjunct represents the projection of the relation R on the corresponding dimension. In our example, the first projection of R contains all the individuals, the same for the second projection, and the third projection contains at least one individual. It should be obvious that under this interpretation $\begin{pmatrix} \forall x_1 \\ \forall x_2 \\ \exists x_3 \end{pmatrix} R(x_1,x_2,x_3)$ expresses a first-order condition.

This type of independence need not be restricted to the standard quantifiers. Here is an example from natural language (due to R. May) which involves two independent cardinality quantifiers:

$$\text{Three frightened elephants were chased by a dozen hunters} \tag{4}$$

We represent it by

$$\begin{pmatrix} 3x_1 \\ 12x_2 \end{pmatrix} E(x_1) \land H(x_2) \land C(x_1,x_2) \tag{5}$$

with truth conditions

$$\begin{matrix} 3x_1 12x_2 [E(x_1) \land H(x_2) \land C(x_1,x_2)] \land \\ 12x_2 3x_1 [E(x_1) \land H(x_2) \land C(x_1,x_2)] \end{matrix} \tag{6}$$

(6) expresses the fact that the first domain of the relation

"x_1 is a frightened elephant chased by a hunter x_2"

contains three, and its second domain 12 individuals.

This sort of independence can be generalized in at least two ways. The first one is to consider independent blocks of quantifiers:

$$\begin{pmatrix} \forall x \exists y \\ \forall z \exists u \end{pmatrix} \Phi(x,y,z,u) \Longleftrightarrow \forall x \exists y \exists z \exists u \Phi(x,y,z,u) \land \forall z \exists u \exists x \exists y \Phi(x,y,z,u) \tag{7}$$

The idea is the same as before: Each row of quantifiers in the prefix is independent from the other rows.

The second generalization arises when allowing a quantifier $Q(x)$ to bind a variable in two formulas, $Q(x)(\varphi(x), \psi(x))$. This gives rise to independent quantifiers of the form

$$\begin{pmatrix} Q_1 x : & \varphi(x) \\ Q_2 y : & \psi(y) \end{pmatrix} R(x, y) \qquad (8)$$

Again, we focus on the case in which $Q_1 x$ and $Q_2 y$ are the standard quantifiers with truth-conditions (in a background model M):

$$\exists_M AB \Longleftrightarrow A \cap B \neq \emptyset, \text{ for } A, B \subseteq M$$
$$\forall_M AB \Longleftrightarrow A \subseteq B, \text{ for } A, B \subseteq M.$$

Here is the second way to generalize independence:

$$\begin{pmatrix} Q_1 x : & \varphi(x) \\ Q_2 y : & \psi(y) \end{pmatrix} R(x, y) \Longleftrightarrow$$
$$Q_1 x [\varphi(x), \exists y (\varphi(x) \wedge \psi(y) \wedge R(x, y))] \wedge$$
$$Q_2 y [\psi(y), \exists x (\varphi(x) \wedge \psi(y) \wedge R(x, y))]. \qquad (9)$$

Thus when $Q_1 x$ is the universal quantifier, and $Q_2 y$ is the existential quantifier, and A and B are the sets which are the extensions of $\varphi(x)$ and $\psi(y)$, respectively, then (9) says that:

> The relation obtained from R by restricting its first domain to A and its second domain to B, has all the universe as its first domain, and at least one individual in B in its second domain.
>
> (10)

We return below to this notion of independence when discussing some applications.

2.1. Independent Quantifiers: The Branching Interpretation (Henkin and Barwise)

There seems to be a sense of independence of quantifiers which does not reduce to Sher's notion. It has emerged from the work of Henkin (1961) but the notion we use here is due to Barwise (1979). Unlike Sher's notion, which considers domains in isolation, Barwise's analysis of independent quantifiers forces us to consider a cluster of domains in the same time in their mutual relationships.

Barwise's notion is formulated for monotone increasing (*most*) or monotone decreasing (*few*) quantifiers, for which it yields essentially a second-order property.

A (monadic) quantifier Q_1 is said to be monotone increasing, if whenever we have $Q_1 x \varphi(x)$ and $\forall x(\varphi(x) \to \psi(x))$, we also have $Q_1 x \varphi(x)$. The notion of a monotone decreasing quantifier is defined analogously. For two monotone increasing quantifiers $Q_1 x$ and $Q_2 y$, Barwise's definition of independent quantifiers is

$$\begin{pmatrix} Q_1 x \\ Q_2 y \end{pmatrix} R(x,y) \iff \\ \exists X \exists Y [Q_1 x X(x) \land Q_2 y Y(y) \land \forall x \forall y (X(x) \land Y(y) \to R(x,y))] \quad (11)$$

(Barwise, 1979, p. 63).

Barwise's definition can be generalized to formulas of the form
$$\begin{pmatrix} \forall x \exists y \\ \forall z \exists u \end{pmatrix} V(x,y,z,u):$$

$$\begin{pmatrix} \forall x \exists y \\ \forall z \exists u \end{pmatrix} V(x,y,z,u) \iff \exists R \exists S [\forall x \exists y R(x,y) \land \forall z \exists u S(z,u) \land \\ \forall x \forall y \forall z \forall u (R(x,y) \land S(z,u) \to V(x,y,z,u))] \quad (12)$$

The notion of independence arising from (11) and (12) is distinct from the one in the previous section. The former does not increase the expressive power of the individual quantifiers, while the present one does so. The distinction may be discussed with the much used Hintikka sentence

> Some relative of each villager and some relative of each townsman hate each other (13)

Hintikka has pointed out the symmetry of quantification in natural language:

> This [example] may ... offer a glimpse of the ways in which branched quantification is expressed in English. Quantifiers occurring in conjoint constituents frequently enjoy independence of each other, it seems, because a sentence is naturally thought of as being symmetrical semantically vis-à-vis such constituents. (Hintikka, 1973, p. 344)

Now if (13) were to be represented by

$$\begin{pmatrix} \exists y : \forall x (V(x) \to R(x,y)) \\ \exists u : \forall z (T(z) \to R(z,u)) \end{pmatrix} H(y,u) \land H(u,y) \quad (14)$$

and quantifiers were taken to be independent in the sense of the previous section, then the truth-conditions of (14) would be given by (9): the first domain of the relation $H(y,u)$ of mutual hatred includes at least one relative for each villager and its second domain includes at least one relative of each townsman.

This account of the truth-conditions of (14) is not the one favoured by Hintikka, however. In order to obtain Hintikka's favourite interpretation, we have to go for the notion of independence expressed by (12). (12) ascribes a so-called *massive nucleus* interpretation to (14): a set containing at least one relative of each villager and one relative of each townsmen such that each villager-relative in the nucleus hates all the townsman-relatives in it, and *vice versa*. Sher's independent quantifier interpretation requires only that at least one relative of each villager hates at least one relative of each townsman. The present interpretation expresses essentially a second-order property.

3. Game-Theoretical Notions of Independence

3.1. Extensive Games for First-Order Logic (GTS)

On Henkin's account, the interpretation of a sequence of standard quantifiers can be displayed in terms of the strategic interaction of two players in a game:

> Imagine, for instance, a "game" in which a First Player and a Second Player alternate in choosing an element from a set I; the infinite sequence generated by this alternation of choices then determines the winner. If we let τ to denote the class of all those sequences for which the First Player is the winner, then the formula $[\exists v_1 \forall v_2 \exists v_3 \forall v_4 \ldots (\tau v_1 v_2 v_3 \ldots)]$ simply expresses the fact that the First Player has a winning strategy. (Henkin, 1961, p. 179)

Henkin's idea is to give quantifiers a game-theoretical content. The universal and the existential are now identified with two players in a semantic game. Their moves consist in choosing individuals from a universe of discourse. An occurrence of $\exists x$ prompts the existential player to find an individual a to be the value of 'x' and $\forall x \varphi$ prompts a similar move by the universal player. In this new setting, the game associated with the sentence $\forall x \exists y F(x,y)$ gives rise to a sequence of choices: a choice of an individual a by the universal player followed by a choice of an individual b by the existential player. If a and b that satisfy $F(x,y)$, then the existential player wins; if not, the universal player wins.

The scopal dependences of quantifiers in a formula can be seen to express the knowledge of the players in the relevant games, under the tacit assumption that "being in the scope of a quantifier" is "knowing the move prompted by it." In classical game theory such knowledge is technically encoded in the notion of the information set of a player for a particular move, and finds its way in the definition of a player's strategy: a function defined on all appropriate information sets in the game, whose value, for a given argument, gives the player the choice to be made. Thus a strategy for the existential player in the game $\forall x \exists y F(x,y)$ (played on a model M which interpretes F) is a function $f : |M| \rightarrow |M|$. It is a winning strategy if it always leads to a win, i.e., $(a, f(a))$ satisfies $F(x,y)$. A strategy for the universal player is an individual b. It is winning, if for any choice a of his opponent, it leads to a win: (b,a) satisfies $F(x,y)$.

Finally, we identify truth and falsity in M with the existence of a winning strategy for the existential and respectively the universal player in the appropriate game. When the sentence the game is associated with is in prenex normal form, the functions which are the winning strategy of the existential player (when they exist) are known as Skolem functions; and the strategies which are the winning strategies of the universal player (when they exist) are known as Kreisel counterexamples.

The game-theoretical setting brings in a correlation between (material) truth (falsity), strategies in extensive games and Skolem functions (Kreisel's counterexamples). It construes the dependence between quantifiers as informational dependence in semantical games.

This way of looking at a sequence of interdependent quantifiers is quite natural when one looks at the mathematical vernacular. Here is a standard definition of a continuous function:

A function $y = f(x)$ is continuous, if for every x_0 in the domain of the function we have: *given a* number α however small, we can find ε such that $|f(x) - f(x_0)| < \alpha$, *given any* x such that $|x - x_0| < \varepsilon$.

In agreement with (a), "Given a" and "given any" have the game-theoretical force of the universal quantifier, and "we can find" that of the existential quantifier. The scopal dependences between the quantifiers in the definition are displayed by the first-order formula (we add also an universal quantifier which corresponds to "every x_0"):

$$\forall x_0 \forall \alpha \exists \varepsilon \forall x [(|x - x_0| < \varepsilon) \rightarrow (|f(x) - f(x_0)| < \alpha)] \qquad (15)$$

In agreement with our earlier remarks, the dependence of the existential quantifier on the first two universal quantifiers is spelled out in the specification of the domain of the strategy function of the existential player in the appropriate game: (15) is true if and only if there is a Skolem function g such that

$$(|c - x_0| < g(x_0, \alpha)) \to (|f(x) - f(x_0)| < a) \qquad (16)$$

Symmetrically, the dependences of the universal quantifiers on each other and on the existential quantifier are spelled out in the falsity conditions of (15):

$$\exists x_0 \exists k \exists h \forall \varepsilon [(|h(x_0, \alpha, \varepsilon) - x_0| < \varepsilon) \wedge (|f(h(x_0, \alpha, \varepsilon)) - f(x_0) \geq \alpha)] \qquad (17)$$

where x_0, k and h are Kreisel counterexamples.

3.2. Independence as Uniformity of Strategies in Extensive Games

In Hintikka and Sandu (1989), languages for imperfect information, or Independence-friendly languages, (*IF*-languages, for short) were introduced, with the purpose of having a general framework for expressing in linear notation quantifier dependences and independences. These languages contain quantifiers of the form

$$(\exists y/W)\varphi, (\forall y/W)\varphi$$

where W is a finite set of variables. Thus

$$\forall x \forall z (\exists y/\{z\})(\exists w/\{x,y\}) R(x,z,y,w) \qquad (18)$$

is intended to express the fact that the first existential quantifier does not depend on $\forall z$ (but only on $\forall x$) and the second existential quantifier does not depend on $\forall x$ and on $\exists y$ (but only on $\forall z$). This formula expresses in a linear notation the kind of independence that Henkin (1961) displayed in the branching form:

$$\begin{pmatrix} \forall x & \exists y \\ \forall z & \exists w \end{pmatrix} R(x,z,y,w) \qquad (19)$$

and defined by

$$\begin{pmatrix} \forall x & \exists y \\ \forall z & \exists w \end{pmatrix} R(x,z,y,w) \Leftrightarrow \\ \exists f \exists g \forall x \forall z R(x,z,f(x),g(z)) \qquad (20)$$

Now in order to illustrate the idea of uniformity, let us focus on the simplest pattern of independent quantifiers:

$$\forall x (\exists y/\{x\}) F(x,y)$$

Game theoretically this formula expresses the fact that the choice of a witness for the existential quantifier is independent of (does not depend on) the choice of a value for the universal quantifier. This is implemented by requirement that any strategy function f of the existential player be uniform in (the value of) x: for any individuals a and b in the relevant universe, $f(a) = f(b)$. But then the existence of a winning strategy in the game comes down to the existence a single individual c such that for all $a : F(a,c)$. Thus the above sentence is truth equivalent to $\exists y \forall x F(x,y)$.

The example illustrate a pattern of switching of quantifiers, $\forall \exists \to \exists \forall$ which is well known in mathematics. For instance, one makes a distinction between the distribution on a test (every student has received a score), and its uniform distribution (there is a score such that every student has received it.)

Our earlier example of the definition of a continuous function provides an interesting case of uniformity involving four quantifiers. According to that definition, a function f is continuous if for every x_0 in its domain we have: given any number $\varepsilon > 0$, there exists $\delta > 0$ such that $|f(x) - f(x_0)| < \varepsilon$ whenever $|x - x_0| < \delta$, for any x. In this definition delta depends on both epsilon and on the point x_0. A function in which delta depends only on epsilon is called uniformly continuous. The difference between simply and uniformly continuous comes down to a switching of quantifiers according to the pattern $\forall\forall\exists\forall \to \forall\exists\forall\forall$. But we can also look at it drawing on the idea of uniform strategy functions. In that context the distinction between continuous and uniformly continuous comes down to the distinction between $\forall x \forall y \exists z \forall w$ and $\forall x \forall y (\exists z/\{x\}) \forall w$. In the later game it is required that the existential player's strategy f be uniform in x (but not in y): for any two sequences (a,b), (c,d) of elements in the relevant universe, $f(a,b) = f(c,d)$, whenever $b = d$. But this is equivalent to (the values of) z depending only on (the values of) y, that is, to the strategies of the existential player being functions of y.

In both examples, the independence of quantifiers, analyzed via the notion of the uniformity of the appropriate strategy functions, reduces to the switching of quantifiers. This might give the impression that quantifier independence qua uniformity is reducible to quantifier switching. But this is not always, as witnessed by the following variations on the patterns of four quantifiers:

$$\forall x \forall y (\exists z/\{x\})(\exists w/\{y,z\})\varphi \quad \text{and} \quad \exists w \forall x \exists y (\exists z/\{x\})\psi$$

In the formula on the left, we want the first existential quantifier to depend only on $\forall y$, and the second to depend only on $\forall x$. In the present jargon, the first strategy function of the existential player must be uniform in its first argument, and the second function must be uniform in both its second and third argument. But no switching of quantifiers will have that effect. Similarly, in the formula on the right we want third strategy function of the existential player to be uniform in its second argument. Again, no quantifiers switching will have that effect. Of course it might still be possible to find ordinary (first-order) formulas which are equivalent with the intended interpretation. But this is not always true. There are choices of φ and ψ for which the above formulas express properties which are not first-order definable. The uniformity account is a conservativity extensive of the game-theoretical account, when we treat the standard quantifiers $\exists x$ and $\forall y$ as $\exists x/\emptyset$ and $\forall y/\emptyset$, respectively.

Before moving to the next notion of independence, let us give two applications of independence as uniformity of strategies in extensive games.

3.3. Lewis' Signalling Games

Signaling in game theory is any convention of play whereby one partner informs the others of his holdings or desires. Von Neumann and Morgenstern observe that it occurs in Whist and Bridge. Hodges remarks: "In these games there are two players and each of the players consists of two partners. ... In general a pair of partners knows different things about the state of the play but one of them can use his public moves to signal information about the other partner." (Hodges, 1997, p. 549).

The phenomenon of signalling is typically encountered in Lewis' signalling games. These are games used to illustrate how words acquire meaning. A signalling game is played by two players, a Sender and a Receiver. The former signals to a Receiver that a certain state m obtains, using a message f. The Receiver does not see the state m, but after receiving the message f, he tries to "interpret" it, i.e., to associate it to the initial state.

Let M denote the set of situations, and F the set of messages. Thus

- The Sender's strategy is any function

$$S : M \rightarrow F$$

- The Receiver's interpretation strategy is any function

$$H : F \rightarrow M$$

- A solution to the signaling game is a strategy pair (S,H) which is successful

$$S(m) = f \Longrightarrow H(f) = m$$

INDEPENDENCE OF QUANTIFIERS IN CLASSICAL LOGIC

- A strategy pair (S,H) is a signalling convention if it is successful and in addition
$$m \neq m' \implies S(m) \neq S(m')$$

- If (S,H) is a signalling convention, the message f signals or "means" that m if $S(f) = m$.

We notice that the condition $m \neq m' \implies S(m) \neq S(m')$ requires that there be as many messages ("signals") as there are states. When this condition is fulfilled, the Sender and the Receiver can coordinate their actions so that they arrive at a signalling convention, despite the fact that the Receiver does not see the initial states. (Of course, there are many conventions, hence the choice between them is a tricky matter which is not going to occupy us here.)

Here is an example. Suppose there are 4 situations $\{S_1, \ldots, S_4\}$ and 4 signals $\{f_1, \ldots, f_4\}$. Here are two signalling conventions:

- Convention 1

$$S_1 \Rightarrow f_1 \Rightarrow S_1$$
$$S_2 \Rightarrow f_2 \Rightarrow S_2$$
$$S_3 \Rightarrow f_3 \Rightarrow S_3$$
$$S_4 \Rightarrow f_4 \Rightarrow S_4$$

According to this convention, f_1 "means" S_1, f_2 "means" S_2, etc.

- Convention 2

$$S_1 \Rightarrow f_2 \Rightarrow S_1$$
$$S_2 \Rightarrow f_1 \Rightarrow S_2$$
$$S_3 \Rightarrow f_4 \Rightarrow S_3$$
$$S_4 \Rightarrow f_3 \Rightarrow S_4$$

According to this convention f_1 "means" S_2, f_2 "means" S_1, etc.

This informal, Lewis type of signalling game can be "defined" by an IF-sentence in a straightforward way. We need two existential quantifiers, one for the Sender ($\exists y_1$), and one for the Receiver ($\exists y_2$). The sentence we are looking for has the form

$$\forall x \exists y_1 (\exists y_2 / \{x\})(M(x) \to \theta(x, y_1, y_2) \land x = y_2)$$

where 'M' represents the set of situations in the language, and $\theta(x, y_1, y_2)$ is a Boolean combination of first-order formulas ensuring that $\exists y_1$ ranges over the set of messages and $\exists y_2$ ranges over the set of states. The conjunct $x = y_2$ ensures that the Receiver must be able to identify the initial state. $(\exists y_2 / \{x\})$ expresses the fact that the Receiver does not see the state signalled by the Sender.

3.4. Extended Signaling Games: Inventing the Code (Skyrms)

In Lewis' signalling games the Sender can send as many signals as there are states. In a series of papers Brian Skyrms has asked: What happens if there are less signals than states? Can the Receiver still recover them? The answer is positive but we need more Senders. Skyrms gives an example which involves 4 states, 2 Senders (each of them having two signals), and one Receiver. More exactly, the scenario is like this:

- There are two Senders and 4 situations $\{S_1, \ldots, S_4\}$.

- Sender 1 can send only two signals, a red or a green flag; Sender 2 can send only two signals, a yellow or a blue flag.

- A signalling system in the extended Lewis' signaling game is a combination of strategies of the three players, two Senders and one Receiver, such that the Receiver always does the right thing for the state.

The question is of course how do the three players coordinate their actions so that the Receiver can recover the initial situations. The reasonable thing to do is for one and the same message to be used to signal two distinct states. Here is an example of a signalling convention:

$$
\begin{array}{ccccccc}
S_1 & \Rightarrow & \text{Red} & \Rightarrow & \text{Yellow} & \Rightarrow & S_1 \\
S_2 & \Rightarrow & \text{Red} & \Rightarrow & \text{Blue} & \Rightarrow & S_2 \\
S_3 & \Rightarrow & \text{Green} & \Rightarrow & \text{Yellow} & \Rightarrow & S_3 \\
S_4 & \Rightarrow & \text{Green} & \Rightarrow & \text{Blue} & \Rightarrow & S_4
\end{array}
$$

Unlike the previous signalling game in which one signal "means" one state, here one signal "means" a disjunctive state, e.g. "Red" means S_1 or S_2, "Blue" means S_2 or S_4, etc.

This sort of game illustrates nicely the emergence of the logical connectives. It can be "defined" in IF logic by a sentence of the form

$$\forall x(\theta(x) \rightarrow (\exists y_1 \exists y_2 (\exists y_3/\{x\})(\theta'(x, y_1, y_2, y_3) \wedge x = y_3)))$$

where $\theta(x)$ and $\theta'(y_1, y_2, y_3)$ are the Boolean formulas specifying the relevant ranges for the four quantifiers.

4. Independence as Nash Equilibria in Strategic Games

There is still a more general way to look at the independence between quantifiers in a game-theoretical setting. We prefer to introduce it by way of an example. On the conception of independence as uniformity of strategies, each of the game associated with the sentences

$$\forall x(\exists y/\{x\})x = y$$
$$\forall x(\exists y/\{x\})x \neq y$$

is nondetermined in every domain which contains at least two individuals. This is easy to see: the uniformity requirement reduces any strategy function for the existential player to an individual. Thus the first sentence is true iff there is an individual y which is equal to any x chosen by the opponent. But this is impossible. And it is false if the universal player can choose an element x which is distinct from any element y chosen by the opponent. But this is impossible too. The same reasoning applies to the second example. We conclude that, on the uniformity interpretation, the two formulas are indistinguishable on any universe which has at least two individuals.

However, the two formulas may be distinguished if we adopt a more fine grained way to distinguish semantic values, as suggested by Ajtai (Blass and Gurevich, 1986). Let us think of the two players representing the two quantifiers as choosing simultaneously elements of the relevant domain and of their odds to win it. The odds of the existential player to win the second game must increase, and her odds to win the first game must decrease, as the universe increases. That is, if the universe consists of 3 individuals, the existential player wins the first game with a probability of $1/3$ and the second game with a probability of $2/3$. And if the universe consists of 4 individuals, then the respective probabilities are $1/4$ and $3/4$. More generally, when the universe contains n elements, the first sentence has "truth-value" $1/n$ and the second value has truth-value $n - 1/n$. In order to account for such facts, we shall introduce a new kind of independence. We formulate it in the context of strategic games with mixed strategies and make appeal to solution concepts in game theory (Nash equilibria). This account has been developed by Sevenster (2006).

We start out by some preliminary definitions from game theory, but we shall limit ourselves to the strategic games associated with the two *IF* sentences above. For φ any of the two formulas above, and M a finite model (which reduces to a set), recall the extensive (semantic) game of imperfect information $G(M, \varphi)$ associated with M and φ. We let $S_\exists$ be the set of strategies of the existential player and $S_\forall$ be the set of strategies of the universal player in $G(M, \varphi)$. Given the uniformity requirement, both of them reduce to the set M. For $i \in \{\exists, \forall\}$, we define the utility function $u_i : S_\exists \times S_\forall \to \mathbb{R}$, for player i:

- $u_\exists(s,t) = 1$ if playing s against t in $G(M,\varphi)$ yields a win for the existential player, (i.e., $s = t$) and $u_\forall(s,t) = 0$ otherwise.

- $u_\forall(s,t) = 1$ if playing s against t in $G(M,\varphi)$ yields a win for the universal player, (i.e., $s \neq t$) and $u_\forall(s,t) = 0$ otherwise.

The tuple $((S_i)_{i \in \{\exists,\forall\}}, (u_i)_{i \in \{\exists,\forall\}})$ is called a strategic game. We denote it by $\Gamma(M,\varphi)$. It is a zero-sum game. We shall restrict ourselves to the case in which the strategy sets S_i are finite.

The value of an IF strategic game $\Gamma(M,\varphi)$ is defined in terms of its equilibrium pairs of mixed strategies.

A mixed strategy σ_i for player i is a probability distribution over S_i. That is, σ_i is a function of type $S_i \to [0..1]$ such that $\sum_{s \in S_i} \sigma_i(s) = 1$. A mixed strategy $\sigma_i : S_i \to \{0,1\}$ is *uniform* over $S'_i \subseteq S_i$ if it assigns equal probability to all $s_i \in S'_i$. We say that σ_i is uniform if it is uniform over its domain S_i. Given a mixed strategy σ for Eloise and a mixed strategy τ for Abelard, the *expected utility* for player i is given by

$$U_i(\sigma, \tau) = \sum_{s \in S_\exists} \sum_{t \in S_\forall} \sigma(s)\tau(t) u_i(s,t).$$

Let $\Gamma(M,\varphi)$ be a strategic game as defined above. Let (σ_0, σ_1) be a pair of mixed strategies. We call (σ_0, σ_1) an *equilibrium* in Γ, if

For every mixed strategy σ'_1, $U(\sigma_0, \sigma_1) \geq U(\sigma_0, \sigma'_1)$
For every mixed strategy σ'_0, $U(\sigma_0, \sigma_1) \geq U(\sigma'_0, \sigma_1)$.

In other words, (σ_0, σ_1) is an equilibrium if no player benefits from unilateral deviation.

A well known result shows that every finite zero-sum, two-player game has an equilibrium in mixed strategies. This is von Neumann's well-known *Minimax Theorem*. Moreover, it is well known that every two equilibriums in zero-sum, two-player games have the same expected utility. But then we can talk about the expected utility of the two players in the game: it is, for each player, the utility returned by any equilibrium in the game.

Let $\Gamma(M,\varphi) = ((S_i)_{i \in \{\exists,\forall\}}, (u_i)_{i \in \{\exists,\forall\}})$ be a strategic IF game. We define the *truth value* of φ in M as the expected utility of the existential player in $\Gamma(M,\varphi)$.

We illustrate the new semantics with an example.

Let $\varphi = \forall x (\exists y/x) x = y$ and M be a model with, say, five objects $a_1, \ldots, a_5$. As pointed out, each of the two players has five strategies in $\Gamma(M,\varphi) = ((S_i)_{i \in \{\exists,\forall\}}, (u_i)_{i \in \{\exists,\forall\}})$:

$$S_\exists = S_\forall = \{a_1, \ldots, a_5\}$$

The utility functions $u_\exists$ and $u_\forall$ are conveniently depicted by the following matrix (we indicate only the utilities of the existential player):

	a_1	a_2	a_3	a_4	a_5
a_1	1	0	0	0	0
a_2	0	1	0	0	0
a_3	0	0	1	0	0
a_4	0	0	0	1	0
a_5	0	0	0	0	1

Let σ (resp. τ) be the uniform strategy over $S_\exists$ (resp. $S_\forall$):

$$\sigma(a_i) = \tau(a_i) = \tfrac{1}{5}, \text{ for } i = 1,\ldots,5$$

It may be checked that (σ, τ) is an equilibrium with value $1/5$.

In the same way, we can show that the truth value of $\varphi' = \forall x (\exists y/x) x \neq y$ is $(n-1)/n$.

The above framework can be generalized to arbitrary IF sentences. It can be shown that the truth value an IF sentence φ on a finite structure M is 1 iff the existential player has a winning strategy in the extensive game associated with M and φ. The present interpretation is thus a conservative extension of the uniformity interpretation. Sandu and Sevenster (forthcoming) contains further results of the equilibria interpretation, and Mann, Sandu and Sevenster (forthcoming) develop this account in full details.

References

Abramsky, S. and Väänänen, J. (to appear). From IF to BI: A tale of dependence and separation.

Barwise, J. (1979). On branching quantifiers in English. *Journal of Philosophical Logic*, 8, 47–80.

Blass, A. and Gurevich, Y. (1986). Henkin quantifiers and complete problems. *Annals of Pure and Applied Logic*, 32(1), 1–16.

Dechesne, F. and Janssen, T. M. V. (2006). Signaling in IF games: A tricky business. In J. F. A. K van Benthem et al. (Eds.), *The Age of Alternative Logics: Assessing the Philosophy of Logic and Mathematics Today* (223–242). Kluwer Academic Publisher, Dordrecht.

Henkin, L. (1961). Infinitistic methods. In *Proceedings of the Symposium on Foundations of Mathematics* (167–183). Pergamon Press, Warsaw.

Hintikka, J. (1973). Quantifiers vs. quantification theory. *Dialectica*, 27, 329–358.

Hintikka, J. (1996). *The Principles of Mathematics Revisited*. Cambridge University Press, Cambridge.

Hintikka, J. and Sandu, G. (1989). Informational independence as a semantical phenomenon. In *Logic, Methodology and Philosophy of Science VIII (Moscow, 1987)*, volume 126 of *Stud. Logic Found. Matm* (571–589). North-Holland, Amsterdam.

Hodges, W. (1997). Compositional semantics for a language of imperfect information. *Logic Journal of the IGPL*, 5, 539–663.
Mann, A., Sandu, G., and Sevenster, M. *Logic and Games*. Cambridge University Press, Cambridge. (forthcoming).
May, R. (1989). Interpreting logical form. *Linguistics and Philosophy*, 12, 387–435.
Sandu, G. and Sevenster, M. (to appear in: *Annals of Pure and Applied Logic*). Equilibrium semantic.
Sevenster, M. (2006). *Branches of Imperfect Information: Logic, Games, and Computation*. PhD thesis, ILLC, University of Amsterdam, Amsterdam.
Sher, G. (1990). Ways of branching quantifiers. *Linguistics and Philosophy*, 13, 390–422.
Skyrms, B. (2006). *Signals*. Presidential Address, Philosophy of Science Association.
Väänänen, J. (2007). Dependence logic. In *London Mathematical Society. Student Texts*, volume 70. Cambridge University Press, Cambridge.

Metaphysics

Carving Nature at Our Joints

Achille C. Varzi

1. The Natural/Artificial Distinction

Driving across Europe has become a remarkable experience. You see a sign that says 'Austria,' or 'Slovakia,' or 'Poland,' and just drive through. No customs barrier, no passport control—just a sign. You say 'Ah—Poland!' and carry on; the sign could be a hundred yards further and it would make no difference. Yet, by crossing that line you enter a different world-district, magically separated from its surroundings—you enter a region where people suddenly speak another language, use a different currency, rely on their own authorities, share a different heritage, and struggle to solve *their* problems and to improve the quality of *their* common life. The line *is* there, even if you don't see it. That sign conceals a long history, perhaps even a thread of blood, though all you see today is a spread of asphalt, souvenir shops, money exchange boots, gas stations, abandoned customs houses. It is more difficult to get that feeling as you drive across the United States of America. Most drivers feel nothing at all as they pass the border between Wyoming and Idaho, for instance—a line whose embarrassing geometric straightness says very little about its history (or says it all). Yet even there something remarkable happens, and Idahoans are proud of their license plates just as Wyomingites are proud of their own. Such is the magic of boundary lines: they are thin, yet powerful; they separate, and thereby unite; they are invisible, yet a lot depends on them, including one's sense of belongingness to a country, a people, a place; they are abstract, in a way, yet people take them seriously and some states expend huge sums of money and sacrifice their soldiers' lives to protect them, or to re-draw them properly.

Not all boundaries are so magic, though—are they? As I was flying over Europe on my way from the States, I did not in fact see any national boundary. Nor did I see the boundaries of Paris, for instance, or those of Switzerland. But I did have the clear and distinct impression of seeing *other* boundaries: the coastline of France along the Atlantic, for instance, or the shoreline of the Geneva Lake. I saw the edges of Long Island when my flight took off. And I saw the boundaries of the Ticino river as I was landing near Milan, Italy. In his celebrated Romanes Lecture of 1907,[1] the British Viceroy of India, Lord Curzon of Kedleston, introduced an important distinction in this regard, a distinction that is in many ways part of common sense, and that geographers have officially embraced ever since. And it fits the bill. It's the distinction between *artificial* boundaries, or frontiers, on the one hand, and *natural* boundaries, or frontiers, on the other.

The boundaries I didn't see would be of the first sort. Most national and state borders are artificial insofar as they are our own making, the product of human decisions and stipulations, an expression of collective intentionality that translates into political, social, and legal agreements whereby it is determined where a certain territory begins and where it ends. So, too, are the boundaries of many other geographic entities, such as plateaus or wetlands or areas of a given soil type, though these may be induced by cognitive or cultural processes, or by scientific stipulation, rather than by legal or political practices. Such artificial boundaries may be drawn with great accuracy (a national border) or left somewhat vague, fuzzy, underspecified (the boundaries of a desert); it depends on the importance we attribute to the relevant demarcations, on the role they play in our lives. But whether sharp or vague, they all qualify as "artificial" precisely because the demarcations are human-induced; they need not correspond to any genuine, physical or otherwise objective differentiations in the underlying territory. They are *de dicto*, so to speak, not *de re* boundaries.

Geographic boundaries of the second sort—the natural, or *de re* boundaries—would by contrast be characterized precisely by their apparent independence from our organizing activity. We can stipulate that a certain portion of the Geneva Lake belongs to Switzerland and the rest to France, and the dividing line will be an artifact. But the shoreline—the border of the whole lake—does not seem to depend on us. It's there regardless, it exists "on its own." Ditto for the boundaries of certain political or administrative entities, such as the Region of Sicily, whose limits are for the most part iden-

---

[1] Lord Curzon of Kedleston (1907). *Frontiers. The Romanes Lecture.* Clarendon Press, Oxford.

tified with the limits of the Sicilian *island*; or such as Spain, which although connected to the continent, is separated from it by the admirably fashioned Pyrenean wall ("the most obvious of features—wrote Joseph Calmette—the plainest of lines, designed by nature in her boldest manner").[2] Artificial boundaries may be subject to controversy. They can be ignored or deleted, and thereby go out of existence; they can be drawn anew, and thereby come into being. Not so with natural boundaries. We are free to ignore them for certain purposes, but we cannot ask a cartographer to omit them from a map of the world. In a physical map we may omit all political boundaries; but a political map will perforce include all physical boundaries—at least, physical natural boundaries that are visible at the relevant scale.

Now, it is, of course, an open question whether Lord Curzon's intuitive distinction is well grounded. That is precisely the question I want to address. First, however, let me emphasize that the question does not only arise in relation to the geographic world that we find depicted in ordinary maps and atlases. It also arises, for instance, in the smaller-scale world featured in a cadastre. Here, too, the parceling of land into real estate is not simply a geometrical affair. In some cases it would seem to rely on natural, pre-existing physical discontinuities, such as creeks, rocks, cliffs, or ditches; in other cases, it is crucial that people *believe* that whoever fenced off a plot of land is the person who actually owns it, so collective intentionality appears to be necessary to explain the difference between landed property and raw land. And what goes for the cadastre goes for everything. Boundaries play a central role at *any* level of representation or organization of the world around us, and so does the relevant artificial/natural distinction. We think of a boundary *every* time we think of an object as of something separated from its surroundings. There is a boundary (artificial) separating my part of the desk from my colleague's, my head from the rest of my body, or the sirloin from the ramp on a butcher's beef chart; there is a boundary (natural) demarcating the interior of this apple from its exterior, the hole from the donut, or shadow from light. Events, too, have boundaries, including temporal ones, and the distinction appears to apply equally well: the end of the war or my turning 18 years old would be examples of artificial boundaries; my birth and death or the point in the cooling process when water begins to solidify would be obvious candidates for natural boundaries. Even abstract entities, such as concepts or properties, may be said to have boundaries of their own. Those expressed by disjunctive predicates such as 'emerose' and 'grue,' or

---

[2] Calmette, J. (1913). La frontière pyrénéenne entre la France et l'Aragon. *Revue des Pyrénées*, 25, 1–19.

by phase-sortals such as 'student' and 'jobless,' would have *de dicto* boundaries. Those expressed by substance sortals and so-called natural kind terms, such as 'cow' or 'water,' would have genuine, *de re* boundaries.

It is not an exaggeration to say that boundaries are at work in articulating every aspect of the reality with which we have to deal. They stand out in every map we draw of the world—not only the world of geography, but the world of nature at large, as well as the secular world that emerges through the weaves of our social and individual practices. And this ubiquitousness of boundaries goes hand in hand with that of the artificial/natural distinction, the apparent contrast between merely *de dicto* and genuinely *de re* demarcations, the opposition—in Barry Smith's more recent terminology—between *fiat* articulations and *bona fide* joints of reality.[3] It is not, therefore, an exaggeration to say that our question bites deeply: How tenable is the distinction? And how does the answer affect our overall metaphysical picture of the world? How does it affect our understanding of the identity and survival conditions of the very things that boundaries demarcate? For obviously, once the *de dicto* / *de re* distinction has been recognized, it can be drawn across the board: not merely in relation to boundaries but also in relation to all those entities that may be said to *have* boundaries. If a certain entity enjoys natural boundaries, it is reasonable to suppose that its identity and survival conditions do not depend on us; it is a *bona fide* entity of its own. By contrast, if its boundaries, or some of its boundaries, are artificial—if they reflect the articulation of reality that is effected through human cognition and social practices—then the entity itself is to some degree a *fiat* entity, a product of our worldmaking. This is not to say that *fiat* entities of the second sort are imaginary or otherwise unreal. As Frege put it, the objectivity of the North Sea "is not affected by the fact that it is a matter of our arbitrary choice which part of all the water on the earth's surface we mark off and elect to call the 'North Sea.'"[4] It does, however, mean that such entities would only enjoy an individuality as a result of our cognitive and/or social practices, like the cookies carved out of the dough: their *objectivity* is independent, but their *individuality*—their being what they are, including their having the identity and survival conditions they have—depends on the

---

[3] Smith, B. (2001). On Drawing Lines on a Map. In A. U. Frank and W. Kuhn (Eds.), *Spatial Information Theory. Proceedings of the Third International Conference.* Springer, Berlin, pp. 475–484. Extended version as: Smith, B. (2001). Fiat Objects. *Topoi,* 20, 131–148.

[4] Frege, G. (1884). *Die Grundlagen der Arithmetik.* Köbner, Breslau (Eng. trans. by J. L. Austin (1950): *Foundations of Arithmetic.* Basil Blackwell, Oxford), at §26.

baker's action. (In the terminology of John Searle, they are "social objects": from a God's eye point of view you don't see a cookie just as you don't see a nation or the North Sea; you only see *us* treating certain chunks of reality as cookies, nations, or the North Sea.)[5]

2. A Tenable Distinction?

So, how tenable is the distinction? Well, it seems to me that the existence of *de dicto* boundaries, hence of *fiat* entities, is uncontroversial. And we can be more fine-grained, if we wish. We can, for instance, draw a further distinction between *fiat* entities that owe their existence to collective intentionality, or to the beliefs and habits of a community, as with geopolitical or social entities at large, and *fiat* entities that emerge instead from the cognitive acts of single individuals, beginning with perception, which as we know from our experience of Seurat paintings has the function of articulating reality in terms of continuous boundaries even when such boundaries are not genuinely present. Individual *fiats* are more ephemeral than social *fiats* because they are rigidly dependent (on *these* acts, taking place *now*) rather than generically dependent (on the existence of relevant acts of a certain kind). We can also distinguish between those cases where a *fiat* entity is carved out as a proper part of a larger entity, as with our initial examples, and those cases where a *fiat* entity is obtained by circumcluding a number of smaller entities within larger wholes, as with Polynesia, the constellation Orion, or a baseball game, or as when we describe the world as consisting of forests, physaliae, schools of fish, swarms of bees, fleets of ships, pairs of shoes.[6] Regardless of all such refinements, it is clear that in each such case *de dicto* boundaries are at work in articulating the reality with which we have to deal, and the entities obtained thereby are themselves of the *fiat* sort. Nor will such entities acquire a *bona fide* status upon further work on our side. Administrative or political borders may in course of time come to involve boundary-markers (barriers, walls, barbed-wire fences, or electronic devices) that will tend in cumulation to replace what is initially a pure *de dicto* boundary with something more substantial. Yet, this is not a process of ontic transformation. Such markers can be very robust, but not so robust as to turn the artificial into the natural. The Chinese Great Wall has survived

---

[5] Searle, J. R. (1995). *The Construction of Social Reality*. Free Press, New York.

[6] On all this, see B. Smith, *op. cit*, as also Smith, B. (1999). Agglomerations. In C. Freksa and D. M. Mark (Eds.), *Spatial Information Theory. Cognitive and Computational Foundations of Geographic Information Science* (267–282). Springer Verlag, Berlin.

for centuries, but the Berlin Wall lasted fifty years and Israel's "Separation Wall" in the West Bank has not been *recognized* by the International Court of Justice. Even Romulus's plowshare could not make natural what natural is not; the blade cuts the soil, tears the grass, uproots all that lies on its *fiat* path.

Let us rather ask whether, and to what extent, it is appropriate to countenance the existence of natural, *de re* boundaries, hence of natural, *bona fide* entities that do not depend on our deeds. And here I am less optimistic. By itself, the concept is perfectly intelligible, and we should be thankful to Lord Curzon for emphasizing its political significance. It is one thing to take up arms and fight in defense of natural borders that are supposed to be "fixed by God," as Lucien Febvre famously put it, quite another to sacrifice your life for a straight Jeffersonian line. Still, even in relation to the geopolitical world, the notion of a natural boundary is all but robust. It is true that I had the impression of *seeing* the shoreline of Long Island from my plane. But it is also true that when you actually go there, things look very different. What looked like a sharp line turns out to be an intricate disarray of stones, sand, algae, piers, boardwalks, cement blocks, musk sediments, marshy spots, decayed fish. And it's not just a matter of our disrespect for Nature. The worry would not change significantly if we took a close look at the coast of a virgin island in the middle of the ocean. Suppose we locate the boundary of the island at the water/sand interface. That boundary is constantly in flux, and it is only by filtering it through our cognitive apparatus—it is only by interpolating objects and concepts—that a clear-cut line may emerge. Even if both water and soil were perfectly still and each were materially homogeneous, it would be hard to locate the boundary with precision. One is reminded here of a question familiar from the early literature on fractals: Just where, and how long, is the coastline of Britain? As it turns out, there seems to be no fact of the matter answering the question. For when measured with increased precision, the coastline would furnish lengths ten, hundred, thousand times as great as the length read off a school map. Cartographers know this well: one works with callipers, but their opening is not fixed by Nature. And if there is no fact of the matter answering the question, "How long is the coastline?," one wonders whether it is even meaningful to think of the coast as of a natural, objectively determined boundary. One begins to wonder whether the island itself might not in some sense be the product of our subjective and approximate worldmaking. And what goes for islands goes for all *prima facie* natural geographical or celestial entities: lakes, rivers, craters, glaciers, mountain chains, whole planets. "Even stars?" Yes, even stars, answered Nelson Goodman: "As we make constellations by picking and putting together certain stars rather than others, so we make stars by drawing certain boundaries rather than others."[7]

It is a short step, now, to extend such doubts to all those boundaries that intuitively belong to the *de re* category, hence all those entities that would seem to enjoy a *bona fide*, mind-independent reality. Take an ordinary material body, say, an apple. On closer look, as we know, an apple is not a solid, continuous object. On closer look, material objects are just swarms of subatomic particles frantically dancing in an otherwise empty space (the "material" volume of an apple is really only one billionth of what we commonly measure), and speaking of their surfaces is like speaking of the "flat top" of a fakir's bed of nails, in Peter Simons' words.[8] To put it differently, on closer look the spatial boundaries of common material bodies involve the same degree of arbitrariness as those of any mathematical graph smoothed out of scattered and inexact data, the same degree of idealization of a drawing obtained by "connecting the dots," the same degree of abstraction as the figures' contours in a Seurat painting. On closer look, therefore, it makes little sense to speak of an apple as of a natural object demarcated by a *de re* boundary: if it has a boundary at all, it is the result of a *fiat* demarcation, a *de dicto* boundary that exists in virtue of our cognitive acts but that is not genuinely present in the autonomous (which is to say mind-independent) physical world.

Nor do we need to resort to microscopes to realize that the ordinary notion of a natural object is far from being clear and distinct. Take this cat, Tibbles—a paradigm example of a living creature, hence a perfect candidate to the status of a natural, *bona fide* individual. Regardless of its subatomic structure, the question of what counts as *it* has no clear answer. Tibbles is eating a chunk of tuna. When it was in the plate, that chunk was definitely not part of Tibbles. But now it is in Tibbles's mouth: Is it part of Tibbles? Will it be part of Tibbles only after some chewing? Only when Tibbles swallows it? Only at the end of the digestive process? Surely, whatever meowing portion of reality we mark off and elect to call 'Tibbles' is something that exists in its own right; as with the North Sea, its objectivity is not affected by the fact that our stipulation is a matter of arbitrary choice.

---

[7] Cfr. Scheffler, I. (1980). The Wonderful Worlds of Goodman. *Synthese*, 45, 201–209, at p. 205, and Goodman, N. (1983). Notes on the Well-Made World. *Erkenntnis*, 19, 99–108, at p. 104 (see also Goodman's original reply in Goodman, N. (1983). On Starmaking. *Synthese*, 45, 210–215, at p. 213).

[8] Simons P. M. (1991). Faces, Boundaries, and Thin Layers. In A. P. Martinich and M. J. White (Eds.), *Certainty and Surface in Epistemology and Philosophical Method. Essays in Honor of Avrum Stroll* (87–99). Edwin Mellen Press, Lewiston, at p. 91.

Yet surely the stipulation adds a *fiat* element to its individuality. Tibbles is not entirely a product of our own making; yet its identity and survival conditions will obtain only relative to us. And this would remain true even if our stipulation were not quite "arbitrary"—even if it fully returned what Jack Wilson calls the "biological individuality" of our cat.[9] Biology is a science, and as such it involves its own stipulations. A workshop held at the University of Utah a few months ago, on the topic *"Edges and Boundaries of Biological Objects,"* focused precisely on the thesis that "delimiting biological objects cannot be determined by empirical facts alone; which facts are salient, and what counts as evidence, often depend on theoretical and conceptual context."[10]

And what goes for objects goes for events. Take the boundary between rest and motion, by itself a source of philosophical puzzlement since Aristotle.[11] On closer look, as we know, a body's being at rest amounts to the fact that the vector sum of the motions of the trillions of restless atoms of which the body is composed, *averaged* over time, equals zero, hence it makes no sense to speak of the *instant* at which an object begins to move, even less to speak of it as of a natural boundary.[12] Or consider biological processes. Earlier I mentioned a person's birth and death as obvious examples of *de re* temporal boundaries, yet the controversies on abortion and euthanasia seem to push for a different treatment. Sometimes it *is* a matter of our deciding whether a person is still alive. *We* decide whether her "vital" functions are still in force, and the criteria for such decisions give expression to our beliefs, our principles, our theories. Similarly, on closer look the initial boundary of a person's life is hardly fixed by Nature alone. Surely that boundary does not coincide with the person's birth (by itself a rather intricate, messy process), except for the registry office. But neither is there an earlier moment that fits the bill comfortably. The moment of *fertilization*? Upon formation of the *zygote*? Upon formation of the *morula*? The beginning of the *implantation* process? Of the *gastrulation* process? The candidates are many. We can base our decision on as many factors we

---

[9] Wilson J. (1999). *Biological Individuality. The Identity and Persistence of Living Entities*. Cambridge University Press, Cambridge.

[10] University of Utah, Department of Philosophy (2008). *Edges and Boundaries of Biological Objects, Annual Colloquium, March 13–15, 2008*. The quote is from the electronic flyer at http://www.phylosophy.org/eb.

[11] Aristotle. *Physics*, VI, 234ff.

[12] Here I concur with Galton. See Galton, A. (1994). Instantaneous Events. In H. J. Ohlbach (Ed.), *Temporal Logic: Proceedings of the ICTL Workshop* (4–11). Max-Planck-Institut für Informatik, Technical Report MPI-I-94-230, Saarbrücken.

like, including up-to-date scientific findings, but a decision it is. And if it is a matter of our (arbitrary or informed) decision, then the boundary is not genuinely *de re* after all and even a person's life becomes, to some extent at least, a *fiat* process.

What about natural properties? Natural kinds? Natural taxa? Here things get more complicated, of course, since the relevant notion of a boundary—especially the notion of a *de re* boundary—is less clear. In fact, when it comes to such entities, the natural/artificial distinction intertwines with the whole realism/nominalism controversy, and our geographic metaphor is bound to sound naïve and dismissive. Still, even here it's obvious and well known that, on closer look, our parochial concerns—historical and cultural circumstances, practical interests and limitations, theoretical priorities—tend to play a major role in the maps we draw of the world. Surely quadrupeds do not form a "natural" kind. But it would be quite remarkable, to use Catherine Elgin's example, if a taxonomy that draws the distinction between horses and zebras where we do aligned at all well with categories fitting the cosmos as a whole, but indifferent to our human faculties and ends (including our interest in domesticating animals).[13] Surely emeroses do not form a natural kind. But neither do roses. Why settle on *Rosa chinensis*? Even in physics, our microscopic categories seem to suffer from a variety of human contingencies. If we construe different isotopes as variants of the same type of atom, it is because of certain reasonable interests that predominated in the development of our best theories. One could as well construe oxygen-17 and oxygen-18 as different types of atoms altogether, hence as "natural" kinds of their own. Here the problem is not that there are no differences in the physical world, as with the Idaho-Wyoming boundary; the problem is that there are *too many* differences, and to privilege some over the others is to draw a *fiat* line—like the dotted line demarcating this apple from its exterior, or Tibbles from the rest of the world. Besides, it is a fact that even within specific domains of inquiry, our scientific practices are not uniform. The thought that the taxa countenanced by biology (for instance) are *fiat* entities seems to clash with "the certainty of biologists on the objective reality of evolution," as David Stamos put it.[14] Yet even today many taxonomists base their classifications more on phenetic than on phylogenetic criteria, and in phenetics a natural classification is simply one in which the members of

---

[13] See Elgin, C. Z. (1995). Unnatural Science. *Journal of Philosophy* 92, 289–302, at p. 297.
[14] Stamos, D. N. (2003). *The Species Problem. Biological Species, Ontology, and the Metaphysics of Biology*. Lexington Books, Lanham (MD), at p. 131, n. 35.

each taxon are on the average more similar to each other than they are to members of other taxa "at the same level" (by itself a problematic notion). Maybe such taxonomists are being sloppy. Maybe the phylogenetic criterion is better. But that's enough to cast the doubt: a criterion is a criterion. We should not shudder if, in the end, we read in the *Annual Review of Ecology and Systematics* that "taxa are human constructs" and "natural taxa are those that are natural to humans."[15] And we should not complain if non-human animals have different taste. After all, there are horses and zebras, but also zorses and hebras. (Didn't Locke even see "the issue of a cat and a rat"?)[16]

3. Conventionalism and Realism

It's pretty clear where all this is going. In the *Phaedrus*, Socrates famously recommends that we should carve the world along its natural joints, trying "not to splinter any part, as a bad butcher might do,"[17] and we know that science and common sense alike have taken this advice very seriously. If all boundaries were the product of some cognitive or social *fiat*, if the lines along which we "splinter" the world depended entirely on our *cognitive* joints and on the categories that we employ in drawing up our maps, then our knowledge of the world would amount to neither more nor less than knowledge of those maps. The thesis according to which all boundaries—hence all entities—are of the *fiat* sort would take us straight to the brink of precipice, to that extreme form of conventionalism according to which "there are no facts, just interpretations." On the other hand, to posit the existence of genuine, *bona fide* boundaries—to think that the world comes pre-organized into natural objects and properties—reflects a form of naïve realism that does not seem to stand close scrutiny.

We know how the compromise solution goes. Perhaps all boundaries are, on closer look, *de dicto* boundaries. It doesn't follow that they must be utterly *arbitrary*, i.e., lack any foundation in reality. It is like beef or veal, says Umberto Eco: "In different cultures the cuts vary, and so the names of certain dishes are not always easy to translate from one language to another, yet it would be very difficult to conceive of a cut that offered at the same moment the tip of the nose and the tail."[18] In other words, perhaps there are

---

[15] Sokal, R.R. (1986). Phenetic Taxonomy: Theory and Methods. *Annual Review of Ecology and Systematics*, 17, 423–442, at p. 424.

[16] Locke, J. (1690). *An Essay Concerning Human Understanding*. Bassett, London, §III.vi.23.

[17] Plato. *Phaedrus*, 265d.

[18] Eco, U. (1997). *Kant e l'ornitorinco*. Bompiani, Milano, at p. 39 (Eng. trans. by

no obligatory paths, no one-way streets in the realm of Being; it still does not mean that anything goes. Some paths will still display a "no entry" sign, some constraints or "lines of resistance" may still be there, making it hard to cut the beast as we like. Out of metaphor, there may still be objective limits to our freedom to carve the world, and it is precisely in this spirit that the realism/conventionalism dichotomy is supposed to be handled. If it is presumptuous to think that the boundaries depicted on our physical maps are perfectly accurate, it is also implausible to think that they are completely off the mark. If it is implausible to think that biology can identify the exact moment at which a person's life begins, it is also implausible to suppose that life begins before fertilization, or at kindergarten. The very notion of "natural kind" to which scientists refer would not betray a commitment to naïve realism but, rather, a form of scientific realism whose cash value is first and foremost *pragmatic*. Just as the maps in our atlases have become more and more accurate, so will the maps drawn by the sciences. And just as cartographers are often forced to redraw their maps as a result of unexpected geo-political changes (the artificial boundary of Israel, but also the prima facie natural boundary of sand between Lybia and Egypt, which keeps drifting under the wind), so biologists and other scientists will not refrain, if necessary, from updating their maps of nature, in an effort to achieve greater accuracy and truthfulness. (The taxonomic misadventures of the platypus would be a good illustration of this fact. What sort of beast is that? Not a MAMMAL, for it lays eggs. Not a REPTILE, for its blood is warm. Not a BIRD, for it's got four legs. For over eighty years, the naturalists were baffled: as René Lesson observed in 1839, that damned beast had set itself "athwart the path of taxonomy to prove its fallaciousness."[19] Yet the beast was there, and eventually the category MONOTREME was created *ex novo*.)

Now, I have no intention of denying the pragmatic reasonableness of this stance. But I do not share its fundamental optimism, as I do not recognize the distinction between metaphysical realism and scientific realism if not, indeed, on a deflated understanding of the term 'realism.' That nobody cuts the beef in funny ways does not mean that the laws of nature prevent it. It means that in spite of our cultural diversities, the culinary taste and aesthetic sense of human beings display surprisingly robust regularities, literally as well as out of metaphor. Consider the debate on unrestricted composition. There is no question that we feel more at ease with certain

---

A. McEwen (1999): *Kant and the Platypus*. Harcourt Brace, New York, at p. 53).

[19] I got the quotation second-hand from Eco, *op. cit.*, at p. 216 (Eng. trans. at p. 250), though I was not able to locate the source.

mereological composites than with others. We feel at ease, for instance, with regard to such things as the fusion of Tibbles's parts (whatever they are), or even a platypus's parts; but when it comes to such unlovely and gerrymandered mixtures as Lewisean trout-turkeys, consisting of the front half of a trout and the back half of a turkey, we feel uncomfortable.[20] Such feelings may exhibit surprising regularities across contexts and cultures. Yet, arguably they rest on psychological biases and *Gestalt* factors that needn't have any bearing on how the world is actually structured. Even if we came up with a formula that jibed with *all* ordinary judgments about what counts as a natural fusion and what does not, it wouldn't follow that there may exist in nature only such objects as answer the formula.[21] If anything, it would follow that the factors that guide our judgments of unity impose systematic constraints on our *fiat* articulations.

Besides, the controversies on biotechnology demonstrate that even such factors are less robust than one might think. We feel horror and disgust for such unlovely and gerrymandered mixtures as chimeras and genetically modified organisms, but we have long learned to feed on orange-mandarins, yogurt, peppermint, and seedless grapes, and we didn't have many scruples when it came to forcing zoological categories to make room for mules and poodles. Either we are adamant that DNA is our model for an organism's individuality—and this is a metaphysical thesis that demands argument—or we must recognize that even the "no entry" signs on the way of Being are on closer look an expression of our contingent biases, reasonable as they might be. Of course, we are free to fight for or against such biases and to study their network in the spirit of honest descriptive metaphysics. After all, the world as *we* represent it is the only world we really care about, for it is the world on which we bet everything, including our happiness. That's why Husserl called it the *Lebenswelt*.[22] Nonetheless, that would not be a way of solving the dilemma between realism and conventionalism; it would be a reasonable way of getting rid of it altogether.

For those who think that the dilemma ought to be addressed explicitly, however, I am going to conclude with three remarks aimed at deflating, at least partly, the sense of collapse that usually comes with the conventionalist

---

[20] Lewis, D. K. (1991). *Parts of Classes*. Blackwell, Oxford, at p. 7.

[21] See van Cleve, J. (1986). Mereological Essentialism, Mereological Conjunctivism, and Identity Through Time. *Midwest Studies in Philosophy*, 11, 141–156, at p. 145.

[22] Husserl, E. (1936). Die Krisis der europäischen Wissenschaften und die transzendentale Phänomenologie. *Philosophia*, 1, 77–176 (Eng. trans. by D. Carr (1970): *The Crisis of European Sciences and Transcendental Phenomenology*. Northwestern University Press, Evanston (IL)), at §9h.

hypothesis, understood in the radical sense according to which there would be no *de re* boundaries whatsoever, whether "one way" or "no entry."

To begin with, one should not mistake the conventionalist hypothesis with the ghost of Berkeleyan *idealism*. As I have described it here, the notion of a *de dicto* boundary is intelligible only to the extent that we acknowledge an appropriate real basis for the sorts of demarcation that are effected by our pencils, trenching tools, and cookie cutters. Even assuming that all boundaries are of this sort, and wholly arbitrary, it does not follow that *everything* is the product of a *percipemus*. This is why I stressed that the cognitive dependence of a *fiat* entity affects its individuality but not its *objectivity*. This may well be seen as a "last-ditch effort" to save realism, as Andrew Cortens dubbed it,[23] but so be it. That the factual material onto which we project our categories should itself be a cognitive construct is a different, stronger thesis, which I do not even understand except in the figurative sense made popular by the thought experiments of rational skepticism (Cartesian demons, brains in a vat, the Matrix). From this perspective, conventionalism is not to be mistaken with Goodmanian *irrealism*, either. For Goodman—and for Richard Rorty—all we learn about the world is contained in right versions of it; and "while the underlying world, bereft of these, need not be denied to those who love it, it is perhaps on the whole a world well lost."[24] For a conventionalist, the world is boneless, impoverished, almost bankrupt, but our love for it is not at stake. For Goodman, a world-version need not be a version of *the world*, just as a Pegasus-picture need not be a picture of Pegasus. For a conventionalist, all the maps we draw are indeed maps of one and the same reality. Putnamian *relativism*, then? Even less. For Hilary Putnam, the "cookie cutter" metaphor founders on the question, "What are the 'parts' of the dough?"[25] No neutral description is available—he says—to compare a Leśniewskian world (with x, y, and the fusion of x and y) and a Carnapian world (with only x and y), and assigning a univocal meaning to 'exists' is already "wandering in Cloud Cockoo Land."[26] For the conventionalist, the metaphor holds and 'exists' corresponds to the standard existential quantifier. The number of *fiat* enti-

---

[23] Cortens, A. (2002). Dividing the World into Objects. In W. P. Alston (Ed.), *Realism and Antirealism* (41–56). Cornell University Press, Ithaca (NY), at p. 54.

[24] Goodman, N. (1975). Words, Works, Worlds. *Erkenntnis*, 9, 57–73, at p. 60. Compare Rorty, R. (1972). The World Well Lost. *Journal of Philosophy*, 69, 649–665.

[25] See Putnam, H. (1987). *The Many Faces of Realism* (Lecture 1). Open Court, LaSalle (IL).

[26] Putnam, H. (2004). *Ethics without Ontology*. Harvard University Press, Cambridge (MA), at p. 85.

ties is up for grabs; but the parts of the dough, which provide the appropriate real basis for our *fiat* acts, are whatever they are and the relevant mereology is a genuine piece of metaphysics. (As far as I am concerned, composition is unrestricted, hence the mereology is Leśniewskian; so either Carnapians are not speaking with their quantifiers wide open, as when we say 'There is no beer' meaning 'There is no beer in the refrigerator,' or else they are objectively wrong.)

Second, what's so bad with conventions being arbitrary? We have just seen that, from a pragmatic perspective, it is not the putative *de re* structure of the world that drives our ways of carving it up, in our folklife as well as in science, but the robustness and utility of certain ways against the ephemeralness and futility (if not the absurdity) of others. If we just replaced Socrates's natural joints with Jefferson's pencil lines, then it would be a disaster, and unfortunately that's exactly what happens in some cases. The decision to annex Alaska to the US echoes the chef's unlikely resolution to prepare a dish with a veal's tail and the tip of its nose, and the thought of classifying people on the basis of their skin color or their intelligence quotient isn't much better than the idea of drawing the Dutch-Belgian border through the houses of Baarle. Nonetheless, in most cases the arbitrariness of our conventions, those that govern our social interactions as well as those that are accorded scientific dignity, epitomizes a democratic reasonableness that treasures experience and cooperation. Conventionalism, just like pre-Kantian empiricism, does obliterate any substantive differences between the laws of nature and train timetables, as some will complain.[27] But timetables are not drawn up at random. They ensue from the necessity to *solve*, in an arbitrary but efficient way, coordination problems that can be extremely complex and that could seriously impair our daily deeds. If we come up with a timetable that works poorly, we change it. If a convention fails to measure up to our expectations, we replace it with a new, hopefully better one. Ditto with the laws of nature. For a conventionalist, not all biological taxonomies (for instance) are on a par. Some are better than others, because they support the "laws" that govern biology's coordination game (laws of variation, selection, organic evolution, population growth, etc.) better. One may object that this sort of pragmatic efficiency calls for more than arbitrary, *fiat* demarcations. But the burden of proof is on the objector, not on the friend of conventionalism. Linnaeus's *Systema Naturae*, the bible of all classical taxonomies, was soaked with essentialism—and the platypus didn't fit in.[28]

---

[27] See e.g. Ferraris, M. (2004). *Goodbye Kant!* Bompiani, Milano, at p. 17.
[28] Linnaeus, C. (1735). *Systema naturae, sive, Regna tria Naturae systematice proposita*

Darwin, by contrast, was adamant that the term 'species' is one "arbitrarily given for the sake of convenience to a set of individuals closely resembling each other"[29]—and his theory is much better.

Indeed, the arbitrariness of conventions does not even rule out that in some cases there is a single best way to go, a single best theory. On David Lewis's classic account, conventions are arbitrary insofar as they always admit of "equally good alternatives."[30] If a problem has a unique solution, then the solution is not conventional. By contrast, as I understand the term here, conventions are arbitrary insofar as they do not depend on the *bona fide* structure of the world. This need not imply that there always be at least two equally good choices that we could make; it simply implies that it is up to us—*in nostro arbitrio*—to make the choice. Here I agree with Mark Heller: in some cases, the adopted convention may be the single best choice, even the obvious best choice.[31] We may adopt a convention precisely *because* it is the best choice, but an arbitrary choice it is. It follows, therefore, that when it comes to the monism/pluralism debate concerning the status of scientific taxonomies (hence: theories), a conventionalist may even side with the monist, at least relative to a certain domain of inquiry. A conventionalist may be a monist, not insofar as there is a unique *correct* way of carving up the world, but insofar as there may be a unique *best* way of doing that.

Finally, for those who, like myself, believe in the significance of so-called prescriptive (or: revisionary) metaphysics, it is worth emphasizing that even a radical conventionalist stance across the board need not yield the nihilist apocalypse heralded by post-modern propaganda. The absence of *de re* boundaries does not coincide, for instance, with the death of the individual. On a Putnamian metaphysics, there *are* no individuals except in a relative sense. On a Goodmanian metaphysics, we *make* individuals by drawing boundaries as we like, and this goes "all the way down."[32] Not so for my conventionalist, or not necessarily. For a conventionalist, the identity of a cat, like the identity of a cookie, a people, a nation, or a constellation, turns out to lack autonomous metaphysical thickness. But other individu-

---

per classes, ordines, genera & species. De Groot, Leiden.

[29] Darwin, C. (1859). *The Origin of Species by Means of Natural Selection*. Murray, London, at p. 52 (repr. (1985): Penguin Classics, London, at p. 108).

[30] See Lewis, D. K. (1969). *Convention. A Philosophical Study*. Cambridge University Press, Cambridge.

[31] Heller, M. (1990). *The Ontology of Physical Objects. Four-Dimensional Hunks of Matter*. Cambridge University Press, Cambridge, at p. 44, n. 44.

[32] See Goodman, N. (1983). Notes on the Well-Made World. *Erkenntnis*, 19, 99–108, at p. 107, n. 6.

als may present themselves. For instance, on a Quinean metaphysics, there is an individual corresponding to "the material content, however heterogeneous, of any portion of space-time, however disconnected and gerrymandered."[33] That the content of some such portions of space-time have *de re* boundaries is a possibility, but it is equally possible that the only boundaries are those warranted by geometry. Either way, the corresponding notion of an individual is perfectly intelligible. The relevant identity conditions are perfectly determinate, and one may suppose that it is perfectly determinate, for any property, whether any given individual possesses it. Such individuals are perfectly nonconventional, yet the overall picture is one that a conventionalist is free to endorse. The conventionalist stance simply entails that *which* of them come to play a role in our life is up to us. From a God's eye point of view, they are all equally real. It's just that only some are salient for us, only some make us "feel comfortable," and only some are selected by us through the imposition of more or less precise and official *de dicto* boundaries.

Evidently, a metaphysics of this sort presupposes the existence of a large, all-embracing four-dimensional "dough." But we have already seen that such a presupposition is not incompatible with a conventionalist stance. Nor is reference to the boundaries warranted by geometry vetoed by that stance, as though it surreptitiously reintroduced *de re* demarcations. As we have outlined it here, the *de dicto / de re* distinction does not apply to the geometric boundaries of space-time any more than it applies to the boundaries presupposed by set theory. Those are bare boundaries, so to speak: they are not "artificial," but neither are they "natural." As for the overall plausibility of the theory, this is not the place to embark on an articulated defense. Surely the intuitive plausibility is pretty low, and perhaps also its scientific tenability. Yet, philosophically the deflationary purism of a theory of this sort would have some advantages, including the extermination of the essentialist cancer that besets those metaphysical theories that try to save common sense against the paradoxes of persistence, vagueness, and material constitution.

But, as I said, this is not the place to embark on an articulated defense of such metaphysics. (And it is but an example. A Sidellean metaphysics of "pure stuff" would do just as well.)[34] The relevant point is that, also in relation to the third worry, the conventionalist hypothesis need not result

---

[33] Quine, W. V. O. (1960). *Word and Object*. MIT Press, Cambridge (MA), at p. 171.

[34] See Sidelle, A. (1989). *Necessity, Essence, and Individuation. A Defense of Conventionalism*. Cornell University Press, Ithaca (NY).

in a delegitimization of all philosophical inquiry; only a redistribution of the relative tasks and concerns of the different fields of inquiry, including metaphysics as well as physics, psychology, or sociology. The costs are obvious, for epistemology and also for ethics. But so are the advantages. There are, on this view, no obligatory or forbidden paths; conventionalism is as liberal as it gets, and it is up to us to erect the "one way" or "no entry" signposts that we find appropriate, just as it is up to us to remove them when things take a turn for the worse. As Michael Dummett put it, the picture of reality as an amorphous lump, not yet articulated in discrete objects, is a good one "so long as we make the right use of it."[35] Most importantly, given that even common sense is far from being an all or nothing affair, except for certain surprising regularities, it is up to us to acknowledge our parochial limits without camouflaging them as "natural." At least we would stop pretending that boundary wars have a "just" solution. To conclude on a rhetorical note, if there is a solution, let's face it, it lies in the reciprocal and democratic—and hard to achieve—agreement among all interested parties.[36] Metaphorically speaking as well as out of metaphor.

---

[35] Dummett, M. (1973). *Frege. Philosophy of Language*. Duckworth, London, at p. 577.

[36] I am thankful to the participants in the *Sixth European Congress of Analytical Philosophy* for their comments on an earlier version of this paper. The paper itself is an abridged version of a longer essay entitled "Boundaries, Conventions, and Realism," which is due to appear in a volume edited by Joseph Campbell, Michael O'Rourke, and Matthew H. Slater (*Carving Nature at its Joints. Topics in Contemporary Philosophy*, Volume 8, Cambridge, MA, MIT Press). I am grateful to the publisher and to the editors for their permission to reprint portions of that essay in the present form.

Philosophy of Science

On Propensity-Frequentist Models for Stochastic Phenomena with Applications to Bell's Theorem

Tomasz Placek

ABSTRACT

The paper develops models of statistical experiments that combine propensities for the occurrence of results with frequencies, the underlying theory being the branching space-times (BST) of Belnap (1992). The models are then applied to analyze Bell's theorem. We prove the so-called Bell-CH inequality via the assumptions of a BST version of Outcome Independence and of (non-probabilistic) No Conspiracy. Notably, neither the condition of probabilistic No Conspiracy nor the condition of Parameter Independence is needed in the proof. As the Bell-CH inequality is most likely experimentally falsified, the choice is this: contrary to the appearances, experimenters cannot choose some measurement settings, or some transitions, with space-like related initials, are correlated, or both.

1. Introduction

The objective of this paper is twofold.[1] First, we will construct models that accommodate both propensity aspects and frequentist aspects of statistical experiments. A view underlying this task has it that on a given run of the experiment, a measurement is capable of producing alternative outcomes, the weights of such capacities being propensities. The observed probabilities result from a certain distribution of types of measurements and propensities of individual measurements to produce outcomes. The frequentist aspect consists in a desideratum that observed frequencies of results should reflect underlying propensities and a distribution of measurement types. The second aim is to analyze in the constructed models a particular variety of Bell's theorem, the so-called Bell-CH theorem.

One might ask at this point, however, why yet another analysis of Bell's theorem? The answer is that derivations of Bell's theorem have thus far been either fully informal, or have left some aspect of the reasoning informal. This resistance to formal treatment stems from the multifarious nature of Bell's theorem: it appeals to modal aspects (e.g., "other results might have occurred"), spatiotemporal aspects, with some rudiments of relativity ("space-like related events should not influence one another"), and probabilistic aspects ("the common past should screen off the correlations"). Thus, to adequately analyze Bell-type theorems, a framework is needed which includes all three features: it is modal, spatiotemporal, and probabilistic.

A brief survey of Bell-type derivations, with respect to the use of formal methods, reveals that physicists (incl. Bell, Mermin, and Shimony) have all done this informally. Starting with their (1999) paper, the so-called Budapest school of Rédei, Szabó and Hofer-Szabó has constructed rigorous models accounting for probabilistic aspects. Belnap and Szabó (1996) and Kowalski and Placek (1999) have presented models of the GHZS theorem (a non-probabilistic Bell-type theorem)[2] with modal aspects rigorously analyzed, but the spatiotemporal aspect left informal. Müller and Placek (1999–2002) have created models accommodating modal and probabilistic aspects of the Bell-Aspect set-up (Aspect, Dalibard, and Roger, 1982), the spatiotemporal aspect being left informal.

---

[1] I would like to thank Thomas Müller, Leszek Wroński, and in particular—Balázs Gyenis for a challenge to construct BST models for experiments with runs. Support by MNiSW research grant 3165/32 is gratefully acknowledged.

[2] See Greenberger, Horne, Shimony, and Zeilinger (1990).

In recent years, however, the branching space-times (BST), as conceived by Belnap (1992), has been developed in a series of papers (by Belnap and his collaborators) into a theory encompassing all the three aspects required in analyses of Bell's theorems.[3] This is an axiomatic second order theory, with probabilities assignable to the so-called transitions of events; the theory has a class of models (the so-called Minkowskian Branching Structures) in which possible histories are isomorphic to the Minkowski space-time. The aim of this paper is thus to harness the formal machinery of BST in order to analyze the Bell-Aspect experiment, with the hope of determining, once and for all, what the theorem's implications actually are.

Although in the technical sections of this paper we work in the rigorous possible-worlds framework of BST, let us begin by asking how a statistical experiment is to be represented in a possible-worlds theory. As a preliminary illustration intending to convey the underlying idea, consider the statistical experiment of tossing a given coin. Let us suppose that, after sufficiently many tosses, the experimenter estimates that the probability of heads, written as $p^{ex}(H)$, tends to 0.6. The experimenter can take two stances in interpreting what $p^{ex}(H) = 0.6$ means. First, she might take the experiment as deterministic, i.e., claiming that, on each toss of the coin, there is exactly one possible outcome, either heads or tails, and the numbers 0.6 vs. 0.4 stem from there being a few *types of runs*, the particular numerical values following from the frequencies of types of runs within the set of all runs of the experiment. Using symbol Λ^H for the set of those types of runs, whose tokens yield Heads, and w_λ for the frequency of runs of type λ in the collection of all runs of the experiment, it must be that

(†) $p^{ex}(H) = \Sigma_{\lambda \in \Lambda^H} w_\lambda,$

and analogously for the probability of tails. At an extreme, only two types of runs may be posited: one in which Heads are determined to occur, and the other in which Tails are determined to occur. In this case, experimental probabilities can be identified with frequencies of corresponding types of runs. A deterministic model of the experiment will consist of a single history, with a chain of "toss, result" pairs, the former preceding the latter.

On the other hand, our experimenter can take an indeterministic stance in interpreting the probability $p^{ex}(H) = 0.6$, acknowledging that a given run can have more than one possible outcome (of which, of course, exactly one occurs at a given time). Types of runs are then thought of as differentiated

---

[3] After Belnap's (1992) paper, the essential development of the theory is reported in Belnap and Szabó (1996), Weiner and Belnap (2006), Belnap (2005), Müller (2002), Müller (2005), Müller, Belnap and Kishida (2008), Wroński and Placek (2009), and Placek and Wroński (2009).

by propensities: if two runs belong to one type, for each possible result, the propensity of a result on one run must agree with the propensity of that result on the other run.[4] In other words, runs of a given type are uniform with respect to propensities. Note a peculiarity of the notion of run: it is a modally thick notion since, to represent a run of a given type, it is not sufficient to specify its (actual) outcome, but all the possible outcomes and propensities thereof. In what follows, we will put forward a certain representation of this notion, understood as modally thick.

For propensities to make contact with experimentally given probabilities, a faithfulness assumption is needed: the number of a given (type-level) result (say, Heads) produced in runs of a given type should approximate the propensity for the corresponding (token-level) result in a run[5] of the given type of runs. The experimental probability $p^{ex}(H)$ is then a weighted average of propensities for H, as obtaining in the types of runs:

$$p^{ex}(H) = \sum_{\lambda \in \Lambda} pr_\lambda(H) w_\lambda, \qquad (1.1)$$

where Λ is the set of types of runs in our coin tossing experiment, $pr_\lambda(H)$ is the propensity for obtaining result H in any run of type $\lambda \in \Lambda$ and w_λ is the frequency of runs of type λ in the set of runs of the experiment.

On this propensity approach, our model of coin tossing must have many histories, as it contains "forks" representing possible outcomes of tosses, and contains all the information concerning types of runs, and propensities for results. In short, it is like a branching structure stretching indefinitely long. We turn now to the task of defining propensity BST models for statistical experiments.

2. Propensity-Frequentist Models of Statistical Experiments

We are now going to explain propensity-frequentist BST models without providing any technical details on the BST theory. The required notions are explained in Section (4), hence, in order to fully grasp the models, the reader is advised, after having become acquainted with the technical notions of that section, to jump back to the present section.

---

[4] Note that we are talking type-results here, like Heads. In the formal treatment we will have outcome tokens, and, in that context, it is more accurate to say that in two runs propensities of (different) outcome-tokens agree.

[5] Equally well we could say "in every run" as we assumed that corresponding results attain the same propensity in every run of a given type.

For the present purpose of introducing propensity-frequentist models of statistical experiments, it suffices to know that BST is a possible-world theory, in which the role of possible worlds is played by histories. Chanciness is represented by branching of histories, so that, for instance, an initial event with two alternative possible outcomes is represented by two branching histories that share an initial event but contain different outcomes. Various types of events are definable in BST of which two are relevant for our purpose here: initial events (which are simply upper bounded subsets of histories) and outcome events (which are more complex set-theoretical constructions). There is a natural notion of consistency: two events are consistent iff there is a history in which the two occur. We also say that two events are space-like related (SLR) iff they are distinct and every element of one and every element of the other are consistent, but none is above the other. A pair consisting of an initial event and an outcome event above it is called a transition. Importantly, transitions are objects on which propensities are defined. Finally, we assume that histories are isomorphic to Minkowski space-time; as a result, events have spatiotemporal locations and relations of special relativity are definable on them. Technically, we work in the framework of the Minkowskian Branching Structures (MBS's), which form a class of BST models (for the explanation, see Section 4).

Statistical experiment idealized Let us suppose that our statistical experiment consists of measuring $\tilde{A}$, which has possible outcomes $\tilde{a}_1, \tilde{a}_2, \ldots, \tilde{a}_k$. Be aware that we are talking types now: in our illustration, $\tilde{A}$ is a generic toss of a particular coin, and $\tilde{a}_1, \tilde{a}_2$ are heads up and tails up on this coin, respectively. To keep clear about the types vs. tokens distinction, we use the tilde to signify type-level notions. The statistical aspect has it that an experiment has runs, and in each run the measurement $\tilde{A}$ and some one of its outcomes, $\tilde{a}_i$, is instantiated. What makes runs is experiment-sensitive: in the simplest cases, instantiations of measurement $\tilde{A}$ is a good guide as to when a run begins. In more sophisticated cases, to be considered later, we need to appeal to some other phenomena. For simplicity, let us assume that on each run n of the experiment a token A^n of $\tilde{A}$ and a token $\mathbf{a}_i^n$ of a single $\tilde{a}_i$ occur. This is not a stringent requirement, as one can always add "a failure outcome" to the list of outcomes. More controversially, we assume that on each run n of the experiment, all outcomes $\mathbf{a}_1^n, \mathbf{a}_2^n, \ldots, \mathbf{a}_k^n$ are in a sense possible. That is, all these outcomes are supposed to be in a BST model and situated in such a way that each sentence "It is possible that $\mathbf{a}_i^n$ will occur" is true if evaluated at A^n. On the other hand, we allow for some $\mathbf{a}_i^n$ to have propensity zero. This means that at a time, $\mathbf{a}_i^n$ is possible, but impossible propensity-wise. This is similar to a dart, idealized to have a point-like head, landing exactly at a given point: this is possible, though has zero probability.

The aim of our next assumption is to circumvent the problem of estimating probabilities on the basis of finitely many runs of the experiment: we assume that the experiment has (countably) infinitely many runs. Clearly, as the experiment may finish at this or that stage, our BST models will have histories with a finite number of runs. Yet we take such histories to be irrelevant for our modeling of a statistical experiment. Thus, what we really require is a possible history with infinitely many runs of the experiment. We will define what it takes for the history of a BST model to adequately represent a statistical experiment.

An objection might be raised at this point that an adequate representation of a statistical experiment, with propensities and the idealization of infinitely many runs, should incorporate infinitely many possible histories, representing varying distributions of the measurement outcomes. Among these histories, there should be a majority of those in which frequencies of results are close enough to propensities. Anything short of accounting for this aspect of probabilities does not deserve the name of a "model of statistical experiments, with propensities"—the objection continues. We agree, yet we leave this ambitious task for some future project. Since the import of this paper is negative, as it boils down to a claim that "a propensity-frequentist model for statistical experiment, subject to such-and-such conditions" is not possible, a simplified representation in terms of one history is enough. That is, if a single history of a simplified representation cannot have some desired feature, then none of the (possibly infinitely many) histories of a full-fledged representation can posses that feature, either.

Turning next to spatiotemporal aspects, we assume the weak gravitational fields condition ensuring that Minkowski space-time is an adequate representation of the spatiotemporal aspects of the phenomena considered.

As for spatiotemporal locations, we naturally require that every possible outcome $\mathbf{a}_i^n$ of A^n belongs to the future of possibilities of A^n. We next assume the same spatiotemporal location of alternative outcomes, that is, the following statement:

Condition 1. *Suppose that outcome $\mathbf{a}_i^n$ occurred in spatiotemporal region R on run n; if on this run $\mathbf{a}_j^n$ ($i \neq j$) occurred instead, it would occur in the same spatiotemporal region R.*

Note that the condition concerns possible outcomes of a measurement in a fixed run. As for possible outcomes of different runs, they are assumed to happen in different spatiotemporal regions: we assume that there is a minimal time interval between measurements in any two consecutive runs.

Condition 2. *There is $\Delta \in \mathfrak{R}$ such that for every run l of the experiment, if A^l occurs in spatiotemporal location L^l, then A^{l+1} occurs at a time-like separated location L^{l+1} such that the Lorentz distance between these two locations is at least Δ. A sequence $A^1, A^2, \ldots$ that satisfies this condition will be called "shifted in time" sequence.*

A statistical experiment in which the above idealizing conditions are satisfied, will here be called an idealized statistical experiment.

MBS models for idealized statistical experiments The task of specifying a BST model with propensities for an idealized statistical experiment will be split into two parts. We first focus on the modal and spatiotemporal aspects of the phenomena. With the idealizations assumed above, these aspects can be adequately captured by MBS's. In the next part, we will add propensities to the picture, and with propensities there will come *types* of runs.

Definition 3 (MBS weakly represents)
In an MBS model, history σ weakly represents an idealized statistical experiment in which measurement event $\tilde{A}$ has k possible outcomes $\tilde{a}_1, \tilde{a}_2, \ldots, \tilde{a}_k$ iff

1. there is in σ a sequence $A^1, A^2, A^3, \ldots$ of shifted in time initial events,

2. for every $n \in \mathcal{N}$ there are k pairwise inconsistent, occurring in the same spatiotemporal region, outcomes $\mathbf{a}_1^n, \mathbf{a}_2^n, \ldots, \mathbf{a}_k^n$,

3. for every $n \in \mathcal{N}$, A^n and the set $\{\mathbf{a}_1^n, \mathbf{a}_2^n, \ldots, \mathbf{a}_k^n\}$ yields the set $\mathfrak{A}^n := \{A^n \to \mathbf{a}_i^n \mid 0 < i \leqslant k\}$ of transitions,[6]

4. for every history h, if A^n occurs in h, then for some $i \leqslant k$, $\mathbf{a}_i^n$ occurs in h as well.

We say that an MBS model weakly represents the experiment in terms of history σ and the sequence $\mathfrak{A}^1, \mathfrak{A}^2, \mathfrak{A}^3, \ldots$ of sets of transitions.

A central idea is that the experiment is analyzable in terms of transitions: these are transitions from a (token-level) measurement event to its possible outcomes (token-level as well). For simplicity, we assume here that only one measurement (type-level) is performed in the experiment. This limitation is removed in the definitions given in the Appendix. Observe the subtle

---

[6] In the formal treatment, these outcome events will be identified with scattered outcome events and accordingly the transitions will be from initial events to scattered outcomes—cf. Definition 9.

interplay between the notion of run and the notion of history σ. By conditions (1) and (4), an infinite sequence of measurements and, after each measurement, one of its possible outcomes occur in history σ. Thus, σ codes a particular way in which the statistical experiment (idealized as infinitely long) can go. In and of itself, history σ (as any other history) does not contain any information as to what and where was/is/will be possible. Yet, this modal information is coded in the model, by means of various transitions. Accordingly, the notion of the run codes the measurement and its possible outcomes. We may thus say that the notion of the run is modally thick.

MBS models with propensities for idealized statistical experiments
A BST model can be equipped with probability measures, with each probability measure μ_T defined on some particular set T of basic transitions. Given that a probability measure is defined on a particular set of basic transitions, this measure imposes probabilities on larger (i.e., not necessarily basic) transitions. The probabilities in question are naturally interpreted as propensities, i.e., degrees of possibility of the transitions in question. Yet, in order for a BST model to have propensities for transitions, a μ_T probability must be defined on an appropriate set T of basic transitions—cf. Definition (19). In modeling an (idealized) statistical experiment we require that the relevant propensities be defined, which in turn means that the relevant μ-probabilities exist.

Definition 4 (MBS has propensities)
An MBS model $\mathfrak{M}$ has propensities for an idealized statistical experiment with measurement event $\tilde{A}$ and its k possible outcomes $\tilde{a}_1, \tilde{a}_2, \ldots, \tilde{a}_k$ iff $\mathfrak{M}$ represents (modally) the experiment in terms of some history σ, a sequence $A^1, A^2, A^3, \ldots$ of initial events, and the sequence $\mathfrak{A}^1, \mathfrak{A}^2, \mathfrak{A}^3, \ldots$ of sets of transitions and propensity is defined on every $t \in \bigcup_{n \in \mathcal{N}} \mathfrak{A}^n$.[7]

Now, even if a BST model has propensities for all the relevant transitions, one might still be unable to relate the propensities to the experimentally estimated probabilities. What we need is a kind of faithfulness assumption, which requires that the numerical values of propensities be exhibited as frequencies of results in sufficiently long series of appropriate runs. Since we assumed that the set of all runs is countably infinite, we straightforwardly

---

[7] It follows from the Definition (19) of propensities in terms of causal probabilities, that pr behaves like probability. For instance, since $\mathfrak{A}^n$ is an exhaustive set of alternative transitions of Definition 20, we have: $\sum_{t \in \mathfrak{A}^n} pr(t) = 1$.

require that in our history a numerical value of propensity for a result occurring in a run of a given type be equal to the limiting frequency of the corresponding results occurring in the type of runs in question.

To illustrate, consider the n-th run. The set $\mathfrak{A}^n$ of transitions has some propensity assignment, that is, for every transition $t \in \mathfrak{A}^n$, its propensity $pr(t)$ is defined. We first require that in history σ there is a non-empty subset Ω of the set of all runs that are "exactly like" the n-th run. We mean by this that a transition on the n-th run and the corresponding transition on each run from Ω have the same propensity. Second, for any $i \leqslant k$, the frequency in Ω of transitions of the form $A^n \to \mathbf{a}_i^n$ must tend to propensity $pr(A^n \to \mathbf{a}_i^n)$.

How can it be interpreted that two sets $\mathfrak{A}_i$ and $\mathfrak{A}_k$ of transitions have the same propensity assignment? Uncontroversially, propensities are ontologically grounded, so that if a measurement process on one run and a measurement process on another run are very much alike, the propensity assignment on the first run should agree with the propensity assignment on the other run. Now it is natural to reverse this story to use the same propensity assignments to define types of runs, as follows:

Definition 5 (types of runs)
Let $\mathfrak{M}$ be an MBS with propensities that weakly represents an idealized statistical experiment consisting in measuring $\tilde{A}$ with k possible outcomes $\tilde{a}_1, \tilde{a}_2, \ldots, \tilde{a}_k$, in terms of history σ and the sequence $\mathfrak{A}^1, \mathfrak{A}^2, \mathfrak{A}^3, \ldots$ of sets of transitions.
The l-th run and the m-th run belong to the same type of runs iff for every $j \leqslant k$: $pr(A^l \to \mathbf{a}_j^l) = pr(A^m \to \mathbf{a}_j^m)$.

Putting the above ideas and observations together, we arrive at this definition of an MBS representing fully, i.e., modally, spatiotemporally, and probabilistically, an idealized statistical experiment:

Definition 6 (MBS model fully represents)
Let $\mathfrak{M}$ be an MBS model with propensities for an idealized statistical experiment consisting of measurement $\tilde{A}$, its k possible outcomes $\tilde{a}_1, \tilde{a}_2, \ldots, \tilde{a}_k$, and their experimental probabilities $p^{ex}(\tilde{a}_1), \ldots, p^{ex}(\tilde{a}_k)$. Let the experiment be weakly represented in $\mathfrak{M}$ in terms of history σ and the sequence $\mathfrak{A}^1, \mathfrak{A}^2, \mathfrak{A}^3, \ldots$ of sets of transitions. We say that $\mathfrak{M}$ fully represents the experiment in terms of propensity assignment pr, history σ and the set $\mathfrak{A}^1, \mathfrak{A}^2, \mathfrak{A}^3, \ldots$ iff there is a partition Λ of the set $\mathcal{N}$ of natural numbers such that

1. for every $l, m \in \mathcal{N}$: $l, m \in \lambda$ for some $\lambda \in \Lambda$ iff for every $j \leqslant k$: $pr(A^l \to \mathbf{a}_j^l) = pr(A^m \to \mathbf{a}_j^m)$,

2. $p^{ex}(\tilde{a}_i) = \sum_\lambda w_\lambda \, pr(A^{min(\lambda)} \to \mathbf{a}_i^{min(\lambda)})$, where $min(\lambda)$ picks a minimal element of $\lambda \subset \mathcal{N}$, the limit on the right-hand side below exists, and

$$w_\lambda = lim_{n \to \infty} \frac{\#\{m \in \mathcal{N} \mid m \leqslant n \land m \in \lambda\}}{n},$$

3. for every $\lambda \in \Lambda$ and $i \leqslant k$, the limit on the right-hand side below exists and:

$$pr(A^{min(\lambda)} \to \mathbf{a}_i^{min(\lambda)}) = \lim_{n \to \infty} \frac{\#\{m \in \mathcal{N} \mid m \leqslant n \land m \in \lambda \land \sigma \in H_{\langle \mathbf{a}_i^m \rangle}\}}{\#\{m \in \mathcal{N} \mid m \leqslant n \land m \in \lambda\}},$$

where $H_{\langle \mathbf{a}_i^m \rangle}$ is the set of histories in which $\mathbf{a}_i^m$ occurs.

To explain, the model has a history with an infinite sequence of initial events, representing (token-level) measurement events. Thus, by partitioning the set of natural numbers, we partition the set of initial events, each element of the partition being intuitively thought of as a type of runs.[8] Clause 1 ensures that Λ is the set of types of runs in the sense of Definition (5). Clause 2 is an Empirical Adequacy condition, as it draws a bridge between experimentally estimated probabilities and propensities (via weights). A value of the weight of a given type is equated to the limiting frequency of runs of this type in history σ. Observe that a type whose tokens appear only finitely many times in σ receives weight zero. The last clause postulates that history σ exhibits propensities of transitions in terms of limiting frequencies of corresponding outcomes, as occurring in runs of appropriate types.

Why might one want to postulate types of runs and propensities, which are unobserved and controversial entities, to account for a statistical experiment? The aim that comes to the fore in the context of quantum non-locality is to explain correlations between spatially distant results. Having seen that explaining such correlations in terms of a deterministic model is not viable, one may turn to models with propensities, as there seems to be more freedom in propensities than in strict determinism. The crucial task is then to derive correlated experimental probabilities from some "more basic" probabilities without correlations and from a distribution of types of runs in the experiment.

Leaving aside the task of constructing deterministic BST models for Bell-type phenomena, as they have been discussed elsewhere, we focus here on indeterministic models with propensities. We will first analyze a simplified EPR set-up, to finally turn to Bell's theorem, deriving the so called

---

[8] Note that Λ is at most countably infinite.

Bell-Clauser-Horne inequality (in short, Bell-CH), cf. Clauser and Horne (1974).[9]

3. Simple EPR Explained Away

A peculiar feature of Bell-type experiments is that chanciness (or its appearance) occurs at two levels: on each measurement device, there is a selection of a measurement setting (e.g., direction of polarization), and each measurement has more than one possible outcome. As a warm-up, we will focus first on a simplified experiment, however, in which no selection of measurement settings occur. This is, in essence, a set-up with fixed polarizations, for which there is a hidden variable model—see Bell (1987); here we will call this "simple EPR." The description of the setup is as follows: there are two space-like related measurement events $\tilde{A}$ and $\tilde{B}$ (type-level), with alternative possible results (type-level) $\tilde{a}_1, \tilde{a}_2, \ldots, \tilde{a}_I$ of $\tilde{A}$ and $\tilde{b}_1, \tilde{b}_2, \ldots, \tilde{b}_J$ of $\tilde{B}$. Every combination $\tilde{a}_i, \tilde{b}_j$ is possible. Results are correlated, i.e.,

$$p^{ex}(\tilde{a}_i) \cdot p^{ex}(\tilde{b}_j) \neq p^{ex}(\tilde{a}_i \text{ and } \tilde{b}_j).$$

We will hereby construct an MBS, in which some history fully represents this experiment. Yet, as we are troubled by the (experimentally observed) correlations above, we attempt the following explanation: the runs of this experiment are not uniform as perhaps the state of the source of particles vary, or the measurement process might not exactly be alike on every run. Yet, one may consider a division of the set of runs into types of exactly alike runs. Our article of faith, motivated solely by our uneasiness with distant correlations, is that in each set of exactly alike runs there are no distant correlations, i.e., probabilities factor. In the language of the Bell literature, we envisage some "hidden variable" parametrizing hidden states of pairs of particles involved in the experiment—"hidden" as it is not accounted by quantum mechanics. The hidden variable can take on a number of values, and each joint probability mentioned above, if conditioned on each value of the variable, factors into single probabilities (conditional on the value). Now, on the assumption of indeterminism, a measurement in a given run has many possible outcomes, and the idea of propensity comes up naturally. Furthermore, drawing on an argument sketched in the last section, if some runs are exactly alike, the propensity of corresponding transitions in these runs should be the same. By this train of reasoning, we end up with the

---

[9] An important step towards this formula was the argument of Clauser, Holt, Shimony, and Horne (1969).

task of constructing an MBS, which yields experimentally observed distant correlations (the results of the existence of many types of runs) with such values of propensities that in each run the distant results are uncorrelated. In other words, a crucial part of the task is to show that it is possible to recover experimentally estimated probabilities, including the correlations above, on the assumption that for every type λ of runs there is no correlation, i.e., for every $i \leqslant K$ and every $j \leqslant J$ we have, schematically,

$$pr(a_i^\lambda) \cdot pr(b_j^\lambda) = pr(a_i^\lambda \text{ and } b_j^\lambda)? \tag{3.1}$$

Our MBS $\mathfrak{M}$ should fulfill these desiderata:

DESIDERATA (†)

1. There is in $\mathfrak{M}$ a history σ containing a sequence $A^1 \cup B^1, A^2 \cup B^2, A^3 \cup B^3, \ldots$ of temporally shifted initial events and for every $n \in \mathcal{N}$: A^n and B^n are SLR;

2. for every $n \in \mathcal{N}$, there are $I \in \mathcal{N}$ pairwise inconsistent, occurring in the same spatiotemporal region, (scattered) outcomes $\mathbf{a}_1^n, \mathbf{a}_2^n, \ldots, \mathbf{a}_I^n$ such that A^n is below each $\mathbf{a}_i^n$ ($i \leqslant K$), and each such outcome is SLR to B^n;

3. for every $n \in \mathcal{N}$, there are $J \in \mathcal{N}$ pairwise inconsistent, occurring in the same spatiotemporal region, (scattered) outcomes $\mathbf{b}_1^n, \mathbf{b}_2^n, \ldots, \mathbf{b}_J^n$ such that B^n is below each $\mathbf{b}_j^n$ ($i \leqslant J$), and each such outcome is SLR to A^n;

4. if $A^n \cup B^n$ occurs in history h, then for some $i \leqslant K, j \leqslant J$, both $\mathbf{a}_i^n$ and $\mathbf{b}_j^n$ occur in h;

5. $\mathbf{a}_i$ and $\mathbf{b}_j$ are consistent, for every $i \leqslant K$ and $j \leqslant J$;

6. for any history h and any n, A^n occurs in h iff B^n occurs in h, i.e., $H_{[A^n]} = H_{[B^n]}$.

From conditions (1), (2) and (3) it immediately follows that for every $n \in \mathcal{N}$, there are two sets of transitions, induced by A^n and B^n, respectively, namely $\mathfrak{A}^n := \{A^n \to \mathbf{a}_i^n \mid 0 < i \leqslant K\}$ and $\mathfrak{B}^n := \{B^n \to \mathbf{b}_j^n \mid 0 < j \leqslant J\}$. With clause (4) added, they also imply that for every $n \in \mathcal{N}$ there is a third set $\mathfrak{C}^n := \{A^n \cup B^n \to \mathbf{a}_i^n \cup \mathbf{b}_j^n \mid 0 < i \leqslant K \text{ and } 0 < j \leqslant J\}$ of transitions from initials to (scattered) outcomes–see Fact (23).

Having accepted the "enumerating" convention: $(A \cup B)^n := A^n \cup B^n$ and $(\mathbf{a}_i^n \cup \mathbf{b}_j)^n := \mathbf{a}_i^n \cup \mathbf{b}_j^n$, it is immediate to see that an MBS $\mathfrak{M}$ which fulfills the above desiderata weakly represents our statistical experiments in terms of history σ and the sequence $\mathfrak{C}^1, \mathfrak{C}^2, \mathfrak{C}^3, \ldots$ of sets of transitions.

Observe that, in the terminology developed in Section (5), the stipulations above amount to saying that for every $n \in \mathcal{N}$, $\mathfrak{A}^n$ and $\mathfrak{B}^n$ is a set of exhaustive alternative transitions of the same spatiotemporal location, and each such pair induces a third set $\mathfrak{C}^n$ of alternative transitions of the same spatiotemporal location. Further, given that A^n and B^n occur in exactly the same histories, the transitions of the last set are also exhaustive—cf. facts of section (5).

We next need to equip our model with propensities, defined for every relevant transition, that is, for every transition from $\mathfrak{A}^n$, $\mathfrak{B}^n$ and $\mathfrak{C}^n$, for every $n \in \mathcal{N}$. Given the facts of section (5), it suffices to postulate that some particular causally significant sets, called sets of past causal loci (*pcl*), are eligible for producing a causal probability space. We require that the numerical values of these propensities recover experimentally estimated probabilities and that with respect to these propensities, every transition from $\mathfrak{A}^n$ be probabilistically independent from every transition from $\mathfrak{B}^n$.

The existence of a model fully representing the experiment requires a set Λ of types of runs, with each $\lambda \in \Lambda$ having associated weight w_λ. Empirical Adequacy (clause 3 of Definition (6)) then dictates:

$$p^{ex}(\tilde{a}_i \text{ and } \tilde{b}_j) = \sum_{\lambda \in \Lambda} w_\lambda \cdot pr(A^{min(\lambda)} \cup B^{min(\lambda)} \to \mathbf{a}_i^{min(\lambda)} \cup \mathbf{b}_j^{min(\lambda)}) \quad (3.2)$$

for every $n \in \mathcal{N}, \lambda \in \Lambda, i \leqslant K,$ and $j \leqslant J$.

What is, however, a proper formulation of the condition that with respect to propensities, every transition from $\mathfrak{A}^n$ be probabilistically independent from every transition from $\mathfrak{B}^n$? It turns out that, thanks to desiderata (2), (3), (5) and (6) imposed on our MBS, the assumptions of Definition (32) of a special case of probabilistic independence are satisfied, and this independence condition amounts to the satisfaction of these equations:

$$pr(A^{m(\lambda)} \to \mathbf{a}_i^{m(\lambda)}) pr(B^{m(\lambda)} \to \mathbf{b}_j^{m(\lambda)}) = pr(A^{m(\lambda)} \cup B^{m(\lambda)} \to \mathbf{a}_i^{m(\lambda)} \cup \mathbf{b}_j^{m(\lambda)}). \quad (3.3)$$

Above $m(\lambda)$ abbreviates $min(\lambda)$, which works like a "selection function"— for every type of runs it chooses a representative run of this type, namely the first run of this type.

Moreover, $\mathfrak{A}^n$ and $\mathfrak{B}^n$ satisfy the assumptions of Fact 29, which in turn allows us to rewrite the above set of equations as follows:

$$(\sum_{l \leqslant J} pr(A^{m(\lambda)} \cup B^{m(\lambda)} \to \mathbf{a}_l^{m(\lambda)} \cup \mathbf{b}_l^{m(\lambda)}))(\sum_{m \leqslant K} pr(A^{m(\lambda)} \cup B^{m(\lambda)} \to \mathbf{a}_m^{m(\lambda)} \cup \mathbf{b}_j^{m(\lambda)})) = pr(A^{m(\lambda)} \cup B^{m(\lambda)} \to \mathbf{a}_i^{m(\lambda)} \cup \mathbf{b}_j^{m(\lambda)}). \quad (3.4)$$

We will now argue that there is a BST model with finitely many types of runs and propensities that fully represents the statistical experiment described, and in this model space-like related transitions are independent propensity-wise. That is, there is a finite partition Λ of $\mathcal{N}$ such that for every $\lambda \in \Lambda$, its weight w_λ is non-zero and the propensity assignment satisfies Equations (3.2) and (3.4). Our task is to find weights w_λ and propensities pr for any set of experimental probabilities p^{ex} that can arise in this experiment.

To introduce a shorter notation, let us put:

$$q_{ij} := p^{ex}(\tilde{a}_i \text{ and } \tilde{b}_j) \quad p_{ij\lambda} := pr(A^{m(\lambda)} \cup B^{m(\lambda)} \to \mathbf{a}_i^{m(\lambda)} \cup \mathbf{b}_j^{m(\lambda)}) \cdot w_\lambda$$

Then we rewrite the above two sets (3.2) and (3.4) of equations as:

$$q_{ij} = \sum_{\lambda \in \Lambda} p_{ij\lambda} \quad \text{and} \quad (\sum_{j \leqslant J} p_{kj\lambda}) \cdot (\sum_{i \leqslant K} p_{im\lambda}) = w_\lambda p_{km\lambda}. \quad (3.5)$$

We assume that q_{ij}'s are given and solve for $p_{ij\lambda}$'s and w_λ's.

We will argue that there are always quasi-deterministic solutions of Equations (3.5), in the sense that every probability $pr(A^n \cup B^n \to \mathbf{a}_i^n \cup \mathbf{b}_j^n)$ is either zero or one. To see that this is indeed the case, we assume that every probability of the form above is either zero or one, and then define partition Λ of natural numbers into at most $I \times J$ subsets λ_{ij} ($i \leqslant K, j \leqslant J$) such that:[10]

$$\lambda_{ij} := \{n \in \mathcal{N} \mid pr(A^n \cup B^n \to \mathbf{a}_i^n \cup \mathbf{b}_j^n) = 1\}.$$

Then we may easily check that we have these solutions for Eqs. 3.5:

$$p_{ij\lambda} = q_{ij} \text{ if } \lambda = \lambda_{ij} \text{ and } 0 \text{ otherwise,} \quad (3.6)$$
$$\omega_\lambda = q_{ij} \text{ if } \lambda = \lambda_{ij} \text{ and } 0 \text{ otherwise.} \quad (3.7)$$

Observe that with these solutions, types of runs work deterministically: if we are in n-th run and the run is of type λ_{ij}, the propensity for the transition $(A^n \cup B^n \to \mathbf{a}_i^n \cup \mathbf{b}_j^n)$ is one, and the propensity for a transition to any

---

[10] The partition has less than $I \times J$ elements if $q_{ij} = 0$ for some $i \leqslant K, j \leqslant J$.

other outcome of $A^n \cup B^n$ is zero. In the hidden-variable language, this is a determinate-value (or deterministic) hidden variable model.

We have just proved that there is propensity assignment that fulfills clause (3) of our Definition (6) of what it means that a BST model fully represents a statistical experiment. As for the remaining clauses, it is rather straightforward to construct an MBS model that satisfies the desiderata (†)—they guarantee that the model modally and spatiotemporally represents the statistical experiment. We need further ensure that each transition of the form $A^n \rightarrowtail \mathbf{a}_i$ or $B^n \rightarrowtail \mathbf{b}_j$ involves only finitely many finitely splitting points—this will guarantee that the transition is eligible for propensity assignment—see Section (4). As postulated above, the model has at most $I \times J$ types of runs. We need to take care that these types be distributed in history σ, in accordance with clause 2 of Definition (6). This is simple if every ω_λ is rational (which in turn depends on whether every $p^{ex}(\tilde{a}_i$ and $\tilde{b}_j)$ is rational) and it is only a little more complex if the above numbers are irrational (they will be recovered as limiting frequencies). Finally, to satisfy clause 4 of the Definition 3, one may require that in a run of type λ_{ij} outcomes $\mathbf{a}_i$ and $\mathbf{b}_j$ occur in history σ. For scarcity of space, we leave the explicit construction of the model to the reader. For the same reason, we do not enter here into a more exciting topic, namely whether it is possible to construct a full fledged indeterministic model with uncorrelated propensities for the experiment, where "full fledged" means that all (or, most) relevant transitions have neither propensity zero nor propensity one.[11]

4. BST, MBS, and Causal Probability Spaces

The aim of Belnap's (1992) theory of branching space-times (BST) is to combine objective indeterminism and relativity. A BST model is a non-empty partially ordered set W subject to some postulates (listed below). W is called 'Our World' and interpreted as the set of all possible point events. A partial ordering $\leqslant$ on W is interpreted as a pre-causal order between point events.[12] Typically Our World W has many possible scenarios, this last idea being rendered by a technical notion of history:

Definition 7
A set $h \subseteq W$ is upward-directed iff $\forall e_1, e_2 \in h \; \exists e \in h$ such that $e_1 \leqslant e$ and $e_2 \leqslant e$.

---

[11] For a discussion, cf. Placek (2000a).
[12] We read '$e_1 \leqslant e_2$' as 'e_2 is in a possible future of e_1,' or as 'e_1 can causally influence e_2.'

A set h is maximal with respect to the above property iff $\forall g \in W$ such that $h \subsetneq g$, g is not upward-directed.
A subset h of W is a history iff it is a maximal upward-directed set.
Hist is the set of all histories in W.
For $h_1, h_2 \in Hist$, any maximal element in $h_1 \cap h_2$ is called a choice event for h_1 and h_2.

A BST model is then defined as follows (for more information about BST, see Belnap (1992)):

Definition 8
$\langle W, \leqslant \rangle$, where W is a nonempty set and $\leqslant$ is a partial ordering on W, is a model of BST if and only if it meets the following requirements:

1. The ordering $\leqslant$ is dense.

2. W has no maximal elements with respect to $\leqslant$.

3. Every lower bounded chain in W has an infimum in W.

4. Every upper bounded chain in W has a supremum in every history that contains it.

5. For any lower bounded chain $O \subseteq h_1/h_2$ there exists an event $e \in W$ such that e is maximal in $h_1 \cap h_2$ and $\forall e' \in O : e < e'$ (this is the Prior Choice Principle, abbreviated as PCP).

A consequence of PCP is that every two histories overlap and have at least one choice point. The postulates also imply that the relation $\equiv_e$ of undividedness of histories at event e (defined by $h \equiv_e h'$ iff $\exists e' : e < e' \wedge e' \in h \cap h'$) is an equivalence relation on the set $H_{(e)} := \{h \in Hist \mid e \in h\}$. Thus, $\equiv_e$ induces a partition Π_e of the set $H_{(e)}$; we write $\Pi_e \langle h \rangle$ (where $h \in H_{(e)}$) for the unique element of partition Π_e to which h belongs. And we say that event e is finitely splitting if $card(\Pi_e)$ is larger than one but finite. For I a subset of a history, one may define undividedness in I: $h \equiv_I h'$ iff $h \equiv_e h'$ for every $e \in I$. This is an equivalence relation on the set $H_{[I]} := \{h \in Hist \mid I \subseteq h\}$, so it induces a partition of it, written as Π_I.

BST allows one to define a few concepts of events (apart from point events) and relations between them. For our purposes, the most important are initial events and scattered outcome events, and the relation of being space-like related.

Definition 9 (types of events and SLR)
An initial event is an upper bounded subset of a history; an outcome chain

event is a lower bounded non-empty chain in W; a scattered outcome event is a set of outcome chain events all of which overlap some one history.

Two point events are space-like related (SLR) if they are distinct, share a history, and are incomparable by $\leqslant$; $A \subseteq h_1$ and $B \subseteq h_2$ are SLR iff every element of A and every element of B are SLR; Initial event A and scattered outcome $\mathbf{O} = \{O_\delta\}_{\delta \in \Delta}$ are SLR iff A and O_δ are SLR, for every $\delta \in \Delta$.

An important class of structures, as they play causal and probabilistic roles, are transitions. A transition is a pair consisting of an initial event and an outcome event located properly above it. We will only use two kinds of transitions:

Definition 10 (transitions)
A transition from initial event to scattered outcome event is a pair $\langle I, \mathbf{O} \rangle$, where I and $\mathbf{O}$ are, resp., an initial event and a scattered outcome event such that $I <_{\forall \exists} \mathbf{O}$, where the last condition means that $\forall e \in I\ \exists O \in \mathbf{O}\ \forall x \in O\ (e < x)$. A basic transition is a pair $\langle e, H \rangle$, where e is a choice event and $H \in \Pi_e$. We will denote transitions by $I \rightarrowtail \mathbf{O}$ and $e \rightarrowtail H$, resp.

A way of looking at EPR phenomena is that a combinatorially allowable history, say one with spin + on the left and spin + on the right, is not possible. A BST property used to analyze this phenomenon is called Modal Funny Business (MFB). A simplified MFB that is sufficient for present purposes is defined as follows:[13]

Definition 11 (MFB)
Two initial events A and B, and two histories h_A, h_B such that $A \subseteq h_A$ and $B \subseteq h_B$ constitute a case of modal funny business (MFB) iff (1) A SLR B and (2) $\Pi_A \langle h_A \rangle \cap \Pi_B \langle h_B \rangle = \emptyset$.

In what follows, we will always assume No Modal Funny Business.

We need a few "propositional" notions, especially the so-called occurrence propositions, defined in terms of sets of histories. We say that an event occurs in history h iff h belongs to the occurrence proposition for this event.

Definition 12 (propositions)
The occurrence proposition for point event e is the set $H_{(e)} := \{h \in Hist \mid e \in h\}$; the occurrence proposition for initial event I is the set $H_{[I]} := \{h \in Hist \mid I \subseteq h\}$; the occurrence proposition for outcome chain event O is the set $H_{\langle O \rangle} := \{h \in Hist \mid O \cap h \neq \emptyset\}$; the occurrence proposition for scattered outcome event $\mathbf{O}$ is the set $H_{\langle \mathbf{O} \rangle} := \bigcap_{O \in \mathbf{O}} H_{\langle O \rangle}$.

---

[13] For a general definition, cf. Müller et al. (2008).

Occurrence propositions permit one to define generally the notion of consistency. Two objects, X and Y are consistent if their occurrence propositions intersect non-emptily.

We turn next to causal notions. A novelty of Belnap's (2005) analysis is that he asks for causes of transitions. Given indeterminism, the best one can get as a causal analysis of transition $I \rightarrowtail \mathbf{O}$ is the notion of a factor that keeps $\mathbf{O}$ possible rather than prohibits its occurrence. The events relevant to this process are choice events at which some history in which I occurs branches from every history in which $\mathbf{O}$ occurs. These events are called *causal loci* of a transition; some particular basic transitions, whose initials are *past* causal loci, are called originating causes, or *causae causantes* of a transition considered. Belnap proves that they satisfy a version of Mackie's INUS condition. The definitions are as follows:

Definition 13 (pcl and cc)
Let $I \rightarrowtail \mathbf{O}$ be a transition from initial event to scattered event.
Event e belongs to the set $cl(I \rightarrowtail \mathbf{O})$ of cause-like loci of the transition iff $\exists h \in H_{[I]} \ h \perp_e H_{\langle \mathbf{O} \rangle}$.
Event e belongs to the set $pcl(I \rightarrowtail \mathbf{O})$ of past causal loci of the transition iff $e \in cl(I \rightarrowtail \mathbf{O})$ and $\exists O \in \mathbf{O} \forall x \in O \ e < x$.
Basic transition $t = e \rightarrowtail H$ belongs to the set $cc(I \rightarrowtail \mathbf{O})$ of causae causantes of the transition iff $e \in pcl(I \rightarrowtail \mathbf{O})$ and $H = \Pi_e \langle h \rangle$ for some history $h \in H_{\langle \mathbf{O} \rangle}$.

An important theorem, proved by Belnap (2002), says that if there is no MFB, then every element of the set of cause-like loci of a transition to a scattered outcome is in the past of some element of the scattered outcome. We put this theorem in the following form:

Theorem 14 (NB's theorem) *If there is no MFB in a BST model, then for any transition $I \rightarrowtail \mathbf{O}$ to scattered outcome, $cl(I \rightarrowtail \mathbf{O}) <_{\forall \exists} \mathbf{O}$.*

In our modeling of statistical experiments, we assume that the underlying space-times are Minkowski's. Thus, we need a class of BST models, in which histories are isomorphic to the Minkowski space-time. These are the so-called Minkowskian Branching Structures (MBS's).[14]

For our present purposes, it suffices to say that in an MBS point events have spatiotemporal locations; we write $s(e)$ for spatiotemporal location of point event e (for brevity, in what follows we abbreviate this term to "st-location"). In a natural way, we can use st-locations of point events to

---

[14] They were introduced by Placek (2000a), then rigorously defined for finitary cases by (Müller, 2002), and defined generally by Wroński and Placek (2009).

define the concept st-location for initial events, outcome chain events, scattered outcomes, etc.

4.1. Causal Probability Spaces

A (classical) probability space is a triple $\langle A, F, \mu \rangle$, where A is a non-empty set called a sample space, F_E is a Boolean σ-algebra of subsets of A (called event algebra), and μ is a normalized to unity, countably additive measure on F. For Müller's (2005) concept of causal probability space (the generalized version), consider a finite set E of finitely splitting points such that minimal elements of E are consistent (since E is finite, its every element is above, or identical to, some minimal element of E). The elements of a causal probability space $\langle A_E, F_E, \mu_E \rangle$ associated with set E are constructed as follows:

1. Consider the complete set of alternative basic transitions that have initials in E, i.e., set $\tilde{T}_E = \{t \in \mathscr{T}r \mid i(t) \in E)\}$, where $\mathscr{T}r$ is the set of basic transitions and $i(t)$ is the initial of transition t.

2. Identify the sample space A_E with the set of all maximally consistent subsets of $\tilde{T}_E$,

3. Since the case is finite, take for event algebra F_E the set-theoretical algebra of subsets of A_E,

4. Define a normalized to unity measure μ_E on F_E.

To explain clause 2 above, two basic transitions $e_1 \rightarrowtail H_1$ and $e_2 \rightarrowtail H_2$ are consistent iff $H_1 \cap H_2 \neq \emptyset$. As was mentioned before, there are finististic assumptions concerning the construction of causal probability spaces. For the record, we capture them here in this definition:

Definition 15 (eligible for producing a causal probability space)
$E \subset W$ is eligible for producing a causal probability space if E is a finite set of finitely splitting points and the minimal elements of E are consistent.

Many causal probability spaces may "live" in a given BST model, and a set of basic transitions may belong to many event algebras, each associated with a different subset E of W. Thus, a question arises of how to represent an element of a "smaller" probability space in a "larger" probability space so that one could subsequently postulate that an element of the smaller probability space and its representative in the larger probability space are ascribed the same value by the two probability measures. The idea that a set of transitions should be ascribed the same probability in the probability spaces of an appropriate kind is the meaning of the Marginal Property, which (following Müller (2005)) we state and postulate for the causal probability spaces.

First, to say which probability spaces are "appropriate" in the statement of the Marginal Property, we define extensions of probability spaces:

Definition 16 (extension)
Let $E \subseteq W$ and $E' \subseteq W$ be sets of splitting points, each eligible for the construction of a causal probability space, $\langle A_E, F_E, \mu_E \rangle$ and $\langle A_{E'}, F_{E'}, \mu_{E'} \rangle$, resp. Then $\langle A_{E'}, F_{E'}, \mu_{E'} \rangle$ is an extension of $\langle A_E, F_E, \mu_E \rangle$ iff $E \subseteq E'$, and there are no $e' \in E'/E$ and $e \in E$ such that $e' < e$.

This means that each "new" element e' is either above or SLR to elements of E.[15]

We next state when an element of one event algebra is representable in some other event algebra, and what its representative is. We do it in steps, defining first an auxiliary notion of *rep* of an element of one sample space in some other event algebra. With the help of this notion we then state what a representative of an element of one event algebra in some other event algebra is.

Definition 17 (representability)
Let $E \subseteq W$ and $E' \subseteq W$ be sets of splitting points, each eligible for the construction of a causal probability space, $\langle A_E, F_E, \mu_E \rangle$ and $\langle A_{E'}, F_{E'}, \mu_{E'} \rangle$, resp. $x \in F_E$ is said to be representable in $F_{E'}$ iff $E \subseteq E'$.
For $a \in A_E$, the rep $r_{E'}(a)$ of a in $F_{E'}$ is: $r_{E'}(a) := \{a' \in A_{E'} \mid a \subseteq a'\}$.
For $x \in F_E$, the representative $R_{E'}(x)$ of x in $F_{E'}$ is:

$$R_{E'}(x) := \bigcup \{z \in F_{E'} \mid \exists a \in A_E \ a \in x \wedge z = r_{E'}(a)\}.$$

Having the notion of representative, we may state the Marginal Property, which we assume in the rest of this paper. Marginal Property requires that an element of one event algebra and its representative in some other event algebra are assigned the same numerical value by the probability measures defined on the two event algebras:

Postulate (marginal property) Let probability space $\langle A_{E'}, F_{E'}, \mu_{E'} \rangle$ be an extension of probability space $\langle A_E, F_E, \mu_E \rangle$ and $x' \in F_{E'}$ be the representative of $x \in F_E$. Then

$$\mu_{E'}(x') = \mu_E(x).$$

---

[15]This and the next definition are Müller's definitions 13 and 12.

Recall now that an element of the sample space of a causal probability space is a maximal consistent set of basic transitions. One might have a foreboding that basic transitions should be independent. With this intuition it is tempting to postulate that the probability of an atom be the product of probabilities of its elements—basic transitions. As Nature can go her own unaccountable ways, we had better not assume this postulate, but merely single out those causal probability spaces that satisfy this condition.

Definition 18 (uncorrelated probability space)
Let $\langle A_E, F_E, \mu_E \rangle$ be a causal probability space. We say that it is uncorrelated if for every $a \in A_E$:
$$\mu_E(A) = \prod_{x \in a} \mu_{\{i(x)\}}(x),$$
where $i(x)$ is the initial event of basic transition x.

Note that a causal probability space is produced out of a set of basic transitions, subject to some requirements. Can we use it to assign probabilities to other BST structures, in particular to (non-basic) transitions? It seems utterly reasonable that if probability is defined for a certain transition, it should be derivable from probabilities of its *causae causantes*. This is a desideratum of the 'nothing but *causae causantes*' that Weiner and Belnap (2006) assumed (their Postulate 4–4):

> When $pr(I \rightarrowtail \mathbf{a}_i)$ is defined, nothing counts except nature-given stochastic features of its *causae causantes*—including the possibility that one may need to take into account not only probabilities of individual *causae causantes*, but also probabilities of certain sets of them, taken as operating jointly. [...]

In line with the tradition, we will call single-case objective probabilities of transitions propensities, and write them as $pr()$. Allowing that *causae causantes* may work together, the desideratum above boils down to this definition of propensities:

Definition 19 (propensities)
A propensity of transition $I \rightarrowtail \mathbf{O}$, if defined, is the following:
$$pr(I \rightarrowtail \mathbf{O}) = \mu_{pcl(I \rightarrowtail \mathbf{O})}(I \rightarrowtail \mathbf{O}).$$

5. Facts About Structures Occurring in EPR-Like Experiments

The purpose of this section is to justify our (seemingly loose) talk of propensities by relating it to hard facts about the so-called causal probability spaces of BST. The justification relies on showing that structures occurring in our BST models allow for defining causal probability spaces, exactly of the sort that assign propensities to the transitions considered. In our analysis of statistical experiments, we have typically considered finitely many transitions from one initial event to scattered outcome events, where the scattered outcome events are pairwise inconsistent. A set of such transitions represents a measurement event with its alternative possible outcome events (both kinds of events are token-level). We define:

Definition 20 (exhaustive alternative transitions)
Let $Tr = \{I \rightarrowtail \mathbf{a}_i \mid i \leqslant K\}$ be a set of transition from initial event I to scattered outcome events $\mathbf{a}_i$. We say that Tr is a set of exhaustive alternative transitions of the same st-location iff

1. $H_{\langle \mathbf{a}_i \rangle} \cap H_{\langle \mathbf{a}_j \rangle} = \emptyset$ if $i \neq j$;   alternative outcomes

2. $\forall h \exists j: h \in H_{[I]} \rightarrow h \in H_{\langle \mathbf{a}_j \rangle}$,   exhaustiveness

3. all $\mathbf{a}_i$'s have the same st-location.   same st-location

For each transition $A \rightarrowtail \mathbf{a}_i$, its set $pcl(A \rightarrowtail \mathbf{a}_i) =: pcl_i$ is consistent. We will always additionally assume that each pcl_i is finite and its every element is finitely splitting. We aim to build a causal probability space associated with $E := \bigcup_{i \leqslant K} pcl_i$, but for this to be viable, the minimal elements of E should be consistent. The fact below testifies that indeed this is the case.

Fact 21. *Let $Tr = \{I \rightarrowtail \mathbf{a}_i \mid i \leqslant K\}$ be a set of exhaustive alternative transitions of the same st-location such that each $pcl_i := pcl(I \rightarrowtail \mathbf{a}_i)$ is finite and its elements finitely splitting. Then the minimal elements in $pcl := \bigcup_{i \leqslant K} pcl_i$ are consistent.*

PROOF. Assume to the contrary that the set of minimal elements of pcl is not consistent. Then for every $h \in H_{[I]}$ there is a minimal element $e \in pcl$ such that $e \notin h$. Accordingly, one may take an arbitrary $h \in H_{[I]}$ and some $i \leqslant K$ such that $e \in pcl(I \rightarrowtail \mathbf{a}_i)$ but $e \notin h$ and e is minimal in pcl. Pick now some $h' \in H_{\langle \mathbf{a}_i \rangle}$; clearly $e \in h'$. By applying PCP to e, h, and h', we have $\exists e':$ $e' < e \wedge h \perp_{e'} h'$, from which $h \perp_{e'} H_{\langle \mathbf{a}_i \rangle}$ follows. Accordingly, $e' \in pcl(I \rightarrowtail \mathbf{a}_i)$. Hence $e' < e$ contradicts the assumption that e is a minimal element of pcl. □

We will now argue that for a set $Tr = \{I \rightarrowtail \mathbf{a}_i \mid i \leqslant K\}$ of exhaustive alternative transitions of the same st-location, such that $pcl := \bigcup_{i \leqslant K} pcl(I \rightarrowtail \mathbf{a}_i)$ is eligible for producing a causal probability space, the sample space A_{pcl} is exactly the set of all sets of causae causantes of transitions from Tr.

Fact 22. *Let $Tr = \{I \rightarrowtail \mathbf{a}_i \mid i \leqslant K\}$ be a set of exhaustive alternative transitions of the same st-location such that $pcl := \bigcup_{i \leqslant K} pcl(I \rightarrowtail \mathbf{a}_i)$ is eligible for producing a causal probability space. Then the sample space A_{pcl} of the probability space $\langle A_{pcl}, F_{pcl}, \mu_{pcl} \rangle$ is: $A_{pcl} = \{cc(I \rightarrowtail \mathbf{a}_i) \mid i \leqslant K\}$.*

PROOF: To the left first. Suppose for reductio that a certain $cc(I \rightarrowtail \mathbf{a}_i)$ is not a member of A_{pcl}, which means that it is not a *maximally* consistent subset of $\tilde{T}_{pcl}$ (recall that every set of causae causantes is consistent). Accordingly, there is $t \in \tilde{T}_{pcl}$, $t = e \rightarrowtail H$, such that $t \notin cc(I \rightarrowtail \mathbf{a}_i)$ and $cc(I \rightarrowtail \mathbf{a}_i) \cup \{t\}$ is consistent $(\star)$. If $e \in pcl(I \rightarrowtail \mathbf{a}_i)$, then either t belonged or was inconsistent with the set $cc(I \rightarrowtail \mathbf{a}_i)$. It thus must be that $e \in pcl(I \rightarrowtail \mathbf{a}_j)$ for some $j \neq i$, $j \leqslant K$. Accordingly, $e <_\exists \mathbf{a}_j$. But it cannot be that $(\dagger)$ $e <_\exists \mathbf{a}_i$: if $(\dagger)$ were true, $H_{\langle \mathbf{a}_i \rangle} \subseteq H^*$ for some $H^* \in \Pi_e$, and, accordingly, $e \in pcl(A \rightarrowtail \mathbf{a}_i)$. Since $\mathbf{a}_i$ and $\mathbf{a}_j$ have the same spatiotemporal st-location, for $\neg(e <_\exists \mathbf{a}_i)$ to obtain, e and $\mathbf{a}_i$ must be inconsistent, from which it follows that t and $cc(I \rightarrowtail \mathbf{a}_i)$ are inconsistent. Contradiction with $(\star)$.

In the opposite direction, A_{pcl} is the set of maximally consistent subsets of $\tilde{T} = \{t \in \mathscr{B} \mid i(t) \in pcl\}$, where $\mathscr{B}$ is the set of basic transitions and $i(t)$ is the initial event of basic transition t. Since by the premises Tr is a set of exhaustive alternative transitions of the same st-location, every $cc(I \rightarrowtail \mathbf{a}_i)$ is a maximally consistent subset of $\tilde{T}$. It remains to be seen that there is no maximal consistent subset of $\tilde{T}$ that is not $cc(I \rightarrowtail \mathbf{a}_i)$ for some $i \leqslant K$. By the condition of exhaustiveness of Definition 20, every element of $\tilde{T}$ belongs to some $cc(I \rightarrowtail \mathbf{a}_i)$ for $i \leqslant K$. It remains to be seen that no element of A_{pcl} can have two elements, $t_1 : e_1 \rightarrowtail H_1$ and $t_2 := e_2 \rightarrowtail H_2$ such that, for some $i \neq j \leqslant K$:

$t_1 \in cc(I \rightarrowtail \mathbf{a}_i)$ but $t_1 \notin cc(I \rightarrowtail \mathbf{a}_j)$ and $t_2 \notin cc(I \rightarrowtail \mathbf{a}_i)$ but $t_2 \in cc(I \rightarrowtail \mathbf{a}_j)$.

Assume for reductio that the above formula holds. It cannot be that $e_1 \in pcl(I \rightarrowtail \mathbf{a}_j)$; otherwise $t_1 \in cc(I \rightarrowtail a_j)$ or t_1 is inconsistent with t_2. Similarly, $e_2 \notin pcl(I \rightarrowtail \mathbf{a}_i)$. Clearly, $e_1 <_\exists \mathbf{a}_i$. Yet, $\mathbf{a}_i$ and $\mathbf{a}_j$ are in the same st-location. But it cannot be that $e_1 <_\exists \mathbf{a}_j$, since then e_1 would belong to $pcl(I \rightarrowtail \mathbf{a}_j)$. It must thus be that e_1 and $H_{\langle \mathbf{a}_j \rangle}$ are inconsistent, and hence, (since $e_1 <_\exists \mathbf{a}_i$) $H_{\langle \mathbf{a}_i \rangle} \cap H_{\langle \mathbf{a}_j \rangle} = \emptyset$. Hence, t_1 is inconsistent with t_2, which means that the two cannot belong to any one element of A_{pcl}. □

To represent a joint experiment in which two measurements occur in separate "stations," we need two sets of exhaustive alternative transitions of the same st-location. Each set represents a measurement in one station. It is a welcome consequence that such two sets induce a third set of exhaustive alternative transitions of the same st-location, naturally representing the joint measurement and its possible joint outcomes:

Fact 23. *Let $Tr_a := \{A \rightarrowtail \mathbf{a}_i \mid i \leqslant K\}$ and $Tr_b := \{B \rightarrowtail \mathbf{b}_j \mid i \leqslant J\}$ be sets of exhaustive alternative transitions of the same st-location such that $\mathbf{a}_i \cup \mathbf{b}_j$ is a scattered outcome event for every $i \leqslant K, j \leqslant J$. Then $Tr_{ab} := \{A \cup B \rightarrowtail \mathbf{a}_i \cup \mathbf{b}_j \mid i \leqslant K, j \leqslant J\}$ is a set of exhaustive alternative transitions of the same st-location.*

PROOF: Immediate. □

The induced set has an interesting property: the sets of past causal loci of, resp., Tr_a, Tr_b and the induced set TR_{ab}, are nicely related:

Fact 24. *For initial events A, B, scattered outcomes $\mathbf{a}, \mathbf{b}$, and $\mathbf{a} \cup \mathbf{b}$ such that $A <_{\forall \exists} \mathbf{a}$, $B <_{\forall \exists} \mathbf{b}$:*

$$pcl(A \cup B \rightarrowtail \mathbf{a} \cup \mathbf{b}) \subseteq pcl(A \rightarrowtail \mathbf{a}) \cup pcl(B \rightarrowtail \mathbf{b}).$$

PROOF. From the order relations assumed in the premises, $A \cup B <_{\forall \exists} \mathbf{a} \cup \mathbf{b}$. Let $e \in pcl(A \cup B \rightarrowtail \mathbf{a} \cup \mathbf{b})$. Then for some $h \in H_{A \cup B}$: $h \perp_e H_{\langle \mathbf{a} \cup \mathbf{b} \rangle}$ and $e <_\exists \mathbf{a} \cup \mathbf{b}$, and hence $e <_\exists \mathbf{a}$ or $e <_\exists \mathbf{b}$. Suppose the former (the other case is symmetric). Clearly, $h \in H_{[A]}$. Pick now an arbitrary $h' \in H_{\langle \mathbf{a} \cup \mathbf{b} \rangle}$; it follows that $h' \in H_{\langle \mathbf{a} \rangle}$ and $h \perp_e h'$, so $h \perp_e \Pi_e \langle h' \rangle$. Since $e <_\exists \mathbf{a}$ and $h' \in H_{\langle \mathbf{a} \rangle}$, it follows that $H_{\langle \mathbf{a} \rangle} \subseteq \Pi_e \langle h' \rangle$. Accordingly, $h \perp_e H_{\langle \mathbf{a} \rangle}$, which proves that $e \in pcl(A \rightarrowtail \mathbf{a})$. □

Fact 25. *Let $A \rightarrowtail \mathbf{a}$ and $A \cup B \rightarrowtail \mathbf{a} \cup \mathbf{a}$ be transitions from initials to scattered outcomes. Let $H_{[A]} \subseteq H_{[B]}$. Then $pcl(A \rightarrowtail \mathbf{a}) \subseteq pcl(A \cup B \rightarrowtail \mathbf{a} \cup \mathbf{b})$.*

PROOF: Let $e \in pcl(A \rightarrowtail \mathbf{a})$. Then for some $h \in H_{[A]}$: $h \perp_e H_{\langle \mathbf{a} \rangle}$. By the premise ($H_{[A]} \subseteq H_{[B]}$), $h \in H_{[A \cup B]}$. Since $H_{\langle \mathbf{a} \cup \mathbf{b} \rangle} \subseteq H_{\langle \mathbf{a} \rangle}$, it follows that $h \perp_e H_{\langle \mathbf{a} \cup \mathbf{b} \rangle}$. Thus $e \in pcl(A \cup B \rightarrowtail \mathbf{a} \cup \mathbf{b})$. □

Combining Facts (24) and (25) above, we arrive at the following:

Fact 26. *For initial events A, B, scattered outcomes $\mathbf{a}, \mathbf{b}$, and $\mathbf{a} \cup \mathbf{b}$ such that $A <_{\forall \exists} \mathbf{a}$, $B <_{\forall \exists} \mathbf{b}$, and $H_{[A]} = H_{[B]}$:*

$$pcl(A \cup B \rightarrowtail \mathbf{a} \cup \mathbf{b}) = pcl(A \rightarrowtail \mathbf{a}) \cup pcl(B \rightarrowtail \mathbf{b}).$$

This fact has a consequence for sets of causaes causantes:

Fact 27. *For initial events A, B, scattered outcomes $\mathbf{a}, \mathbf{b}$, and $\mathbf{a} \cup \mathbf{b}$ such that $A <_{\forall \exists} \mathbf{a}$, $B <_{\forall \exists} \mathbf{b}$, and $H_{[A]} = H_{[B]}$:*

$$cc(A \cup B \rightarrowtail \mathbf{a} \cup \mathbf{b}) = cc(A \rightarrowtail \mathbf{a}) \cup cc(B \rightarrowtail \mathbf{b}).$$

PROOF: As a hint for the proof, note that if $e <_\exists \mathbf{a} \cup \mathbf{b}$, then $\Pi_e \langle \mathbf{a} \cup \mathbf{b} \rangle$ is identical to $\Pi_e \langle \mathbf{a} \rangle$ or to $\Pi_e \langle \mathbf{b} \rangle$, or to both, depending on whether $e <_\exists \mathbf{a}$, or $e <_\exists \mathbf{b}$, or both. In the opposite direction, if $e <_\exists \mathbf{a}$, then $\Pi_e \langle \mathbf{a} \rangle$ is identical to $\Pi_e \langle \mathbf{a} \cup \mathbf{b} \rangle$, as $\mathbf{a} \cup \mathbf{b}$ is a scattered outcome by assumption. □

Suppose now that we have three sets of exhaustive alternative transitions: $Tr_a := \{A \rightarrowtail \mathbf{a}_i \mid i \leq K\}$, $Tr_b := \{B \rightarrowtail \mathbf{b}_j \mid i \leq J\}$ and $Tr_{ab} := \{A \cup B \rightarrowtail \mathbf{a}_i \cup \mathbf{b}_j \mid i \leq K, j \leq J\}$ (the induced set). Suppose further that their past causal loci yield causal probability spaces, associated respectively with $E_1 := \bigcup_{i \leq I} pcl(A \rightarrowtail \mathbf{a}_i)$, $E_2 := \bigcup_{j \leq J} pcl(B \rightarrowtail \mathbf{b}_j)$, and $E := \bigcup_{i \leq I, j \leq J} pcl(A \cup B \rightarrowtail \mathbf{a}_i \cup \mathbf{b}_j)$. We would like to represent elements of the event algebras associated with E_1 and E_2, resp., in the event algebra associated with E. But is E eligible for producing a causal probability space, given that each E_1 and E_2 is eligible for producing a causal probability space? And is the probability space associated with E an extension of probability spaces associated with, respectively, E_1 and E_2? Under a rather special condition of $H_{[A]} = H_{[B]}$, this is the case indeed.

We need to prove that (1) $E_l \subseteq E$ and (2) $e \in E/E_l \rightarrow (e >_\forall E_l \vee e \text{ SLR } E_l)$, (3) the minimal elements of E are consistent if minimal elements of each E_l are consistent, and (4) E is finite and its elements are finitely splitting, if each E_l is finite and its elements are finitely splitting ($l = 1, 2$).

Given that $H_{[A]} = H_{[B]}$, (1) is a consequence of Fact (25). (3) is a consequence of Fact (21). (4) follows from Fact (24). Finally, (2) is proved in this Fact:

Fact 28. *Suppose that there are three sets of exhaustive alternative transitions: $Tr_a := \{A \rightarrowtail \mathbf{a}_i \mid i \leq K\}$, $Tr_b := \{B \rightarrowtail \mathbf{b}_j \mid i \leq J\}$ and (the induced set) $Tr_{ab} := \{A \cup B \rightarrowtail \mathbf{a}_i \cup \mathbf{b}_j \mid i \leq K, j \leq J\}$ such that $H_{[A]} = H_{[B]}$. Then: if $e \in \bigcup_{i \leq K, j \leq J} pcl(A \cup B \rightarrowtail \mathbf{a}_i \cup \mathbf{b}_j) =: E$, but $e \notin \bigcup_{i \leq K} pcl(A \rightarrowtail \mathbf{a}_i) =: E_1$, then $e >_\forall E_1 \vee e \text{ SLR } E_1$.*

PROOF: Given Fact 26, the premises imply that $e \in pcl(B \rightarrowtail \mathbf{b}_j)$ for some $j \leq J$, but (†) $\forall i \leq K : e \notin pcl(A \rightarrowtail \mathbf{a}_i)$. For reductio, assume that $e < e'$ for some $e' \in E_1$. It follows that $e' \in pcl(A \rightarrowtail \mathbf{a}_k)$ for some $k \leq K$. Since $e < e'$, it must be that $e \in pcl(A \rightarrowtail \mathbf{a}_k)$. Contradiction with (†). □

Representatives. To investigate representability in these settings, let us assume that $Tr_a := \{A \rightarrowtail \mathbf{a}_i \mid i \leq K\}$ and $Tr_b := \{B \rightarrowtail \mathbf{b}_j \mid j \leq J\}$ are sets of exhaustive alternative transitions of the same st-location, such that $H_{[A]} = H_{[B]}$ and that the two sets induce $Tr_{ab} = \{A \cup B \rightarrowtail \mathbf{a}_i \cup \mathbf{b}_j \mid i \leq K, j \leq J\}$, yet another set of exhaustive alternative transitions of the same st-location.

Consider now four sets of splitting points, with which we wish to associate causal probability spaces:

$C := pcl(A \rightarrowtail \mathbf{a}_i)$ for a fixed $i \leqslant K$,
$D := \bigcup_{i \leqslant K} pcl(A \rightarrowtail \mathbf{a}_i)$,
$E := \bigcup_{i \leqslant K, j \leqslant J} pcl(A \cup B \rightarrowtail \mathbf{a}_i \cup \mathbf{b}_j)$,
$G := pcl(A \cup B \rightarrowtail \mathbf{a}_i \cup \mathbf{b}_j)$ for fixed $i \leqslant K, j \leqslant J$.

As explained in Definition (19), the propensity of $A \rightarrowtail \mathbf{a}_i$ is determined by a measure μ_C, taken on the set of causae causantes of the transition, that is:[16]

$$pr(A \rightarrowtail \mathbf{a}_i) = \mu_C(cc(A \rightarrowtail \mathbf{a}_i)).$$

Given that $pcl(A \rightarrowtail \mathbf{a}_i)$ is eligible for producing a causal probability space (for every $i \leqslant K$), does D have the same property? Clearly, $C \subseteq D$ and D is a finite set of finitely splitting points, if each $pcl(A \rightarrowtail \mathbf{a}_i)$ is. So D has minimal elements. Moreover, the minimal elements are consistent— cf. Fact (19). Is then the event algebra F_C representable in F_D? Obviously, $C \subseteq D$ and if $e \in D/C$, then $\forall e_1 \in C : e \not< e_1$; for, if $e < e_1$ and $e_1 \in C$, it means that $e_1 \in pcl(A \rightarrowtail \mathbf{a}_i)$, from which it follows that $e \in pcl(A \rightarrowtail \mathbf{a}_i)$. Contradiction with $e \in D/C$.

Finally, by Fact (22), $cc(A \rightarrowtail \mathbf{a}_i) \in A_C$, i.e., the singleton of $cc(A \rightarrowtail \mathbf{a}_i)$ is an atom in the event algebra associated with C. This singleton is also an atom in the event algebra associated to D, i.e., $cc(A \rightarrowtail \mathbf{a}_i) \in A_D$. For by the condition of alternative outcomes, adding to $cc(A \rightarrowtail \mathbf{a}_i)$ a transition from $cc(A \rightarrowtail \mathbf{a}_k)/cc(A \rightarrowtail \mathbf{a}_i)$ (where $k \neq i$) would destroy consistency. Thus,

$$\mu_D(cc(A \rightarrowtail \mathbf{a}_i)) = \mu_C(cc(A \rightarrowtail \mathbf{a}_i)) = pr(A \rightarrowtail \mathbf{a}_i).$$

Let us next turn to representability of events from F_D in the event algebra F_E. By Fact (26), F_E is eligible for producing a causal probability space. By Fact (25), $D \subseteq E$. And by Fact (28), if $e \in E/D$, then $e >_\forall D \vee e$ SLR D. Thus, F_D is representable in F_E. It is then immediate to see that the *rep* of $cc(A \rightarrowtail \mathbf{a}_i) \in A_D$ in F_E is the set: $\bigcup \{cc(A \cup B \rightarrowtail \mathbf{a}_i \cup \mathbf{b}_j) \mid j \leqslant J\}$. We thus have:

$$pr(A \rightarrowtail \mathbf{a}_i) = \mu_C(cc(A \rightarrowtail \mathbf{a}_i)) = \mu_D(cc(A \rightarrowtail \mathbf{a}_i)) =$$
$$\mu_E(\bigcup \{cc(A \cup B \rightarrowtail \mathbf{a}_i \cup \mathbf{b}_j) \mid j \leqslant J\}) = \sum_{j \leqslant J} \mu_E(cc(A \cup B \rightarrowtail \mathbf{a}_i \cup \mathbf{b}_j)),$$

(5.1)

---

[16] Since the singleton of $cc(A \rightarrowtail \mathbf{a}_i)$ belongs to F_C, and not $cc(A \rightarrowtail \mathbf{a}_i)$ itself, we should write $\mu_C(\{cc(A \rightarrowtail \mathbf{a}_i)\})$; to avoid eyestrain, we neglect the curly brackets.

where the last equation follows from the fact that the singleton of each $cc(A \cup B \rightarrowtail \mathbf{a}_i \cup \mathbf{b}_j)$ is an atom of F_E. Consider finally the set G: clearly, $G \subseteq E$, so if E is eligible for producing a causal probability space, so is G. By an argument exactly like the one at the beginning of this paragraph, F_G is representable in F_E. And clearly, $\{(cc(A \cup B \rightarrowtail \mathbf{a}_i \cup \mathbf{b}_j))\} \in F_G$ has the representative $\{(cc(A \cup B \rightarrowtail \mathbf{a}_i \cup \mathbf{b}_j))\} \in F_E$. Hence:

$$\mu_E(cc(A \cup B \rightarrowtail \mathbf{a}_i \cup \mathbf{b}_j)) = \mu_G(cc(A \cup B \rightarrowtail \mathbf{a}_i \cup \mathbf{b}_j)) = pr(A \cup B \rightarrowtail \mathbf{a}_i \cup \mathbf{b}_j). \quad (5.2)$$

The last equation follows from the Definition (19) of propensities. Combining equations (5.1) and (5.2) above, we get this rule on calculating propensities in models satisfying rather special conditions:

Fact 29. *Let each $\{A \rightarrowtail \mathbf{a}_i \mid i \leqslant K\}$ and $\{B \rightarrowtail \mathbf{b}_j \mid j \leqslant J\}$ be a set of exhaustive alternative transitions of the same st-location and $\mathbf{a}_i$ consistent with $\mathbf{b}_j$ for every $i \leqslant K, j \leqslant J$. Let each $E_i := pcl(A \rightarrowtail \mathbf{a}_i)$ be eligible for producing a causal probability space and $H_{[A]} = H_{[B]}$. Then*

$$pr(A \rightarrowtail \mathbf{a}_i) = \sum_{j \leqslant J} pr(A \cup B \rightarrowtail \mathbf{a}_i \cup \mathbf{b}_j). \quad (5.3)$$

6. Towards an Adequate Statement of Probabilistic Funny Business

Having seen how to calculate propensities in causal probability spaces satisfying a rather special condition, we now turn towards formulating a condition to the effect that two transitions, with SLR initials, are probabilistically independent, or (in the BST terminology), do not constitute probabilistic funny business (PFB). For our search of a proper condition, it is instructive to note that under the particular circumstances assumed in the last section, sets of causae causantes of two transitions do not overlap.

Fact 30. *Let each $\{A \rightarrowtail \mathbf{a}_i \mid i \leqslant K\}$ and $\{B \rightarrowtail \mathbf{b}_j \mid j \leqslant J\}$ be a set of exhaustive alternative transitions of the same st-location such that every $\mathbf{a}_i$ is consistent with every $\mathbf{b}_j$. Then*

$$pcl(A \rightarrowtail \mathbf{a}_i) \cap pcl(B \rightarrowtail \mathbf{b}_j) = \emptyset.$$

PROOF: Assume to the contrary that there is $e \in pcl(A \rightarrowtail \mathbf{a}_i) \cap pcl(B \rightarrowtail \mathbf{b}_j)$, which entails $e <_\exists \mathbf{a}_i \wedge e <_\exists \mathbf{b}_j$. There are now two cases:
(i) $H_{\langle \mathbf{a}_i \rangle} \perp_e H_{\langle \mathbf{b}_j \rangle}$, from which it follows that $\mathbf{a}_i$ and $\mathbf{b}_j$ are inconsistent, which contradicts the assumption.

(ii) there is some history h such that $h \notin H_{\langle \mathbf{a}_i \rangle}$, $h \notin H_{\langle \mathbf{b}_j \rangle}$, $h \perp_e H_{\langle \mathbf{a}_i \rangle}$ and $h \perp_e H_{\langle \mathbf{b}_j \rangle}$. But, since $e \in pcl(A \rightarrowtail \mathbf{a}_i)$ and $e \in h$, it must be that $A \subset h$, and then by Exhaustiveness, $h \in H_{\langle \mathbf{a}_k \rangle}$ for some $k \leqslant K$. By the condition of the same st-location of Definition 20: $e <_\exists \mathbf{a}_k$, and hence $H_{\langle \mathbf{a}_k \rangle} \perp_e H_{\langle \mathbf{b}_j \rangle}$, from which inconsistency of $\mathbf{a}_k$ and $\mathbf{b}_j$ follows, contradicting the assumption. □

The immediate consequence is no overlap of the respective sets of causae causantes:

Fact 31. *Let each $\{A \rightarrowtail \mathbf{a}_i \mid i \leqslant K\}$ and $\{B \rightarrowtail \mathbf{b}_j \mid j \leqslant J\}$ be a set of exhaustive alternative transitions of the same st-location such that every $\mathbf{a}_i$ is consistent with every $\mathbf{b}_j$. Then*

$$cc(A \rightarrowtail \mathbf{a}_i) \cap cc(B \rightarrowtail \mathbf{b}_j) = \emptyset.$$

What would be a proper statement of the probabilistic independence of two transitions $A \rightarrowtail \mathbf{a}$ and $B \rightarrowtail \mathbf{b}$, each belonging to a different set of exhaustive alternative transitions of the same st-location, subject to the extra condition that $H_{[A]} = H_{[B]}$? By Facts (27) and (31), if the causal probability space associated with $E = pcl(A \cup B \rightarrowtail \mathbf{a}_i \cup \mathbf{b}_j)$ is uncorrelated and $H_{[A]} = H_{[B]}$, the following obtains:

$$\prod_{e \in pcl(A \rightarrowtail \mathbf{a}_i)} \mu_{\{e\}}(e \rightarrowtail \Pi_e \langle \mathbf{a}_i \rangle) \cdot \prod_{e \in pcl(B \rightarrowtail \mathbf{b}_j)} \mu_{\{e\}}(e \rightarrowtail \Pi_e \langle \mathbf{b}_j \rangle)$$
$$= \prod_{e \in pcl(A \cup B \rightarrowtail \mathbf{a}_i \cup \mathbf{b}_j)} \mu_{\{e\}}(e \rightarrowtail \Pi_e \langle \mathbf{a}_i \cup \mathbf{b}_j \rangle) \qquad (6.1)$$
$$= \mu_{pcl(A \cup B \rightarrowtail \mathbf{a}_i \cup \mathbf{b}_j)}(cc(A \cup B \rightarrowtail \mathbf{a}_i \cup \mathbf{b}_j)).$$

Combining this equation with the formula for propensities and Equation (5.3), we get this definition of a special case of probabilistic independence (or of No PFB):

Definition 32 (probabilistic independence, special case)
Let $A \rightarrowtail \mathbf{a}$ and $B \rightarrowtail \mathbf{b}$ be transitions from initial events to scattered outcome events and $\mathbf{a}$ be consistent with $\mathbf{b}$ and $H_{[A]} = H_{[B]}$. Then $A \rightarrowtail \mathbf{a}$ and $B \rightarrowtail \mathbf{b}$ are probabilistically independent iff

$$pr(A \rightarrowtail \mathbf{a}) \cdot pr(B \rightarrowtail \mathbf{b}) = pr(A \cup B \rightarrowtail \mathbf{a} \cup \mathbf{b}).$$

The immediate consequence of this definition and Equation 6.1 is the following fact:

Fact 33. *Let $A \rightarrowtail \mathbf{a}$ and $B \rightarrowtail \mathbf{b}$ be transitions from initial events to scattered outcome events and $\mathbf{a}$ be consistent with $\mathbf{b}$ and $H_{[A]} = H_{[B]}$. Then $A \rightarrowtail \mathbf{a}$ and $B \rightarrowtail \mathbf{b}$ are probabilistically independent if the probability measure $\mu_{pcl(A \cup B \rightarrowtail \mathbf{a} \cup \mathbf{b})}$ is uncorrelated.*

This definition, although useful in analysis of an EPR-style set-up of Section (3), is not applicable to the Bell-Aspect experiment, because of the assumption $H_{[A]} = H_{[B]}$. In the experiment, a measurement event in one wing may occur together with each of a few measurement events in the other wing. Thus, we need to remove this assumption. As a consequence of relaxing the above condition, we need something like a "conditional propensity," since we will now investigate, let's say, the propensity of transition $A \rightarrowtail \mathbf{a}_i$, on the supposition that B occurs. It is not immediately clear how this concept can be related to causal probability spaces, since B is not even a transition, so there is no room for it in an event algebra of a causal probability space.

The propensity of $I \rightarrowtail \mathbf{O}$ conditional on initial J (written as $pr(I \rightarrowtail \mathbf{O} \mid J)$) should be derived from the set of only those elements of the set $\tilde{T}_{pcl(I \rightarrowtail \mathbf{O})}$ of the alternative basic transitions that do not prohibit the occurrence of J. In other words, we need to take into account only those elements of $\tilde{T}_{pcl(I \rightarrowtail \mathbf{O})}$ that are consistent with J. Given No MFB, every transition from $\tilde{T}_{pcl(I \rightarrowtail \mathbf{O})}$ whose initial is not below some element of J is consistent with J. As for $t \in \tilde{T}_{cc(I \rightarrowtail \mathbf{O})}$, whose initial is below some element of J, it is consistent with J only if it belongs to the set $cc(I \rightarrowtail \mathbf{O})$ of causae causantes and J is consistent with $\mathbf{O}$. Now, if all inconsistent alternative basic transitions with the same initial $e <_\exists J$ are removed from the set $\tilde{T}_{cc(I \rightarrowtail \mathbf{O})}$, in the "pruned" set $\tilde{T}$ there will only be a single basic transition with initial e, namely one belonging to $cc(I \rightarrowtail \mathbf{O})$. Clearly, a basic transition unaccompanied by its alternatives in the "pruned" set $\tilde{T}_{cc(I \rightarrowtail \mathbf{O})}$ has no probabilistic consequences. Hence, in constructing a causal probability space for propensity conditional on J, we remove from $pcl(I \rightarrowtail \mathbf{O})$ all its elements that lie below some element of J. Then in the usual way, we build a sample space out of maximal consistent subsets of the set of basic transitions, whose initials lie in the pruned set.

From these considerations, the following recipe emerges for the construction of a causal probability space for conditional propensity $pr(I \rightarrowtail \mathbf{O} \mid J)$:

Consider the set $E = \{x \in pcl(I \rightarrowtail \mathbf{O}) \mid x \not<_\exists J\}$. Construct the complete set $\tilde{T}$ of alternative basic transitions that have initials in E. As the sample space A_E take the set of all maximally consistent subsets of $\tilde{T}$. Then in a familiar way define the event algebra F_E and probability measure μ_E. The sought probability space is now the triple $\langle A_E, F_E, \mu_E\rangle$. Conditional propensity is then defined as follows:

Definition 34
For transition $I \rightarrowtail \mathbf{O}$ from initial I to scattered outcome $\mathbf{O}$, its propensity conditional on initial J, $pr(I \rightarrowtail \mathbf{O} \mid J)$, where J is consistent with $\mathbf{O}$, is the following:
$$pr(I \rightarrowtail \mathbf{O} \mid J) = \mu_E(cc_J(I \rightarrowtail \mathbf{O})),$$
where $E = \{x \in pcl(I \rightarrowtail \mathbf{O}) \mid x \not<_\exists J\}$ and $t \in cc_J(I \rightarrowtail \mathbf{O})$ iff $t \in cc(I \rightarrowtail \mathbf{O})$ and $i(t) \not<_\exists J$.

Next we have these facts relevant to conditional propensities:

Fact 35. *Let $A \rightarrowtail \mathbf{a}$ and $B \rightarrowtail \mathbf{b}$ be transitions to scattered outcomes such that $\mathbf{a}$ and $\mathbf{b}$ are consistent. Then:*
$$\text{if } t \in cc_B(A \rightarrowtail \mathbf{a}), \text{ then } t \in cc(A \cup B \rightarrowtail \mathbf{a} \cup \mathbf{b}), \tag{6.2}$$
where $cc_B(A \rightarrowtail \mathbf{a}) = \{t \in cc(A \rightarrowtail \mathbf{a}) \mid i(t) \not<_\exists B\}$.

PROOF: For the implication above to fail, there must be $t = \{e\} \rightarrowtail H$ such that for some $h \in H_{[A]}$: $h \perp_e H_{\langle \mathbf{a}\rangle}$ and $e \not<_\exists B$, but (†) for every $h' \in H_{[A]}$ such that $h' \perp_e H_{\langle \mathbf{a}\rangle}$, $h' \notin H_{[B]}$. Since $\mathbf{a}$ is consistent with $\mathbf{b}$, e and B are consistent as well. Given this consistency, $e \not<_\exists B$ means: $\forall x : (x \in B \rightarrow (e \text{ SLR } x \vee x \leqslant e))$. Consider now two subsets that partition B: $B_1 = \{x \in B \mid x \text{ SLR } e\}$ and $B_2 = \{x \in B \mid x \leqslant e\}$. Clearly, e SLR B_1. By No MFB, there is $h^* \in \Pi_e\langle h\rangle \cap H_{[B_1]}$. Hence $B_1 \subseteq h^*$ and (since $e \in h^*$) $B_2 \subseteq h^*$, so $B \subseteq h^*$. From $e \in h^*$, it also follows that $A \subseteq h^*$. And since $h^* \equiv_e h$, it must be that $h^* \perp H_{\langle \mathbf{a}\rangle}$. Contradiction with (†). □

The immediate consequence of this implication and Fact (24)[17] is that, for $cc_B(A \rightarrowtail \mathbf{a}) = \{t \in cc(A \rightarrowtail \mathbf{a}) \mid i(t) \not<_\exists B\}$ and symmetrically for cc_A, the following is true:
$$cc_B(A \rightarrowtail \mathbf{a}) \cup cc_A(B \rightarrowtail \mathbf{b}) = cc(A \cup B \rightarrowtail \mathbf{a} \cup \mathbf{b}). \tag{6.3}$$

Accordingly, if we frame the notion of probabilistic independence in terms of conditional propensities as defined above, and require further that $cc_B(A \rightarrowtail \mathbf{a}) \cap cc_A(B \rightarrowtail \mathbf{b}) = \emptyset$, we get that $A \rightarrowtail \mathbf{a}$ and $B \rightarrowtail \mathbf{b}$ are probabilistically independent if the relevant causal probability measure is uncorrelated—see the fact below. This motivates the following definition of probabilistic funny business.

---

[17] Additionally one needs to observe that if $e \in pcl(A \cup B \rightarrowtail \mathbf{a}_i \cup \mathbf{b}_j)$, then $\neg(e <_\exists A \cup B)$.

Definition 36 (probabilistic funny business)
Transitions $A \rightarrowtail \mathbf{a}$ and $B \rightarrowtail \mathbf{b}$ (where A and B are initial events and $\mathbf{a}$, $\mathbf{b}$ scattered outcomes) are a case of probabilistic funny business (PFB) iff

1. A SLR B;

2. $\mathbf{a} \cup \mathbf{b}$ is a scattered outcome event;

3. $cc(A \rightarrowtail \mathbf{a}) \cap cc(B \rightarrowtail \mathbf{b}) = \emptyset$ (causal separation);

4. $pr(A \rightarrowtail \mathbf{a} \mid B) \cdot pr(B \rightarrowtail \mathbf{b} \mid A) \neq pr(A \cup B \rightarrowtail \mathbf{a} \cup \mathbf{b})$.

With this definition, we have a desired fact: if the basic transitions are uncorrelated, then (if two transitions are causally separated, then they do not constitute a case of PFB).[18]

Theorem 37. *Let $A \rightarrowtail \mathbf{a}$ and $B \rightarrowtail \mathbf{b}$ be causally separated transitions from initial events to scattered outcomes such that A SLR B and $\mathbf{a} \cup \mathbf{b}$ is a scattered outcome. Then if the measure $\mu_{pcl(A \cup B \rightarrowtail \mathbf{a} \cup \mathbf{b})}$ is uncorrelated, $A \rightarrowtail \mathbf{a}$ and $B \rightarrowtail \mathbf{b}$ are not a case of PFB.*

PROOF. Assume this notation: $G = pcl(A \cup B \rightarrowtail \mathbf{a} \cup \mathbf{b})$, $F = \{e \in pcl(A \rightarrowtail \mathbf{a}) \mid e \not<_\exists B\}$, $E = \{e \in pcl(B \rightarrowtail \mathbf{b}) \mid e \not<_\exists A\}$, $cc_A(B \rightarrowtail \mathbf{b}) = \{t \in cc(B \rightarrowtail \mathbf{b}) \mid i(t) \not<_\exists A\}$, and $cc_B(A \rightarrowtail \mathbf{a}) = \{t \in cc(A \rightarrowtail \mathbf{a}) \mid i(t) \not<_\exists B\}$. Our theorem follows from this sequence of equations:

$$\mu_G(cc(A \cup B \rightarrowtail \mathbf{a} \cup \mathbf{b})) \stackrel{1}{=} \prod_{t \in cc(A \cup B \rightarrowtail \mathbf{a} \cup \mathbf{b})} \mu_{(i(t))}(t) \stackrel{2}{=}$$

$$\prod_{t \in cc_A} \mu_{(i(t))}(t) \cdot \prod_{t \in cc_B} \mu_{(i(t))}(t) \stackrel{3}{=} \mu_F(cc_A) \cdot \mu_E(cc_B),$$

where $\stackrel{1}{=}$ and $\stackrel{3}{=}$ hold because basic transitions are uncorrelated, and $\stackrel{2}{=}$ because of Fact (35) (cf. Equation (6.3)) and causal separation. □

Now, in rather special circumstances, which however are most likely satisfied in experimental tests of Bell's theorem, causal separation of transitions is true.

Fact 38. *Let $\mathfrak{A} = \{A \rightarrowtail \mathbf{a}_i \mid i \leq K\}$ be a set of exhaustive transitions of the same st-locations, and $B \rightarrowtail \mathbf{b}$ be a transition to a scattered outcome such that $\mathbf{a}_i$ is consistent with $\mathbf{b}$ for every $i \leq K$. Then $A \rightarrowtail \mathbf{a}_i$ is causally separated from $B \rightarrowtail \mathbf{b}$ for every $i \leq K$.*

---

[18] This is an analogue of Müller's (2005) Theorem 1.

PROOF. For $A \rightarrowtail \mathbf{a}_i$ and $B \rightarrowtail \mathbf{b}$ *not* to be causally separated, there must be some $e \in pcl(A \rightarrowtail \mathbf{a}_i) \cap pcl(B \rightarrowtail \mathbf{b})$, from which it follows that for some $h_1 \in H_{[A]} : h_1 \perp_e H_{\langle \mathbf{a}_i \rangle}$. By exhaustiveness condition, for some outcome event $\mathbf{a}_k$ of some transition from $\mathfrak{A}$: $h_1 \in H_{\langle \mathbf{a}_k \rangle}$. Hence e is consistent with $\mathbf{a}_k$. Hence, by same st-location, since $e <_\exists \mathbf{a}_i$: $e <_\exists \mathbf{a}_k$. We will now argue that, contrary to the premise of the fact, $\mathbf{a}_k$ and $\mathbf{b}$ are inconsistent. Pick a "witness" history $h^* \in H_{\langle \mathbf{a}_i \rangle} \cap H_{\langle \mathbf{b} \rangle}$. Observe that for $h \in H_{\langle \mathbf{a}_k \rangle}$, since $e <_\exists \mathbf{a}_k$, it must be that $h \equiv_e h_1$, and hence $h \perp_e h^*$, whereas for $h \in H_{\langle \mathbf{b} \rangle}$, (since $e <_\exists \mathbf{b}$) it must be that $h \equiv_e h^*$. Contradiction. □

6.1. Parameter Independence

A crucial premise in Bell's theorem is that an outcome of a measurement performed in one wing is to be independent of what is measured in the other wing, this being the content of Parameter Independence (PI). The present framework affords it a precise reading: the (conditional) propensity of a transition $A \rightarrowtail \mathbf{a}$ is the same, no matter whether conditioned on event B or on B', where each B and B' are SLR to A. But is Parameter Independence true in the present framework?

The answer is that, given certain assumptions, naturally thought to be satisfied in a Bell's setup, Parameter Independence is true. To prove a theorem to this effect, let us first observe this fact:

Fact 39. *Let A, B and B' be initial events and B have the same st-location as B'. Let also $A \rightarrowtail \mathbf{a}$ be a transition to scattered outcome $\mathbf{a}$ and $\mathbf{a}$ SLR B as well as $\mathbf{a}$ SLR B'. Then $cc_B(A \rightarrowtail \mathbf{a}) = cc_{B'}(A \rightarrowtail \mathbf{a})$.*

PROOF: We need to show that for every $e \in pcl(A \rightarrowtail \mathbf{a})$: $e \not<_\exists B$ iff $e \not<_\exists B'$. But since $\mathbf{a}$ is SLR to B as well as to B', by No MFB there are histories $h_1 \in H_{\langle \mathbf{a} \rangle} \cap H_{[B]}$ and $h_2 \in H_{\langle \mathbf{a} \rangle} \cap H_{[B']}$. Since $H_{\langle e \rangle} \subseteq H_{\langle \mathbf{a} \rangle}$, the two histories testify that e is consistent with B and e is consistent with B', respectively. The equivalence above follows then from same st-location. □

From the fact it follows that Parameter Independence is true. That is, if an outcome $\mathbf{a}$ of measurement A in one wing of the experiment is SLR to each alternative measurement selectable in the other wing, then this outcome is independent, propensity-wise, from measurements in the other wing. More precisely, we have: $pr(A \rightarrowtail \mathbf{a} \mid B) = pr(A \rightarrowtail \mathbf{a} \mid B')$.

As it stands, the proof of the theorem depends on the same st-location of B and B', which means of course that the two events are inconsistent. This suggests a possibility of proving the same conclusion from a different set of premises, in which same st-location of B and B' is replaced by the

7. Bell-Aspect Experiment

The experiment that Clauser, Shimony, Horne and Holt (1969) envisaged was carried out by Aspect and his team in 1980–1981. It seems to involve two levels of chanciness: The production of results, a_1, a_2 on the left and b_1 and b_2 on the right, seem to be chancy. And there is (apparently) freedom in the selection of settings on the apparatus to the left as well as on the apparatus to the right. We will denote these selection events by S_L and S_R, respectively. As a result of the selection on the left, there is either the measurement event Li or the measurement event Li', read as the measurement in the left wing with the setting put to i (i').[20] Similarly on the right, after selection event S_R there is either measurement Rj or measurement Rj'. To keep track of the wings and settings, we write Lia for result a of the measurement on the left, with setting i, and similarly for $Li'a, Rjb$ and $Rj'b$. As for spatiotemporal features, a selection event in one wing is space-like related to results occurring in the other wing, and, consequently, a measurement event performed in one wing is space-like related to results occurring in the other wing.

Let us now focus on a statistical aspect of the experiment described above. We assume the idealization that the experiment has infinitely many runs, and the results obtained on the runs permit one to estimate probabilities of "joint results," like $p^{ex}(Lia \wedge Rjb)$, and probabilities of "single" results, like $p^{ex}(Lia)$. (In total, there are 16 probabilities of the former kind, and 8 of the latter kind.) But how are the runs of this experiment differentiated? It is natural to say that the experiment is 'timed' by emissions for pairs of particles from the source. Yet, we will not lose on generality, and avoid a commitment to ontology of particles, if we postulate that runs of the experiment are differentiated by the occurrences of pairs of selection events.

To recall, run of the experiment is a modally thick notion, as it involves measurement events with all possible settings, and all possible results. As for types of runs, they are differentiated by propensity assignments to the

---

[19] More precisely, choice events between histories comprising B and histories comprising B' should be SLR to **a**.

[20] To add some physics, the measurement on the left of the electron's spin projection on the direction i (or i').

transitions involved. This has a consequence that although some two occurrences of, for instance, S_L look perfectly similar, they may still belong to different types of runs, as transitions to some corresponding results occurring in two runs obtain different propensities. In a similar vein, two runs are subsumed to different types because some transition that has not occurred in either of these runs (this being said from some later point of view), receives different propensities on these runs.

We list now some properties, idealizations, and desiderata on a run of the Bell-CH experiment.

REQUIREMENTS:

1. Li is inconsistent with Li' and Rj is inconsistent with Rj';[21]
2. selection events S_L and S_R are consistent and $S_L <_\forall Li \cup Li'$, and $S_R <_\forall Rj \cup Rj'$;
3. there are exactly two setting at each wing: $\forall h: S_L \subset h \Rightarrow Li \subset h \lor Li' \subset h$ and $\forall h: S_R \subset h \Rightarrow Rj \subset h \lor Rj' \subset h$;
4. same st-location of alternative measurements: $s(Li) = s(Li')$, $s(Rj) = s(Rj')$, etc.;
5. same st-location of alternative outcomes of each measurement: $s(Lia) = s(Lia')$, $s(Li'a) = s(Li'a')$, $s(Rjb) = s(Rjb')$, and $s(Rj'b) = s(Rj'b')$;
6. SLR relations: Lia SLR S_R, $Li'a$ SLR S_R, Rjb SLR S_L, and $Rj'b$ SLR S_L;
7. exhaustiveness: if $h \in H_{[Li]}$, then there is a result a such that $h \in H_{\langle Lia \rangle}$, and analogously about Li', Rj and Rj';
8. every measurement has three possible results: $+$, $-$, and f, where f is a "failed outcome"; that is: $a, b \in \{+, -, f\}$;
9. every outcome of every measurement on the left is consistent with every outcome of every measurement on the right.

We take it that selection events as well as measurement events are represented by BST initial events, whereas outcomes are interpreted as scattered outcome events. Then, as a result of the above conditions, in each run n of the experiment there are four sets of exhaustive transitions of the same st-location (keep in mind that a run is a modally thick notion, and runs are

---

[21] Since $Lia \cup Rjb$ and the like are scattered outcomes, and $Li \cup Rj$ is below it, the latter is consistent.

differentiated by occurrences of pairs of selection events):

$$\{Li^n \cup Rj^n \rightarrowtail Lia^n \cup Rjb^n \mid a,b \in \{+,-,f\}\} \quad (7.1)$$
$$\{Li'^n \cup Rj^n \rightarrowtail Li'a^n \cup Rjb^n \mid a,b \in \{+,-,f\}\} \quad (7.2)$$
$$\{Li^n \cup Rj'^n \rightarrowtail Lia^n \cup Rj'b^n \mid a,b \in \{+,-,f\}\} \quad (7.3)$$
$$\{Li'^n \cup Rj'^n \rightarrowtail Li'a^n \cup Rj'b^n \mid a,b \in \{+,-,f\}\}. \quad (7.4)$$

That is, above the selection $S_L^n \cup S_R^n$, there are the above four sets of transitions in the model. Yet, every history passing through this selection passes through exactly one of the initials $Li^n \cup Rj^n, Li'^n \cup Rj^n, Li^n \cup Rj'^n$ and $Li'^n \cup Rj'^n$. Hence, if a history σ in an MBS is to fully represent this experiment, we need to ensure that each of the four types of initials above occurs in σ sufficiently many times. This follows from conditions (3) and (4) of Definition (6). Note, however, that the first clauses of the above definition need to be modified, to accommodate *alternative* measurements, the gist of the Bell-CH set-up. The general definition of an MBS model fully representing a statistical experiment adequate for the Bell-CH setup is provided in the Appendix.

In the literature, the Bell-CH inequality (which is violated by quantum mechanics and most likely, by Nature as well) is derived from the three "independence" premises: Outcome Independence, Parameter Independence, and No Conspiracy. The first says that outcomes in two wings are probabilistically independent, conditional on the measurement settings and a value of the hidden variable. The second is the claim that the outcome in one wing is probabilistically independent of the measurement setting selected in the other, conditional on a value of the hidden variable. The third amounts to assuming that the measurement settings and values of hidden variables are probabilistically independent.

As one might expect, a similar bunch of conditions will figure in our BST derivation of the Bell-CH inequality; there will, however, be some significant alterations. To draw a link between the two approaches, a value of a hidden variable is here understood as a type of runs of the experiment. Outcome Independence (recall, it is conditional on a hidden variable) amounts to No Probabilistic Funny Business: observe that the latter is supposed to hold in each type of runs. Parameter Independence presently takes on the following form:

$$pr(Li^{m(\lambda)} \rightarrowtail Lia^{m(\lambda)} \mid Rj^{m(\lambda)}) = pr(Li^{m(\lambda)} \rightarrowtail Lia^{m(\lambda)} \mid Rj'^{m(\lambda)}), \quad (7.5)$$

where $m(\lambda)$ abbreviates $min(\lambda)$. Observe that the set-up considered satisfies the premises of Parameter Independence: points (6) and (9) of the description of the set-up imply that every measurement in one wing is SLR to every outcome in the other wing. And, by clause (4), measurements in one wing have the same st-location. Hence Parameter Independence is provably true—cf. Theorem (39). This is the first surprise.

As for the condition of No Conspiracy, it is essential for our proof that in each run n of the experiment, each setting—Li^n, Li'^n, Rj^n and Rj'^n—is possible. It follows that in every run *of every type* each setting is possible. Note that this independence of selected settings and types of runs is modal and not probabilistic. This is the second surprise.[22]

The proof of Bell-CH inequalities takes advantage of the arithmetical fact which states that for any real numbers u, u', v, and v' from the $[0,1]$ interval,

$$-1 \leqslant uv + uv' + u'v' - u'v - u - v' \leqslant 0. \tag{7.6}$$

Make then these substitutions:

$$u = pr(Li^{m(\lambda)} \rightarrowtail Li+^{m(\lambda)} \mid Rj^{m(\lambda)}) = pr(Li^{m(\lambda)} \rightarrowtail Li+^{m(\lambda)} \mid Rj'^{m(\lambda)}) \tag{7.7}$$

$$v = pr(Rj^{m(\lambda)} \rightarrowtail Rj+^{m(\lambda)} \mid Li^{m(\lambda)}) = pr(Rj^{m(\lambda)} \rightarrowtail Rj+^{m(\lambda)} \mid Li'^{m(\lambda)}) \tag{7.8}$$

$$u' = pr(Li'^{m(\lambda)} \rightarrowtail Li'+^{m(\lambda)} \mid Rj^{m(\lambda)}) = pr(Li'^{m(\lambda)} \rightarrowtail Li'+^{m(\lambda)} \mid Rj'^{m(\lambda)}) \tag{7.9}$$

$$v' = pr(Rj'^{m(\lambda)} \rightarrowtail Rj'+^{m(\lambda)} \mid Li'^{m(\lambda)}) = pr(Rj'^{m(\lambda)} \rightarrowtail Rj'+^{m(\lambda)} \mid Li^{m(\lambda)}), \tag{7.10}$$

where the right-hand side equations rest on Parameter Independence. With the substitution made, multiply then the sides of Inequality (7.6) by w^λ, employ No PFB and sum over $\lambda \in \Lambda$. Finally, use the Empirical Adequacy condition, that is, clause (4) of Definition (6). The result is the constraint on observable probabilities— the Bell-CH inequality:

$$-1 \leqslant p^{ex}(Li+ \wedge Rj+) + p^{ex}(Li+ \wedge Rj'+) + p^{ex}(Li'+ \wedge Rj'+) - \\ p^{ex}(Li'+ \wedge Rj+) - p^{ex}(Li+) - p^{ex}(Rj'+) \leqslant 0. \tag{7.11}$$

The failure of this inequality means that there is no MBS fully representing the (idealized) statistical Bell-CH experiment, subject to the condition that outcomes are independent (in the sense that the pairs of corresponding transitions are not cases of PFB). More precisely, given the above condition, one cannot satisfy "probabilistic" conditions (1) and (3) of Definition (6).

---

[22] In recent papers of Hofer-Szabó (2008), Grasshoff, Portmann and Wüthrich (2005), and Portmann and Wüthrich (2007) Bell-type inequalities are derived from variously weakened versions of (probabilistic) No Conspiracy. One might thus have a premonition that *probabilistic* No Conspiracy is not needed for the proof.

8. Discussion

The main result of this paper is this: given (1) the assumptions needed to construct causal probability spaces and (2) the idealizations made in the construction of propensity-frequentist models for Bell-CH experiment, a frequentist-propensity model for Bell-CH that satisfies (i) (non-probabilistic) No Conspiracy and (ii) No Probabilistic Funny Business (No PFB), is not possible. No Conspiracy requires that in each run n of the experiment, it is possible to select each Li and Li' on the left and each Rj and Rj' on the right. No PFB is a BST rendition of Outcome Independence. Notably, Parameter Independence, which is another premise of Bell's theorem, is unnecessary, as it is derivable from the assumptions (1) and the idealizations (2), mentioned above. Thus, given that our world satisfies the assumptions (1) and the idealizations are "reasonably close to the reality," there is conspiracy in our world, or probabilistic funny business, or both. By Theorem 37, probabilistic funny business implies that the corresponding μ measure is correlated, which means that some *basic* transitions, with SLR initials, are correlated.

Since the main result is conditional on the assumptions (1) and the idealizations (2), let us examine them, starting with the former. First, there are finitistic assumptions in Müller's (2005) construction of causal probability spaces: the underlying set of choice events must be finite, and each element thereof—finitely splitting. As Müller himself indicates, there are ways to remove these limitations. Also, an important Fact 21 has an infinitistic version, i.e., with no finitistic assumptions in the premises. Another important assumption is No MFB. This means that although a joint occurrence of some SLR results is not observed in an experiment, a BST model of the phenomenon nevertheless has histories with the joint occurrence of the two results, but it has probability zero.

A similar move occurs in the idealizations (2). It is assumed that if some type-level measurement result $\tilde{a}$ of measurement event $\tilde{A}$ is observed in a statistical experiment, then in a model, if in the n-th run a corresponding (token-level) measurement event A^n occurs, then there is a history in which both A^n and $\mathbf{a}^n$ occur. This sounds like a proliferation of possibilities: if a result is possible, then it is possible in every run in which the corresponding measurement event occurs. The proliferation is mitigated by probabilities, however, as the results might have probability zero assigned. Other idealizations concern spatiotemporal locations of events, yet as we indicated above, they can be weakened, if one wants, but in my opinion they are still quite reasonable as they stand.

Thus, as the assumptions are rather weak, and the idealizations either realistic or capable of being appropriately weakened, we are facing a choice:

> there is conspiracy in our world, or some *basic* transitions, with SLR initials, are correlated.

Long live experimental metaphysics!

9. Appendix: General Propensity Models

By comparing a simplified EPR experiment of Section (3) with the Bell-CH experiment, we have seen that runs of experiments might be counted differently. Also, we have seen that, since the measurements considered might be incompatible, it should not be required that all measurement events occur in a run. This suggests a simplified and abstract approach to be sketched here.

To begin with, we assume that an experimenter somehow knows how to count runs of the experiment; suffice it to say that she counts the runs by natural numbers. Suppose next that our experimenter believes that there is a finite set $\mathfrak{X}$ of measurements (type-level) that can be carried out (instantiated), separately or jointly, at each run of the experiment. Let also be the case that every $\tilde{X} \in \mathfrak{X}$ has a finite set $\{\tilde{x}_1, \tilde{x}_2, \ldots, \tilde{x}_k\}$ of possible outcomes. Now, to capture this structure of possible measurements and their possible outcomes, we postulate that for every $n \in \mathcal{N}$ there is a set T^n consisting of transitions from possible measurements (token-level) to their possible outcomes (also token-level) that are available at the n-th run. In the context of Bell-CH, T^n should contain usual "measurement" transitions, like $(Li \cup Rj)^n \rightarrowtail (Lia \cup Rjb)^n$ as well as less standard "selection" transitions, like $S_L^n \rightarrowtail Li^n$. First we define what it means when we say that a history in an MBS weakly (i.e., modally and spatiotemporally) represents a statistical experiment:

Definition 40 (MBS weakly represents)
Consider a statistical experiment in which there is a set $\mathfrak{X}$ of measurements such that each is possible at every run and for every $\tilde{X} \in \mathfrak{X}$ there is a finite set $\mathfrak{P}_X := \{\tilde{x}_1, \tilde{x}_2, \ldots, \tilde{x}_k\}$ of possible outcomes. An MBS $\mathfrak{M}$ weakly represents this statistical experiment in terms of mapping Y, history σ and sequence $T^1, T^2, T^3, \ldots$ of sets of transitions iff

1. Y is an injective function from the set of triples $\langle n, \tilde{X}, \tilde{x}_i \rangle$, where $n \in \mathcal{N}$, $\tilde{X} \in \mathfrak{X}$ and $\tilde{x}_i \in \mathfrak{P}_X$, to the set of transitions from initials to scattered outcomes in $\mathfrak{M}$;

2. $\mathfrak{J}_X^n := \{Y(n,\tilde{X},\tilde{x}_i) \mid \tilde{x}_i \in \mathfrak{P}_X\}$ is the set of alternative exhaustive outcomes of the same st-location;

3. the sequence $T^1, T^2, \ldots$ is shifted in time, where $T^n = \bigcup_{\tilde{X} \in \mathfrak{X}} \mathfrak{J}_X^n$;

4. for every $n \in \mathcal{N}$, there is at least one $t \in T^n$, whose outcome occurs in σ.

We say $t \in T^m$ and $t' \in T^n$ are corresponding transitions iff there is $\tilde{X} \in \mathfrak{X}$ and $\tilde{x}_i \in \mathfrak{P}_X$ such that $t = Y(\langle m, \tilde{X}, \tilde{x}_i \rangle)$ and $t' = Y(\langle n, \tilde{X}, \tilde{x}_i \rangle)$.

Then we define types of runs.

Definition 41
The m-th run and the n-th run belong to the same type of runs iff for every $t \in T^m$ and $t' \in T^n$, if t and t' are corresponding transitions, then $pr(t) = pr(t')$.

Definition 42 (MBS model fully represents)
Let $\mathfrak{M}$ be an MBS that weakly represents, in terms of mapping Y, history σ and sequence $T^1, T^2, T^3, \ldots$, a statistical experiment characterized by sets $\mathfrak{X}$ and $\{\mathfrak{P}_X\}_{\tilde{X} \in \mathfrak{X}}$. We say that $\mathfrak{M}$ represents fully the experiment in terms of Y, σ and propensity assignment pr iff there is a partition Λ of the set $\mathcal{N}$ of natural numbers such that

1. for every $l, m \in \mathcal{N}$: $l, m \in \lambda$ for some $\lambda \in \Lambda$ iff for every $t \in T^l$ and $t' \in T^m$, if t and t' are corresponding transitions, then $pr(t) = pr(t')$,

2. $p^{ex}(\tilde{x}) = \sum_\lambda w_\lambda pr(Y(\langle min(\lambda), \tilde{X}, \tilde{x} \rangle))$, and

$$w_\lambda = \lim_{n \to \infty} \frac{\#\{m \in \mathcal{N} \mid m \leq n \land m \in \lambda\}}{n},$$

3. for every $\lambda \in \Lambda$:

$$pr(Y(\langle min(\lambda), \tilde{X}, \tilde{x} \rangle)) = \lim_{n \to \infty} \frac{\#\{m \in \mathcal{N} \mid m \leq n \land m \in \lambda \land \sigma \in H_{\langle x \rangle}\}}{\#\{m \in \mathcal{N} \mid m \leq n \land m \in \lambda\}}.$$

Let us write $X^n \rightarrowtail \mathbf{x}^n$ for $Y(\langle n, \tilde{X}, \tilde{x} \rangle)$. The transitions to consider in constructing a model of the Bell-CH experiment are the transitions from the sets of Equations (7.1)–(7.4), as well as "selection" transitions: $(S_L \cup S_R)^n \rightarrowtail (Li \cup Rj)^n$, $(S_L \cup S_R)^n \rightarrowtail (Li' \cup Rj)^n$, $(S_L \cup S_R)^n \rightarrowtail (Li \cup Rj')^n$, and $(S_L \cup S_R)^n \rightarrowtail (Li' \cup Rj')^n$. Because selection event $(S_L \cup S_R)^n$ is below any other event in any transition from T^n, by clause (4) of Definition (40) we get that $(S_L \cup S_R)^n$ occurs in history σ. As the experimenter strives to achieve a non-zero experimental probability of any combination of settings, in history σ there must be an infinite number of runs with each combination of settings.

References

Aspect, A., Dalibard, J., and Roger, G. (1982). Experimental test of Bell's inequalities using time-varying analyzers. *Physical Review Letters*, 49, 1804–1807.
Bell, J. S. (1987). *Speakable and Unspeakable in Quantum Mechanics*. Cambridge University Press, Cambridge.
Belnap, N. (1992). Branching space-time. *Synthese*, 92, 385–434. Postprint archived at PhilSci Archive, http://philsci-archive.pitt.edu/archive/00001003.
Belnap, N. (2002). EPR-like "funny business" in the theory of branching space-times. In T. Placek and J. Butterfield (Eds.), *Nonlocality and Modality* (293–315), NATO Science Series. Kluwer Academic Publisher, Dordrecht.
Belnap, N. (2005). A theory of causation: causae causantes (originating causes) as inus conditions in branching space-times. *British Journal for the Philosophy of Science*, 56, 221–253.
Belnap, N. and Szabó, L. (1996). Branching space-time analysis of the GHZ theorem. *Foundations of Physics*, 26(8), 982–1002.
Clauser, J., Holt, R., Shimony, A., and Horne, M. (1969). Proposed experiment to test local hidden-variable theories. *Physical Review Letters*, 23, 880–884.
Clauser, J. and Horne, M. (1974). Experimental consequences of objective local theories. *Physical Review D*, 10, 526–535.
Grasshoff, G., Portmann, S., and Wüthrich, A. (2005). Minimal assumption derivation of a Bell-type inequality. *British Journal for the Philosophy of Science*, 56, 663–680.
Greenberger, D., Horne, M., Shimony, A., and Zeilinger, A. (1990). Bell's theorem without inequalities. *American Journal of Physics*, 58(12), 69–72.
Hofer-Szabó, G. (2008). Separate- versus common -common-cause-type derivations of the Bell inequalities. *Synthese*, 163(2), 199–215.
Hofer-Szabó, G., Rédei, M., and Szabó, L. (1999). On Reichenbach's common cause principle and Reichenbach's notion of common cause. *British Journal for the Philosophy of Science*, 50, 377–399.
Kowalski, T. and Placek, T. (1999). Outcomes in branching space-time and GHZ-Bell theorems. *British Journal for the Philosophy of Science*, 50, 349–375.
Müller, T. (2002). Branching space-time, modal logic and the counterfactual conditional. In T. Placek and J. Butterfield (Eds.), *Nonlocality and Modality* (273–291), NATO Science Series. Kluwer Academic Publisher, Dordrecht.
Müller, T. (2005). Probability theory and causation: a Branching Space-Times analysis. *British Journal for the Philosophy of Science*, 56(3), 487–520.
Müller, T., Belnap, N., and Kishida, K. (2008). Funny business in branching space-times: Infinite modal correlations. *Synthese*, 164(1), 141–159.
Placek, T. (2000a). *Is Nature Deterministic? A Branching Perspective on EPR Phenomena*. Jagiellonian University Press, Cracow.
Placek, T. (2000b). Stochastic outcomes in branching space-time. An analysis of the Bell theorems. *British Journal for the Philosophy of Science*, 51(3), 445–475.
Placek, T. and Wroński, L. (2009). On infinite EPR-like correlations. *Synthese*, 167(1), 1–32.
Portmann, S. and Wüthrich, A. (2007). Minimal assumption derivation of a weak Clauser-Horne inequality. *Studies in History and Philosophy of Modern Physics*, 38(4), 844–862.
Weiner, M. and Belnap, N. (2006). How causal probabilities might fit into our objectively indeterministic world. *Synthese*, 149(1), 1–36.
Wroński, L. and Placek, T. (2009). On Minkowskian branching structures. *Studies in History and Philosophy of Modern Physics*, 40, 251–258.

Philosophy of Mind

Independent Intentional Objects

Katalin Farkas

Intentionality is customarily characterised as the mind's direction upon its objects. This characterisation allows for a number of different conceptions of intentionality, depending on what we believe about the nature of the objects or the nature of the direction. Different conceptions of intentionality may result in classifying sensory experience as intentional and non-intentional in different ways. In the first part of this paper, I present a certain view or variety of intentionality which is based on the idea that the intentional object of a sensory experience must be Independent; that is, an intentional object must be such that its existence doesn't depend on being experienced (except in some very special cases). This means, for example, that sense-data understood as mind-dependent objects are not intentional objects, because their existence depends on the occurrence of an experience. In the second part of the paper, I will sketch a view of how sensory experiences can acquire an Independent object.

1. Moore and the 'Act-Object' Model

When I declare my interest in intentional objects whose existence does not depend upon being experienced, I want to set aside an alternative suggestion that detecting some 'objects' in experience is in itself sufficient for intentionality. I shall now look at a certain theory, which holds that we can discover an 'act-object' structure in (a certain class of) experiences prior to any commitment about the nature of these objects—the objects can turn out to be mind-dependent, mind-independent, mental, physical or neither.

A philosopher who famously defended a version of the act-object model of conscious states is G. E. Moore. I am going describe this model based mainly on Moore's famous paper "The Refutation of Idealism" (1903).[1] There he writes:

> We have then in every sensation two distinct terms, (1) 'consciousness,' in respect of which all sensations are alike; and (2) something else, in respect of which one sensation differs from another. It will be convenient if I may be allowed to call this second term the 'object' of a sensation: this also without yet attempting to say what I mean by the word. (Moore, 1903, p. 444)

For example, the sensation of green and the sensation of blue differ in their objects—green and blue, respectively—but similar in that they are both acts of conscious awareness of these objects. Moore's particular examples involve visual sensations, and subsequent discussions of these arguments take place mostly in the context of the philosophy of perception. It is important to keep in mind, however, that Moore believes that the same analysis applies to all conscious mental events, including thoughts.

Of course, in saying that every conscious episode has an object, Moore doesn't claim that all conscious mental states are intentional. But in order to see what else would be needed to endow conscious states with intentional objects, it will be useful to get a better understanding of the act-object model. To proceed further in the argument, we need to introduce a distinction. When we say that something has mind-dependent existence, this may mean that its existence depends on there being a mind or minds in general, or it may mean that it depends on a particular mental act. We can illustrate this difference on a Kantian view: according to this view, the phenomenal world is mind-dependent in the sense that it has a structure and properties that make sense only with reference to our minds. If my teacup is next to my notebook, this state of affairs—since it includes, for example, spatial relations—is mind-dependent in this general sense. However, the state of affairs is not mind-dependent in the sense that it would depend on someone's particular mental act of perceiving it or thinking about it. I shall call this latter kind of mind-independence 'act-independence.'

---

[1] Perhaps this isn't the most charitable way of proceeding given that, in the preface of the collection *Philosophical Studies* (1922) where the "The Refutation of Idealism" was republished, Moore himself says the following: "This paper now appears to me to be very confused, as well as to embody a good many down-right mistakes" (Moore, 1922, p. viii). I cannot but agree with this judgement, but my reason for relying on it is the same reason why Moore decided to include the paper in the collection after all: namely, that many references have been made to the paper by 1922 and ever since, and it will be instructive to see how certain ideas have been shaped in the course of this discussion.

It may seem that in the "The Refutation of Idealism," Moore is not entirely neutral about the nature of the object of experiences since he insists that "we can and must conceive the existence of blue as something quite distinct from the existence of the sensation. We can and must conceive that blue might exist and yet the sensation of blue not exist" (Moore, 1903, p. 446). The dialectic is a bit tricky here. Moore's main goal in the paper is to undermine the Idealist thesis that "*esse* is *percipi*," which he understands as the claim that an experience and what is experienced are the same: that a sensation of blue is the same as blue. The distinction between the two elements—the conscious act and its object—is part of the attack on the Idealist thesis. In one possible view, the argument against the Idealist thesis proceeds in two steps: first, we identify the act-object structure in the experience; second, we argue that the object doesn't depend on the act for its existence. In another understanding, the very distinction between act and object is sufficient to show the possible independent existence of the object. Certain paragraphs suggest that Moore promotes something like the latter line in "The Refutation of Idealism": he seems to question the very coherence of the claim that the object is not identical to the act, and yet necessarily connected to it (Moore, 1903, p. 446). Why such a claim should be self-contradictory is far from clear, and in a later paper, "The Status of Sense-Data" Moore clearly opts for the former understanding when he says that an *a priori* discoverable necessary connection between object and act is at least conceivable, even if he doesn't think that there is such a connection, because of further considerations (Moore, 1914, p. 366). Given the fact that Moore himself declared that "The Refutation of Idealism" contains mistakes (see footnote 1 above), I think we can safely take the attitude of the latter paper as a guide.

Therefore I will consider Moore's position as the position that we can detect an act-object structure in all conscious events even before forming any commitment about the nature of the object. This is consistent with a number of Moore's subsequent writings (including "The Status of Sense-Data") where the project is to find out what sort of things sense-data are. Sense-data are a subclass of the objects of conscious events, namely, the objects of sensory experiences. As it is familiar, Moore thinks that the existence of sense-data should be generally accepted before we try to decide whether they are mental, physical, or other kinds of entities.

What is the argument for the act-object model, understood in this noncommittal way?[2] The few sentences quoted above may suggest the follow-

---

[2] For an illuminating analysis of Moore's argument, see Hellie (2007).

ing line: different types of conscious events have something in common—their being conscious events—but they are also different, obviously. Let us call the respect in which they differ their 'object.' If this were the argument, the existence of objects in conscious acts would be uncontroversial. Of course, one could reasonably complain that calling the dimension of difference 'object' is tendentious; if we are just introducing terminology, why don't we call it, for example, 'quality'? Or even better, 'quale'? Indeed I do think that the introduction of 'objects' at this stage is tendentious, though to be fair to Moore, he doesn't rely on this simple argument. In "The Refutation of Idealism," he considers some alternative ways of accounting for the difference among various types of conscious events, but he declares that the object-view is superior to them.

The main competitor to the object-view is what Moore calls the 'content' view. This terminology is confusing when viewed from the contemporary perspective, but what Moore means is very simple: on the 'content' view of experiences, a blue sensation and a green sensation differ in a *quality*, namely, the former is blue and the latter is green. A blue sensation is like a blue bead or a blue beard: "The relation of the blue to the consciousness is conceived to be exactly the same as that of the blue to the glass or hair: it is in all three cases the *quality* of a *thing*" (Moore, 1903, p. 448).

So is blue a quality or the object of a sensation of blue? Is the sensation itself blue, or is it *of* blue? Moore claims that we can decide between these two options by appealing to introspection. Introspection reveals nothing to support the content- (that is, quality-) view, though it doesn't reveal anything to refute it either. However, introspection does offer a positive support for the object-view, and hence the object-view emerges as the overall winner.

> Whether or not, when I have the sensation of blue, my consciousness or awareness is thus blue, my introspection does not enable me to decide with certainty: I only see no reason for thinking that it is. But whether it is or not, the point is unimportant, for introspection *does* enable me to decide that something else is also true: namely that I am aware of blue, and by this I mean, that my awareness has to blue a quite different and distinct relation. It is possible, I admit, that my awareness is blue *as well* as being *of* blue: but what I am quite sure of is that it is *of* blue . . . (Moore, 1903, p. 450)

Moore warns that this exercise is not easy: it is actually quite difficult to discover through introspection that a sensation involves awareness of a blue object. The difficulty lies not in finding the *blue*; that's the easy part. What is hard to detect is the other element: the 'act,' that is, the awareness relation to the blue object. This is where we find the famous sentences frequently quoted in recent discussions on the 'transparency of experience':

... the moment we try to fix our attention upon consciousness and to see what, distinctly, it is, it seems to vanish: it seems as if we had before us a mere emptiness. When we try to introspect the sensation of blue, all we can see is the blue: the other element is as if it were diaphanous. Yet it can be distinguished if we look enough, and if we know that there is something to look for. My main object in this paragraph has been to try to make the reader see it; but I fear I shall have succeeded very ill. (Moore, 1903, p. 450)

He had good reason to fear: I do not think that Moore made a convincing case for the introspective evidence in support of the object-view. I shall try to explain this in the next section.

2. Transparency and Independent Objects

I will now put aside the case of a blue sensation for a moment, and propose to consider another example: feeling dizzy, or dizziness. It is usual to classify dizziness not as a perceptual experience—since it doesn't involve one of the five external senses—but as a bodily sensation. But Moore claims the act-object model to be valid for all conscious episodes, and so it should apply to dizziness too. On what Moore calls the 'content' view, dizziness is an experience with a certain dizzy quality, which distinguishes it from another experience with, say, a nauseous quality. In contrast, on the object-view favoured by Moore, dizziness involves awareness of a—dizzy?—object, and this distinguishes it from another experience involving awareness of a nauseous (?) object.

What I find difficult here is to decide, through introspection, between these competing views of the ontological structure of dizziness. I have met people who honestly assured me that when they focus on a bodily sensation—dizziness, or pain, or nausea—they can clearly detect an *object* of which they are aware. I'm afraid I simply don't see this. Of course, I grant that in the introspective act, one is in a second-order mental state, whose object is another mental state. When I am reflecting upon an experience, I am in a state which has an object: the whole of the original experience. This, however, doesn't show that the original experience itself involved a separate act of awareness and object. I myself am much more inclined to say that when I reflect on dizziness, I simply find an experience with a certain quality. But—not unlike Moore—I don't feel very optimistic that I can convince the doubtful just by encouraging them to reflect more on their experiences. The debate here reaches an impasse.

To move things further, I suggest that we compare Moore's discussion of the sensation of blue with more recent considerations on the issue of the 'transparency' of experience. Michael Tye also shares with us the results of his introspecting a perceptual experience of blue:

> Standing on the beach in Santa Barbara a number of summers ago on a bright sunny day, I found myself transfixed by the intense blue of the Pacific Ocean. . . . It seems to me that what I found so pleasing in the above instance, what I was focusing on, as it were, were a certain shade and intensity of the colour blue. I experienced blue as a property of the ocean not as a property of my experience. My experience itself certainly wasn't blue. Rather it was an experience that represented the ocean as blue. What I was really delighting in, then, was a quality *represented* by the experience, not a quality *of* the experience. It was the color, blue, not anything else that was immediately accessible to my consciousness and that I found so pleasing. This point, I might note, seems to be the sort of thing G. E. Moore had in mind when he remarked that the sensation of blue is diaphanous (see Moore, 1922, p. 22). When one tries to focus on it in introspection one cannot help but see right through it so that what one actually ends up attending to is the real colour blue. (Tye, 1992, p. 161)

Tye thinks he makes the same point as Moore, but after what's been said in the previous section, we can see that there are a number of differences. For one thing, we saw that Moore didn't think that introspection shows that blue is not a quality of the experience (though it doesn't show the contrary either). Further, Moore thought that one can, after all, bring into focus the act of awareness, if only one tries hard enough.[3] But the most important difference is the following: Tye proceeds by first identifying an act-independent object for his experience, the Pacific Ocean; he then observes that he experienced blue *as a property of the ocean*, not as a property of his experience. This move is missing from Moore's procedure. Moore seems to suggest that we can first identify blue as the object and then ask whether this is part of the physical object or a mental particular (or something else).

This is a crucial difference, because, in my view, this feature of Tye's procedure makes his introspective exercise much more convincing than that of Moore. I complained above that we may find it hard to choose between a dizzy quality and a dizzy object on the basis of introspection. But the case of blue is different. In the case of dizziness, there is no easily identifiable object such that we experience that object, and experience dizziness *as* the property of that object.[4] In the case of blue, there is: for example, the ocean.

---

[3] See also Hellie (2007) for the difference between Moore's argument and the more recent discussions of transparency.

[4] It may be suggested that the 'object' we experience to have the dizzy quality is the *subject* herself. Perhaps there is a development of this idea that eventually works, but at a first

It seems to me very clear that the way things seem in my experience, my experience is not blue, because if there is anything that's blue, it's the ocean. It is in this contrast—experience *versus* ocean—that it becomes manifest that something characterising the phenomenal character of this episode, namely an awareness of blue, is presented to me as an awareness of a feature of an object, rather than as an awareness of a feature of my experience. And to anticipate the conclusions of the second part of the paper, this is what convinces me that seeing the Pacific Ocean is an intentional episode, whereas feeling dizzy isn't.

It has been objected that Tye is too quick in moving from the observation that he experiences blue as a property of the ocean to the claim that the experience *represents* this property. Maybe other accounts are available for the initial observation (see Martin, 2002). In any case, it seems to me that all phenomenologically convincing appeals to something like the transparency observations proceed in the same way as Tye does: first they identify an act-independent object that one seems to perceive, and then note that the features that we would offer in a characterisation of our experience are experienced as properties of this act-independent object.

In Gilbert Harman's example, the object is a tree: "When Eloise sees a tree before her, the colors she experiences are all experienced *as features of the tree* and its surroundings" (Harman, 1990, p. 39, my emphasis). In M. G. F. Martin's example, the object is a lavender bush: "I attend to what it is like for me to introspect the lavender bush through perceptually attending to the bush itself while at the same time reflecting on what I am doing. So it does not seem to me as if there is any object apart from the bush for me to be attending to or reflecting on while doing this" (Martin, 2002, pp. 380–381). In all these cases, the phenomenological observations are committed to the apparent act-independence of the object and its properties.

How does this all relate to the issue of the intentionality of sensory experiences? Intentionality, as I said, is the mind's direction upon an object, and this characterisation may suggest that once we find an *awareness of an object*, we have intentionality. This isn't the sense of intentionality I am after when I raise a question about the intentionality of sensory experiences. The alternative in which I am interested is a special variety of directedness,

---

glance, it won't do. The defender of the 'content'-view (that is, the 'quality'-view) can agree that dizziness is a quality of the subject, by being a quality of her experience. In a reflective second-order mental state, we may see that the experience involves an exemplification of a quality by the subject. But there is still no introspective evidence for thinking that the subject somehow doubles in the original experience, by being both the subject and the object of the experience at the same time.

which is directedness at something else, at something *beyond*. The point is not simply that the object is not identical to the experience—this is true on theories which regard sense-data as mental particulars, because the object is only a constituent of the experience, not identical to it. Rather, as I shall say, in a genuine act of directedness, the object is 'Independent,' in the sense that its existence doesn't depend on the occurrence of the experience. This also applies to nonexistent objects: they don't exist, and hence *a fortiori* their existence doesn't depend on being experienced. The formulation here is not equivalent to saying that the object could exist independently of the experience, because we may want to accommodate impossible objects. The following holds for impossible objects: it is not the case that whenever they are experienced, they exist.[5]

Of course, to a certain extent, the point is terminological. 'Intentionality' has been used in different senses, and no sense is given independently of what we make of it in particular theories, nor can I claim a special right over the term 'intentional.' However, there are fruitful and less fruitful questions that one can ask about the nature of sensory experiences. As I tried to explain, there is a question that I don't find particularly fruitful: this the question of whether the phenomenology of sensory experience reveals an *object* in the case of every sensation—even when there is no mind-independent object—or whether the phenomenal character is better characterised in these cases as the experience having some quality. But I do find another question fruitful: this is the question of whether we are inclined to characterise the phenomenal character of some of our experiences with reference to properties that we experience as mind-independent properties of mind-independent objects. I think the answer to this question is yes, for some experiences, though not for all, and this is very significant in accounting for the nature of sensory episodes.

3. The Independence of the Experienced Situation

Intentionality is the mind's direction upon its object, and as I explained in the previous sections, I am interested in a specific variety: when the mind is directed upon an Independent object, i.e., when an experience has an ob-

---

[5] There are exceptions in some very special cases. Suppose physicalism is true, and consider the experience of looking at some device which shows the image of your brain having precisely this visual experience. But here the act-dependence of the object is not due to some general connection between experiences and their objects (like in the case of sense-data), but simply to the coincidence between the object and the brain-state that realises the experience.

ject whose existence doesn't depend on being experienced. Which sensory experiences are intentional in this sense? By 'sensory experiences' I mean perceptual experiences associated with the five senses, and bodily sensations like pains, itches, hunger, dizziness and all sorts of introceptive experiences. Under 'perceptual experiences' I mean to include both veridical and non-veridical (illusionary or hallucinatory) experiences.

In a perceptual experience, things appear as being in a certain way. Let's call things being in a certain way a 'situation.' A perceptual experience is directed at a situation, or we might say that the situation is the intentional object of the experience, where 'object' in this instance is not meant to be a specific ontological category, but simply whatever the experience is directed at. If the situation is the object of the experience, and intentionality requires Independence, then the *whole* situation has to be Independent. This means that not only *what* is experienced—the object in the narrower sense, that is, the individual or particular—but also *the way* it is experienced as being, must be Independent. The requirement of Independence entails that neither the existence of the individual, nor its having the experienced properties depend on being experienced.

I take it that the nature of a sensory experience is given by its phenomenal character. Now the question is: how much of the nature of a sensory experience is given by its intentional properties? And this question now translates as: how much of the phenomenal character of a sensory experience is given by features that are experienced as act-independent properties of act-independent particulars?

The most plausible version of the view that pain is an intentional episode holds that the object of, say, a pain in the ankle, is the ankle. As Tim Crane (2003) points out, the main argument for the intentionality of bodily sensations like pain is that they have a felt location in one's body: the experience appears to be directed at a certain region of one's body, for example, the ankle. The ankle is Independent: its existence doesn't depend on the occurrence of the pain experience, it could exist even if didn't hurt. However, there is a further element in the phenomenal character of the experience: the ankle is not merely experienced, but it's experienced as *hurting*. And this aspect of the situation, I claim, is act-dependent. Something couldn't hurt without being experienced: there are no unfelt pains. If you don't feel it, it doesn't hurt. Once the experience ceases, so does the hurting.

One option for analysing the structure of the pain-experience would be to say that the experienced situation consists of a body part hurting, that is, the body part is experienced as hurting. This seems to be, for example, Alex Byrne's view: he says that the content of a certain pain experience is "that there is a pain in the toe" or that the toe is hurting (Byrne,

2001). This situation is not entirely Independent as we have just seen: though the ankle or the toe are Independent, the way they are experienced as being—i.e. hurting—is not. This would contrast with the case of seeing the blue ocean, where both the ocean and the way it is seen as being—blue—are Independent. This is compatible with the view that colours are response-dependent properties. The relevant sense of mind-dependence in Independence is act-independence, and this would be granted by response-dependence theories as well.

Crane has a somewhat different view. He holds that an intentional episode has intentional content, which is a matter of some object being presented under certain aspects. Besides, there is a further element that he calls the 'intentional *mode*,' which refers to the relation of the subject to the intentional content. Believing, desiring, visually experiencing, for example, are intentional modes. In the visual experience of the blue Pacific Ocean, the blue ocean—or the ocean being blue—is the content, and the visual experiencing is the mode. In Crane's view of bodily sensations, the *hurting* in pain can be regarded not as part of the content of the intentional episode, but rather as the mode. So if we compare the experience of the ankle with that of the Pacific Ocean, the hurting would be analogous not to the blue, but to the visual experiencing. This analysis means an explicit acknowledgement of the act-independence of the hurting aspect of the experience: intentional modes cannot be exemplified outside mental episodes.

There is a further alternative intentionalist account of pain: that pain is the experience of damage in a body part (Tye, 2002). If this were right, then both the experienced particular (the ankle) and the way it is experienced as being (damaged) would be Independent, because one could have damage in a body part that is not experienced. However, I find this account of pains completely implausible. If we want to do justice to the phenomenal character of pains, one cannot leave out the essentially experiential feature of *hurting*.

The phenomenal character of a pain episode is best described with reference to a combination of Independent and non-Independent elements. In pain, an intentional object makes its appearance: the body part that hurts. But the way the object is experienced as being—that it's hurting—is essentially dependent on the subject's having this mental episode. Hence it is not intentional: it isn't directed at something beyond the experience, but rather it is a matter of the experience having a certain quality. Now I want to say that once we realise that some experiences are not entirely intentional, this opens the way to asking whether there are some that are entirely non-intentional. It seems to me that the answer is yes, and a certain type of dizziness can plausibly be regarded as such a case. Sometimes dizziness is described as one's 'head spinning,' but I don't think that in every case, the sensation has to have an obviously locatable region in the body: sometimes it really is just the way one feels, that is, the qualities that characterise the experience are all attributed to the experience or to the subject of the experience, rather than to any object. In other words, dizziness is a pure *quale*; simply a modification of the subject's consciousness.[6]

There are two separate questions here. One is whether we agree on the analysis of the particular case of dizziness. The other is whether there is any general reason to think that an experience couldn't be a pure quale, without an intentional object; because if there is no such reason, then even if I am wrong about dizziness, there may be other cases. I'm not sure who has the burden of proof here, but I don't see such a general reason. Here again it is important that we are interested in the Independent variety of intentional objects. Famously, Brentano makes the following claim:

> Every mental phenomenon includes something as object within itself, although they do not all do so in the same way. In presentation, something is presented, in judgement something is affirmed or denied, in love loved, in hate hated, in desire desired and so on. (Brentano, 1924/1995, p. 88)

Superficially, this applies to sensations as well. One could say that in every feeling, something is felt—one feels pain, dizziness or nausea, and so on. But whatever kind of object 'dizziness' is, it is certainly not Independent. Just pointing out that a feeling of dizziness is a feeling *of something*—i.e., of dizziness—is not sufficient to show that the sensation has an Independent object: dizziness couldn't exist without being experienced. In fact, at the time of writing the above lines, Brentano himself held that intentional objects were immanent to the intentional act, which can arguably be understood as the denial of the Independence of the object.[7]

Alex Byrne (2001) defends the claim that all bodily sensations, as well as all perceptual experiences, are intentional, but Byrne has a much more

---

[6] I want to put aside a certain issue: by claiming the existence of a pure quale, I don't mean to take sides in the dualism/physicalism debate. For all I say here, the characteristic modification of the subject's consciousness which constitutes the phenomenal character of dizziness may be a completely physical event.

[7] As well known, Husserl criticised Brentano for holding that intentional objects are immanent, arguing instead that they should be regarded as *transcendent* (for Husserl's notion of transcendence, see Zahavi 2003, pp. 16ff.). I am tempted to say that my proposal is along the same lines and in the talks which preceded this paper, I talked about the transcendence of the intentional situation. However, without wanting to go into exegetical details, I started to suspect that my notion of Independence is not quite the same as Husserl's notion of 'transcendence,' and hence I decided that I better avoid using the same term.

liberal conception of intentional objects than I do. He holds that if afterimages, phosphenes or pains-as-objects turn out to be mental particulars which cannot exist independently of being experienced, they can still serve as intentional objects. He argues that even if objects are conceived in this way, the sensations can be characterised as 'things seeming in a certain way'—it seems to one that her toe is hurting, or that there is an afterimage before her—and this means the sensations have intentional content. There is indeed a sense in which when one feels dizzy, 'things seem in a certain way' or rather 'feel in a certain way,' but the 'things' in this case need not be Independent objects at all, they may just be placeholders for the grammatical subject of the sentence. As I said before, we are hard pressed in these cases to say exactly which Independent objects are the objects of the experience. In any case, the point is this: even if Byrne is right that every sensation has some object, Independent or not, it would require a separate argument to show that every sensation must have an Independent object. And my point here is that I don't see a general reason to assume that they do.

4. The Acquisition of an Intentional Object and the Projection of a Property

Some sensations may lack any intentional object, but I believe that this feature of a sensation is not stable; it can change. Let me give an illustration of what I mean. Some people are prone to sudden drops in their blood sugar level. The symptoms are very characteristic: sudden dizziness or faintness, feeling hot, weak and shaky. I occasionally have this experience. When it first happened, I didn't realize what it was. In the course of time, I learnt that the cause is low blood sugar level, and that the best thing to do is to eat something sugary, which makes the symptoms disappear quite quickly. Nowadays, if this happens to me, I might say to someone "I feel that my blood sugar is dropping, I have to have a biscuit." And to my mind, this is indeed *what* the feeling is: it is *the feeling of my blood sugar dropping*. The idea of blood sugar has an immediate presence in my mind when I have the experience; there is no apparent inference at all, the very feeling seems to have the character that is best described in these words. The case is similar to other cases when I describe a feeling in terms of its object: the feeling of something cold touching my skin, or the feeling of a vein's throbbing in my leg. And just as the other feelings are *about*, or *directed at*, the cold touch on my skin, or the vein in my leg, the presently discussed feeling is *about* blood sugar. At least this is how it strikes me.

The general possibility suggested by this case is the following. There is a sensation, which is regularly caused by a certain event. A subject may first identify the sensation on the basis of its characteristic feel, without being aware of its cause. When she learns about a cause, she becomes able to infer the presence of this cause from having the experience. After a certain time, the experience in truth becomes a 'sign of the cause' in the consciousness of the subject; that is, the experience starts to feel as a feeling of its cause. We might call this 'the process of an experience acquiring an object through interpretation.' It is clear that being about the level of blood sugar is not an intrinsic feature of the experience (at least not as the subject feels it initially). One needs the interpretative process to endow the experience with this object. However, after the interpretative process, it seems that the cause becomes a genuine intentional object of the experience. It is an object, because its presence in the experience is completely immediate—or in any case, as immediate as the presence of the ankle in the experience of one's hurting their ankle. And it is a genuine intentional object, because it is Independent: it is possible for the blood sugar level to drop without the subject feeling it, for example, during sleep.

The experience, at this stage, is similar to the case of pain, in the following respect: it involves an Independent intentional object—the event of the blood sugar dropping—but the other element that characterises the phenomenal nature of the experience, the way this feels, is not Independent. However, in some cases a further development is possible. Hurting, as I said, is not experienced as a property that the ankle could have without being experienced. But in theory, we could form the idea of a *different* property: a property of being disposed to cause a certain kind of sensation. Wittgenstein considers the following possibility in §312 of the *Philosophical Investigations*: "The surfaces of the things around us (stones, plants, etc.) have patches and regions which produce pain in our skin when we touch them. ... In this case we should speak of pain-patches on the leaf of a particular plant just as present we speak of red patches" (Wittgenstein, 1953/2001, p. 37).[8]

If we conceived the pain-patches analogously to the red patches, then having a pain-patch would be an Independent feature: the leaf would have pain-patches even if no one touched it. 'Pain' in 'pain-patch' would refer to a different property than in 'pain in one's ankle.' The latter is a property that only conscious creatures can exemplify, the former can be a property of inanimate objects. If a certain kind of object or event is seen as a potential

---

[8] Thanks to Alex Byrne for calling my attention to this example.

regular cause of a certain kind of sensation, we can move to the idea that it remains a potential cause of the sensation even if no one is actually having the sensation. An unfelt sensation is a contradiction in terms, but an unfelt regular type of cause of a certain kind of sensation is not. We may call this process 'the projection of a property upon an object,' but we need to keep in mind that the projection results in a different property.[9]

Here, again, we need to remember that the sense of mind-independence relevant to Independence is act-independence. Even if the property of having pain-patches is mind-dependent in the sense that it makes an essential reference to the experiences of sentient creatures, something's exemplifying this property would not depend on a particular act of experiencing it.

Notice that in this case the pain is projected onto an external object, rather than onto the body part where the pain is felt, so in addition to the change of the property through projection, the intentional object of the experience changes. This would be a variation of the process described above: a *change*, rather than acquisition of intentional object through interpretation. In fact, someone may think that the blood sugar case was similar; if the original sensations—feeling dizzy, shaky, etc.—already had an object, then the process of interpretation changed the object, rather than introduce it. The principle is the same though: something becomes an object of the experience by being associated with it as its regular cause.

The pain-patch in Wittgenstein's example is imaginary because, as a matter of fact, we usually don't project pains on external objects. There must be reasons for this which have to do with the conditions of projecting properties. I cannot go into details here, but one likely factor is that pains tend to linger even when their external cause is removed from the presence of the observer. Compare seeing a knife and being cut by a knife. If you throw the knife out the window, the visual sensation of the shape and colour cease, but the pain probably doesn't. This partly explains why we attribute shape and colour to the knife, and pain to ourselves.

If we did project pain on external objects, then hurting would become a form of external perception, in addition to sight, hearing, touch, smell and taste. The characteristic feature of perceptual experiences is that their basic nature is given by an intentional situation: in the normal cases, both the object, and the way it is experienced as being, are Independent. However, there are often non-Independent residues in the experience. What strikes me as the most natural account of blurry vision is the view that the blurriness is

---

[9] This is the kind of distinction that Christopher Peacocke (1983) makes between *red* and *red'*, where the former is a property of objects, the latter is a property of sensations.

not experienced as the property of the mind-independent objects that I perceive; blurriness is something that is brought to existence specifically by the occurrence of the experience. Suppose I have a blurry visual experience, and I close my eyes. The impression is that the individuals, their colours and distances would continue to exist even when I don't have the experience. But for all I know, all blurriness might have ceased from the world when I closed my eyes. Blurriness is not Independent.[10]

5. Pure Sensations

Many sensory experiences involve a combination of Independent and non-Independent elements. But a further important observation is that even when the experience has acquired an intentional object through interpretation, and a property is projected onto an object, the fundamental sensory aspect of the experience is given by the original non-intentional sensation, or quale. It is true that this isn't how perceptual experience usually strikes us. The transparency observations have a significant force: the phenomenal character of a perceptual experience is often most naturally described by mentioning act-independent objects and their act-independent properties. The reason is that the processes of interpretation and projection are very natural and automatic. If we wanted to discover the purely sensory, we should inquire into the conditions that facilitate these processes, because we can expect to find the purely sensory where these conditions are missing.

The main reason for interpreting our sensations as providing a testimony of the mind-independent world is the highly organised and stable structure of experience, which responds in a uniform and predictable way to our movements and other actions. One seldom reflects upon this fact, but it really is very remarkable. If I only think about my present visual experience of the small coloured icons on my text editor program, it is rather amazing what a fine detail it offers, and how reliably these details seem to hang together.

Compare this experience, for example, with the fleeting impression of an afterimage. In an afterimage, several factors are missing that are present in my current visual experience. First, in an afterimage, there is no fine detail. Second, the details don't seem to be available for further investigation: it doesn't seem as if I could go and attend to the various aspects of

---

[10] Tye (2002) has an alternative account of blurriness, which is consistent with an intentional account, but I think it is less plausible than this one. See Crane (2006) for an effective critique of Tye's view on this point.

the image, because it changes with every second, and starts to evaporate before my very eyes. Third, the image has a certain way of moving as I move my eye, which gives the impression that my eye movement somehow drags the picture with itself. All these together give me the distinct impression that the afterimage is not an act-independent entity, and the properties that characterise my experience of the afterimage—colours, shapes—are not experienced as properties of act-independent objects. One has the feeling that once the experience of a blue-and-red afterimage is gone, for all we know, nothing may remain blue or red. Whereas if I close my eyes after seeing a tomato, I don't have the impression that all red has gone out of the world.

Some people say that the colours of the afterimage are experienced as features of the white surface against which the illusion occurs, and hence afterimages are simply hallucinations. But their afterimages must be very different from mine, because mine don't look like that at all. And neither do those of Ned Block, apparently:

> Afterimages—at least the ones that I have tried—don't look as if they are really objects or as if they are really red. They look ... illusory. Try it out yourself. Don't get me wrong. I agree that an image experience and a tomato experience share something that one might call a color property. My point is that when one has an afterimage one has no tendency to think thereby that anything is really red ... (Block, 1996, p. 32)

In the framework developed in this paper, the property of being 'really red'—which we are not inclined to attribute to anything in the afterimage experience—is a projected property, that is, a property that could be instantiated independently of any particular experience. Since the qualities that characterise the afterimage experience are felt as being instantiated only as long as the experience occurs, we are not inclined to attribute such a projected property to anything. Further, in terms of the proposal presented here, what is common to the afterimage experience and the tomato experience is the sensation-property red, which is different from, although has an intimate connection to, 'real red.'[11]

I said that we would have a chance to experience pure sensations when the conditions of interpreting them as giving a testimony of the mind-independent world are missing. So imagine sensations which have even less detail, stability and apparent independence of our actions than afterimages.

---

[11] I argued above that Moore fails to make progress in answering the question of whether a sensation *is* blue or is *of* blue. One reason for this failure is that the formulation of the question seems to presuppose that the *same* property is a candidate for being the property of the experience, and the property of its object. But this may not be the case.

Imagine a chaos of shapes and colours that change every time we blink, random ringing in the ear, fleeting impressions of strange tastes in one's mouth, and so on. These would be the pure sensations which are the building blocks of our perceptual experiences. It's only when impressions are organised into a structure that we form an idea of the experience-independent world that sustains them.[12]

References

Block, N. (1996). Mental paint and mental latex. *Philosophical Issues*, 7, 19–49.
Brentano, F. (1995). *Psychology from an Empirical Standpoint* (A. C. Rancurello, D. B. Terrell, and L. McAlister, Trans., 2nd ed.). Routledge, London. (Original work published 1924).
Byrne, A. (2001). Intentionalism defended. *Philosophical Review*, 110, 199–240.
Crane, T. (2003). The intentional structure of consciousness. In A. Jokic and Q. Smith (Eds.), *Consciousness: New Philosophical Perspectives* (33–56). Oxford University Press, Oxford and New York.
Crane, T. (2006). Is there a perceptual relation? In T. Gendler and J. Hawthorne (Eds.), *Perceptual experience*. Oxford University Press, Oxford.
Harman, G. (1990). The intrinsic quality of experience. *Philosophical Perspectives*, 4, 31–52.
Hellie, B. (2007). That which makes the sensation of blue a mental fact: Moore on phenomenal relationism. *European Journal of Philosophy*, 15, 334–366.
Martin, M. G. F. (2002). The transparency of experience. *Mind and Language*, 17, 376–425.
Moore, G. E. (1903). The refutation of idealism. *Mind, New Series*, 12, 433–453.
Moore, G. E. (1914). The status of sense-data. *Proceedings of the Aristotelian Society, New Series*, 14, 355–380.
Moore, G. E. (1922). *Philosophical Studies*. Reprinted: Routledge, London, 2000.
Peacocke, C. (1983). *Sense and Content*. Clarendon Press, Oxford.
Tye, M. (1992). Visual qualia and visual content. In T. Crane (Ed.), *The Contents of Experience*, (158–176). Cambridge University Press, Cambridge.
Tye, M. (2002). Representationalism and the transparency of experience. *Noûs*, 36, 137–151.
Wittgenstein, L. (2001). *Philosophical Investigations: The German Text, with a Revised English Translation*. Blackwell, Oxford. (Original work published 1953).
Zahavi, D. (2003). *Husserl's phenomenology*. Stanford University Press, Stanford.

---

[12] This paper is a written and extended version of the lecture I gave at the ECAP conference in Kraków, August 2008. Previously to the ECAP conference, a predecessor of the paper was presented at a workshop in Canberra on The Representational and Relational Nature of Experience. I am grateful to the audiences in Kraków and Canberra for their comments; especially to Alex Byrne, David Chalmers, Tim Crane, Robert Hopkins and Susanna Schellenberg.

Social Philosophy, Political Philosophy and Philosophy of Law

Market Efficiency and Contractual Justice

Peter Koller

1. Introductory Remarks

In recent decades, neoliberal thinking has experienced a remarkable heyday, within academic discourse as well as in public life. This thinking, which actually is widely in line with traditional economic liberalism rather than being really new, relies on a firm belief in the market to the effect that a free market economy based on private property is the best system for regulating most social affairs, at any rate much better than any centralized rule, apart from the fact that a sort of centralized power is deemed to be necessary to provide the institutional framework of the market. Accordingly, neoliberals take it for granted that the market, provided that it can operate freely under appropriate framing conditions, automatically leads to generally acceptable results. This belief is usually defended by some standard arguments that, at first glance, have some intuitive plausibility, such as the importance of private property and free exchange for individual autonomy, the internalization of external costs through property rights, the virtues of the price mechanism, the need for sufficient incentives, the productive capacity of the market, and the like. At closer glance, however, it turns out that the belief in the market's excellence relies on a number of heroic assumptions some of which are little reflected in the relevant disciplines, in particular economics and philosophy, let alone public discussion.

These assumptions include both theoretical and normative ones, which, however, are closely interconnected. While the first have to do with the real processes and effects of markets under certain framing conditions, the normative assumptions refer to the standards on the basis of which these processes and effects are to be judged. I am mainly interested in the latter, the *normative* assumptions, and will address the theoretical only insofar as it is necessary for my targets. Those who believe in the excellence of the market, obviously assume that a market order is not only more efficient than feasible alternative systems, but also morally acceptable, i.e. by and large just, or at least not significantly unjust. So their normative assumptions contain standards of efficiency as well as requirements of morality and justice. When we seek to learn a bit more about these assumptions by consulting the relevant literature, we face a mixed situation. While there is a lot of useful stuff on the efficiency of markets, because economists have dealt with that matter extensively, the scholarly discussion on market related problems of justice appears somewhat unsatisfactory: Even though issues of *distributive* justice have been widely dealt with both in philosophy and in economics, particularly since the appearance of Rawls' theory of justice (cf. Rawls, 1971; Nozick, 1974; Lucas, 1980; Chapman and Pennock, 1989; Sunstein, 1997; Miller, 1999; Dworkin, 2000), the topic of *contractual* justice has been almost completely neglected, in spite of its obvious relevance for the moral evaluation of market orders. I want to make an attempt to fill this gap and will, therefore, mainly focus on contractual justice.

With this aim in view, I want to begin with a closer glance at the normative assumptions which are relevant for the evaluation of markets, namely the standards of efficiency and morality, including various standards of justice. Then, I am going to recap the most significant insights of the economic theory of the market that relies on efficiency. On this basis, I will try to identify the requirements of contractual justice which define the conditions under which market transactions qualify as just. Finally, I shall point out some implications of my approach to contractual justice which are contradictory to neoliberal thinking.

2. Normative Assumptions

Following Kant's differentiation between technical, pragmatic and moral guide-lines of human conduct (cf. Kant, 1785/1968, pp. 44ff.), the normative standards for evaluating and shaping social affairs, practices, institutions and orders, including markets, may roughly be divided into three sorts: standards of *efficiency*, standards of *common good*, and standards of *morality and justice* (cf. Koller, 2002, pp. 44ff.). In the present context, I shall leave the common good aside and focus only on efficiency, on the one hand, and morality and justice, on the other. Let me briefly explain my understanding of these concepts.

The viewpoint of *efficiency* aims to judge social affairs with regard to their *expediency* on the basis of the *actual preferences* of the individuals concerned against the background of the respective *status quo* of social life. Although there is no complete agreement on how individual preferences are to be translated into a generally acceptable collective preference function, two criteria of social efficiency find wide-spread acceptance in economics: first, the *Pareto principle* according to which a state of affairs is efficient, when it is impossible to change it in a way that is to the benefit of at least one party involved without worsening the situation of others; and, secondly, the *Kaldor-Hicks principle* which qualifies a state of affairs as efficient, if this state, compared with others, benefits some of the parties involved to such an extent that they could compensate all those who fare worse (cf. Buchanan, 1985, pp. 4ff.; Coleman, 1988, pp. 67ff.). Even though these criteria, contrary to the view of many economists, are neither free of theoretical problems nor easily applicable, I shall rely on them in my following considerations, since I deem them to be acceptable in the present context and do not know any better alternative.

By contrast, considerations of *morality and justice* are directed at evaluating social affairs in the light of their *general acceptability* for all people concerned from an *impartial point of view*. So the reference point for the comparative assessment of the feasible alternatives states of affairs is a hypothetical *situation of equal freedom of choice* (rather than the status quo), and the measure of evaluation are the *reasonable interests* of the parties involved, such as their essential needs (rather than their actual preferences). Every comprehensive morality includes two kinds of moral standards that claim to be generally binding: on the one hand, some perfectly universal demands of *general morality* which apply to all people vis-à-vis others, irrespective of their specific social relationships, such as the commandment not to harm others without justification, and, on the other hand, the demands of *justice* which are context-dependent in the sense that they have to do with specific social relationships among the parties involved, such as dividing goods, exchanging assets, awarding achievements, exercising power, compensating torts, etc.

Since, however, there is a great variety of social relationships subject to demands of justice that hardly can be reduced to a common denominator, it is necessary to differentiate between various typical forms of social interaction and check whether there are plausible standards of justice that fit with them, in order to find out whether such standards exist at all. This strategy was already pursued by Aristotle when he distinguished between two sorts of justice that have been named 'distributive' and 'commutative' justice (Aristotle, 1954, V.7, 1130bff.). Even though this distinction leads to the right direction, it is still too simple, for it does not only mix diverse forms of social interaction that raise different problems of justice (as exchanging goods and correcting wrongs), but it also fails to pay attention to some important forms of interaction which obviously are subject to justice (such as exercising power). So I think it requisite to distinguish between four kinds of justice, each of which refers to a particular elementary type of social relationships: (1) distributive justice—communal relationships, (2) contractual justice—exchange relationships, (3) political justice—power relationships, and (4) corrective justice—wrongness relationships. On the basis of this conceptual distinction, it is possible to identify a number of widely shared and reasonable, though rather general and vague demands of justice (Koller, 2001, pp. 19ff.).

Distributive justice applies to *communal relationships*. These are interpersonal constellations whose participants have a common claim to certain goods or are commonly bound to bear certain burdens. In short, individuals maintain a communal relationship, when and insofar as they share some common goods or common burdens or both. Justice demands that common goods and burdens are to be distributed in a way that is reasonably acceptable to all parties involved. I think there is a general principle of distributive justice that works for all distributive problems, although the specific criteria that apply to them vary with their special features. This is the *principle of equal treatment* which runs as follows: The goods and burdens of a communal relationship are to be distributed *equally* among all parties unless an unequal distribution is justified by reasons generally acceptable from an impartial viewpoint. Taken alone, the principle of equal treatment is rather weak and abstract, but it is not completely empty, for it acquires significant strength when it is combined with further assumptions (Rawls, 1971, p. 62; Feinberg, 1973, pp. 99ff.; Miller, 1976, pp. 24ff.).

Contractual justice deals with *exchange relationships*, i.e. contractual transactions through which two (or perhaps some more) independent parties, who are all endowed with transferable assets, voluntarily agree on a mutual exchange of certain assets which they possess. A paradigm case of exchange relationships are market transactions. In order to be just, such transactions must be in the well-considered interest of all parties involved, so that none of them has reason to complain about their results. The requirements of contractual justice will be discussed in detail later on.

Political justice focuses on *power relationships*, which occur when particular individuals or collectives have (or should have) authorized power which enables them to determine the ways of conduct of other people through binding decisions backed by force. There are plausible reasons to assume that, at least in large social unions, some form of authorized power is necessary in order to secure a generally beneficial social order. Since, however, any such power involves the danger of getting misused, it is subject to certain requirements of justice in order to keep it within acceptable limits. It is generally agreed that power is not an end in itself, but a means to achieve an acceptable social order. Accordingly, power is deemed to be just only if it serves *legitimate purposes* and is exercised in an *impartial way*. Its purposes are twofold: first, enforcing the well-founded duties and rights of individuals, particularly human rights, and, second, facilitating projects of cooperation to the benefit of all people involved, such as the provision of public goods (see Lucas, 1980, pp. 154ff.; Höffe, 1987, pp. 62ff.).

Eventually, *corrective justice* applies to *wrongness relationships*. These emerge when people commit wrongdoings by flouting binding social norms or violating the rights of others. This constellation demands for a correction of committed wrongs in order to restore the social order. The ways of such correction are the object of *corrective justice*, which itself includes two issues: compensating the victims of wrongs for their damages, on the one hand, and punishing wrongdoers who are guilty of severe misdemeanours, on the other. Both forms of corrective justice rely on certain requirements of *proportionality*, which need not to be discussed in the present context (cf. Lucas, 1980, pp. 124ff.; Sterba, 1980, pp. 63ff.; Braithwaite and Pettit, 1990; Coleman, 1992).

So much to the basic kinds of justice which, notwithstanding their analytical differences, are actually interlinked, since, in social reality, human interaction usually takes place within complex social networks that often combine all forms of relationships under consideration. So the different kinds of justice and their demands are bound together through manifold interrelations some of which are necessary, while others are contingent. I just want to point out one *necessary* interrelation between contractual and distributive justice that is of particular importance: According to any plausible conception of contractual justice, an exchange relationship can be deemed to be just only under the condition that it starts from a previous distribution of the respective rights and possessions of the parties involved that itself appears acceptable, i.e. by and large just or at least not extremely unjust, a condition which eventually relies on the demands of general morality and distributive justice. For it is obvious that contractual transactions will generate just outcomes only if all parties involved are entitled to dispose of the goods and services which they want to exchange. In this sense, distributive justice takes precedence over contractual justice. And since political and corrective justice also require an acceptable initial distribution of individual rights and assets, distributive justice is in fact fundamental to all other demands of justice.

If the various basic demands of justice, taken together, are applied to complex social orders, they facilitate the formation of more complex ideas of justice in regard to the orders under consideration. If applied to the order of modern societies, they amount to the idea of *social justice* (Koller, 2003), and they suggest the idea of *international* or *global justice*, if applied to international relations or the entire global order (Koller, 2009). It is true, however, that these ideas, even though widely acknowledged in general, are greatly contested in detail as to their particular understanding or conception, where most disputes centre on the scope and impact of distributive justice in the context of the respective complex idea. The conceptions that attach great importance to the communal matters within the social order under consideration and, therefore, have to grant distributive justice a major role for its appropriate regulation generally tend to a more or less egalitarian view, whereas those which regard the respective order as a sort of marketplace where distributive justice has no significant function are usually inclined to a more or less libertarian perspective (see Koller, 1995). Yet, I cannot discuss these controversial issues here in more detail.

Before I turn to market efficiency I should say a word on the relationship between efficiency and justice. Since both apply to the same objects which they only illuminate from different viewpoints, they eventually must be combined in a way that results in a *comprehensive evaluation* of those affairs. This raises the question of how efficiency and justice are related to each other, a question which is particularly important when they are conflicting rather than coincident. In principle, there is the rule, which is widely accepted in general, though not always easily applicable in detail, that *justice has priority to efficiency*. The reason for this rule, I think, is the fact that, in most cases, considerations of efficiency actually have stronger motivational force than the more ideal requirements of justice. This makes it necessary to reverse their ranking order as to their normative force, since, otherwise, justice would always by overruled by efficiency and have no impact at all. In practice, however, things are more complicated, for the priority rule obviously relies on the condition that the relevant reasons of justice and efficiency are both well-founded and of like relevance. But this condition is often not met. It may occur, for example, that there are strong and highly plausible considerations of efficiency for a social arrangement, while only weak and contested reasons of justice speak against it. In such cases, the simple priority rule will not work. So it will be requisite to balance all relevant reasons for and against the alternatives under consideration in order to come to a decision that appropriately pays tribute to efficiency and justice as well.

3. Market Efficiency

The market as a social practice, through which people coordinate their individual activities in a decentralised way via private exchange transactions, requires sufficiently reliable social rules that determine the individual rights to transferable assets and guarantee the keeping of contracts. Therefore, a market needs appropriate property and contract rules, which, in large societies, must be provided by the legal order (cf. Fritsch, Wein, and Ewers, 2001, pp. 6ff.). For a first approximation to the question of how the legal framework of a market is to be arranged in order to make it work, it may be helpful to take a look at the rules through which markets have been actually regulated by modern legal orders. The general contents of these rules, which can be found in various parts of private law, such as person law, property law, contract law and trade law, can be generalized in form of a traditional *legal conception of the market* whose essential features may be summarized as follows (see Graf, 1997).

The domain of *marketable goods* comprehends all those things that can be object of transferable property rights. This domain is limited by norms according to which some assets are either declared inalienable, such as human rights, or deprived of their commercial exploitation, e.g. one's physical organs. Here, I am not interested in these limitations in detail, I just want to point to the fact that virtually all modern legal systems submit the market to various constraints by making the contractual transfer of certain goods illegal from the start. The group of *possible participants in the market* includes all persons, be they individuals or collectives, who possess legal competence, i.e. the right to conclude binding transactions. In general, individuals ought to have some basic intellectual capabilities that enable them to pursue their own interests to a sufficient degree. Accordingly, they acquire full legal competence only with a certain age, and they may lose it again, if they fall short of a certain minimum level of rationality. Furthermore, all modern legal orders have certain rules concerning the *procedural conditions of contracts* in order to make sure that the parties conclude them voluntarily with sufficient information about the relevant properties of the respective goods and services. To this end, the law not only prohibits the use of force and coercion, but also rules out obvious forms of fraud and deception from which a party would gain advantage at the expense of others. Of course, a legal order also has to take care for the *enforcement of binding contracts* by giving the contracting parties an enforceable claim to fulfilment or compensation. Eventually, modern legal orders include some *regulations of market competition* which, at least in theory, aim at securing a fair market process in which no party is able to dictate the terms of trade, e.g. the prices, at the expense of others.

All these legal regulations have come into being in the course of history on the basis of a more or less vague idea of the requisite framing conditions of a functioning market without any proper theory of the market. Rather, such a theory has emerged only with the development of modern economics, i.e. the *economic theory of the market*. The *locus classicus* of this theory, of course, is Adam Smith's famous *invisible hand thesis*. According to it, a free market economy that guarantees private property, the keeping of contracts and open competition automatically generates general wealth to the benefit of all individuals involved, even if all people would pursue nothing else than their own self-interest, for the market would not only regulate the supply of every good in accordance with its demand, but also induce a constant increase of the division and productivity of labour to the benefit of the whole society (see Smith, 1759/1976b, IV.i.10; 1776/1976a, IV.ii; cf. Grampp, 2000; Minowitz, 2004).

This thesis, which belongs to the very core of modern economic thinking, was the starting point of a highly productive industry of economic theorizing that has ramified into several directions. Today, there are essentially two main approaches which both aim to function as a positive theory for the empirical description and explanation of real markets and also as a normative conception of economic policy for the institutional framing of well-functioning markets: on the one hand, the approach of *Neoclassical Economics* that centres around the model of a perfect market (cf. Haslinger and Schneider, 1983), and, on the other hand, the approach of so-called *Austrian Economics* which emphasizes the innovative dynamics of market competition as a driving force of social progress (see Streissler, 1988). I want to take a brief glance at these approaches.

The *model of a perfect market* of *Neoclassical Economics* provides a formal prove that, under certain ideal conditions, the market results in an equilibrium of demand and supply in which the price of every good falls to the lowest possible level of its production costs and a Pareto-optimal allocation of scarce resources is achieved (cf. Samuelson and Nordhaus, 1998, pp. 137ff.; Stiglitz, 1997, pp. 28ff, 171ff, 345ff.). Those conditions include that all participants are in possession of a sufficient stock of resources which allow them free access to the market, that all participants have perfect rationality and complete information on the relevant facts so that their choices always meet their well-considered interests, that market transactions are costless and free of external effects, that the marketable goods and services are fully homogenous and divisible, that their demand and supply are fully elastic, and, last but not least, that the market competition works perfectly in a way that no party is powerful enough to dictate unilaterally the terms of trade (cf. Buchanan, 1985, pp. 14ff.; Gauthier, 1986, pp. 85ff.; Fritsch et al., 2001, pp. 33ff.). Accordingly, a perfect market requires the *irrelevance of social power differentials* among its participants which could distort its outcomes, in order to operate in a fully efficient way.

The *Austrian School of Economics*, which tends to be sceptical about the idealized models of neoclassical economics, conceives of the market as an *engine of social progress*. This view essentially relies on two arguments which have the form of plausible considerations rather than being precise proofs. One argument comes from Joseph Schumpeter who described the market as a *process of creative destruction*: Since market competition forces the producers to meet the consumers' demands as good as possible in order to succeed, it exerts a constant pressure on them to improve their products and ways of production and, thereby, stimulates an ongoing progress of the economy leading to permanent growth (see Schumpeter, 1934/1961, pp. 57ff.; 1942/1976, pp. 81ff.; cf. Baumol, 2002; McCraw, 2007). The second argument is the conception of the *market as a discovery procedure* by Friedrich A. Hayek: As the price mechanism of a free market provides the individuals with appropriate information on the constantly changing facts that are relevant for their economic activities, it automatically manages to conduct vast streams of dispersed knowledge to their most effective use which no centralized institution could ever achieve; therefore, Hayek regards a free market as the paradigm of a spontaneous social order that eventually leads the activities of a great many independently acting individuals to generally beneficial outcomes by submitting them to a constant process of trial and error (see Hayek, 1945; 1976, pp. 107ff.; 1968/1978; 1979, pp. 67ff.; cf. Kirzner, 1979).

It is pretty clear, however, that virtually all markets in real economy do more or less deviate from the ideal of the perfect market. So most economists, at least the exponents of neoclassical theory, admit that real markets, which do not meet the previously mentioned conditions of perfect market competition, may suffer from *market failures*, i.e. generate inefficient results. I just want to name the most significant cases of such inefficiencies: the inability of the market to provide a sufficient supply of public and meritoric goods; the fact that market transactions that have negative external effects may cause social costs that exceed their benefits; the distortions of price fixing in cases of monopolistic and oligopolistic competition; the problems of moral risk and adverse selection; and the cyclical instabilities and inefficient equilibria that frequently occur in real markets (see Fritsch et al., 2001, pp. 95ff.; Sturn, 2002). Even though the adherents of Austrian Economics are more reluctant to concede that markets can fail because of these facts, they would also not deny that some of them work badly, if they fall short of certain framing conditions that guarantee their openness and free operation (cf. Cordato, 1980).

Anyway, the economic theory of the market faces various critical objections, some of which amount to a radical critique on the part of heterodox economists or leftist social theorists, while others represent a more moderate critique within the mainstream of economics. The *radical critique* usually maintains that economic market theory has degenerated into an ideology which glorifies the market system by hushing up the gross deficiencies and injustices of the capitalist economy (see Sherman, 1972; Hunt and Sherman, 1990, pp. 180ff.). In contrast, the *moderate critique*, if addressed to neoclassical economics, mainly challenges the model of the perfect market by raising two objections: first, that this model could neither explain economic reality nor guide economic policy, because it is greatly unrealistic; and, secondly, that it is completely static, since it would only focus on the allocative function of the market without paying attention to its dynamics (cf. Arndt, 1979, pp. 34ff.; Rothschild, 1980; Streissler, 1980). And as far as Austrian Economics is concerned, one may object that its market conception lacks any clear standard for the evaluation of the outcomes of free market processes, because its general reliance on social progress appears much too vague and spongy (cf. Koller, 2000).

Even though these critical points against the ideal of a perfect or well-functioning market appear by and large justified and should be taken seriously, they could not shatter the great fascination und impact of this ideal, and probably they will not succeed in future either. At any rate, this ideal still plays a dominant role in economics. Why is that so? In my opinion, this fact results from a number of reasons which include ideological, but also objective motives. The main motive, I think, is the fact that the market ideal, in spite of its unrealistic assumptions, illuminates the most significant aspects and virtues of a market order in a simple and convincing way. Due to this fact, this ideal may serve the ideological purpose of polishing up the image of capitalism, but it can also help to improve our understanding of economic reality, both its positive description and its normative evaluation. In the context of *positive economics*, it is useful to interpret the ideal of a perfect market as a simplified theoretical starting point from which economic analysis, through a successive approximation of its models to reality, may move to a more adequate understanding of how more or less imperfect markets operate (Stiglitz, 1997, pp. 187ff.). Furthermore, the ideal of a perfect or well-functioning market may also render good service in *normative economics*, because it provides a normative guide-line for the evaluation of real markets as to whether and to what extent they meet the requirements of efficiency. Since it is certainly true that a well-functioning market is a greatly efficient device for the allocation and production of economic goods, even though one may quarrel about the appropriate standards of efficiency and their application in real life, one may expect that real markets tend to become less efficient to the extent in which they deviate from the ideal.

On the basis of these results, which admittedly are not very original, I am now going to defend my main thesis which may be somewhat more interesting. This is the proposition that the ideal conditions of a perfect or well-functioning market also represent demands of *contractual justice*, i.e. moral demands that apply to exchange relationships or contractual transactions.

4. Contractual Justice

The view that a well-functioning market does not only generate efficient outcomes, but, in general, also warrant just exchange relationships may sound surprising against the background of contemporary economics, which does not pay much attention to contractual justice, but it is by no means new. It already appears, though in a rudimentary form, in the writings of medieval and early modern thinkers, and it was also advocated by some distinguished theorists of Neoclassical and Austrian economics. Furthermore, there are good reasons to suppose that this view is implicitly shared by most economists who always refer to the achievements of a well-functioning market, even if they do not explicitly claim its virtue of being in line with contractual justice. Otherwise, their preference for free markets would be hard to explain. Anyway, I want to argue that the ideal of a perfect or well-functioning market is closely related to the idea of contractual justice, a fact which makes that ideal even more appealing.

Contractual justice applies to *exchange relationships*, i.e. contractual transactions of which a market is composed. An old and in a way still influential conception of contractual justice is the *equivalence principle*, according to which an exchange relationship is just, if the goods or services exchanged are of equal value or equivalent (cf. Aristotle, 1954, V.7, 1131b 33ff.). This conception, however, relies on the condition that there is a generally acknowledged objective measure on the basis of which the value of the respective assets can be determined independently from the procedure of the particular transactions. The search for such a value measure, which went on for centuries, from the middle ages until modern times, was a central issue of the *theory of just price*, a matter of manifold speculations and controversies. For example, some thinkers advocated the view that, following Aristotle, the proper value of goods would depend on their *utility value*, i.e. the degree of preference satisfaction that they usually provide to those who use them. By contrast, others argued that the just price of goods would result from their *production costs plus a profit margin* guaranteeing their producers an income befitting their social rank. Eventually, there is a further view that, in a rudimentary form, already appeared in some medieval authors, such as Thomas Aquinas, and then dominated the social philosophy of enlightenment and classical political economy until Marx, namely the view that the value of goods, in the last instance, would flow from the *labour efforts* that, under normal conditions, were necessary to their production (cf. Kaulla, 1904; Wood, 2002, pp. 132ff.).

It is well-known that all endeavours to find an objective measure of the proper value of goods on the basis of which their just price could be determined have failed. And it appears greatly improbable that such a measure can be found at all, since it would hardly fit into the context of a highly differentiated economy based on division of labour and commercial transactions, whose productivity is driven forward by market competition. This result recommends another, a procedural conception of contractual justice that ties the justice of exchange relationships to the conditions under which contractual transactions occur rather than any objectively given value of the respective goods. Accordingly, a contractual transaction appears just, if it takes place under conditions which warrant that it is in the well-considered interest of all parties involved so that each derives sufficient benefit from it. As a well-functioning market appears to meet these conditions to a great extent, it is plausible to conclude that the just price is nothing else than the market price which automatically emerges from the operation of such a market. In other words, the just price is the price in a well-functioning market that guarantees fair exchange relationships. So contractual justice demands a *fair market* (cf. Koslowski, 1988, pp. 227ff.).

This conception of contractual justice, the *fair market conception*, is much older than the subjective value theory of modern economics, for it can be traced back to middle ages and early modernity (cf. Langholm, 1998, pp. 77ff.; Sturn, 2007). There are historical studies which show that a number of medieval scholars, including theologians and jurists, already advocated the view that, under certain conditions, the just price would coincide with the market price (see Höffner, 1953; de Roover, 1958; Trusen, 1967/1997). According to one study, for example, most German scholars in late middle ages agreed on the view "that the value of a commodity would usually be determined in a just way by its general appreciation and the strength of the demand for it. This, however, would be true only, if fraud, coercion and error were excluded. The market price, one thought, would be always just, if it were not distorted through unfair manipulation and speculation or an irresponsible conduct of existing monopolies, in short, if it were the result of a honest competition among responsible merchants" (Trusen, 1967/1997, p. 535, my translation).

A pretty elaborated theory of just price fixing in a well-ordered market was developed by the Spanish late scholastics of the *School of Salamanca* in the 16th century. These thinkers, particularly Francisco de Vitoria and Luis de Molina, declared that the just price would not be determined by the profit and loss on the part of the merchants, but rather by the relationship between supply and demand at the place where the commodities were sold. Consequently, the price of a good could vary from place to place, but would level off to a single price at the same place. They also emphasized the importance of open competition for the emergence of just market prices as well as for the proper functioning of the market in general. Therefore, they regarded regulative measures by the public authorities as justified or even requisite, whenever the market process was distorted by some obstacles, such as private monopolies and other forms of market power that enabled some of the parties involved, e.g. merchants and guilds, to push the prices either above or below the competitive level (see Höffner, 1953, pp. 194ff.; de Roover, 1958, pp. 424ff.; Langholm, 1998, pp. 77ff.).

Even though the great thinkers of classical political economy, to my knowledge, did not deal with the problem of contractual justice explicitly, one may assume that they implicitly shared the fair market view. This view does not only fit well into their theoretical understanding of the market as the most appropriate mechanism of regulating the economy, but also makes their emphatic plea for a free market system even more plausible. Furthermore, it is to observe that those classical thinkers who actually spoke about issues of justice, namely Adam Smith and John Stuart Mill, were mainly concerned with the institutional framing conditions of markets. Smith, for example, conceived of justice as the totality of legal rules and institutions necessary for the effective enforcement of individual rights and duties, particularly those that were necessary for the protection of property and the keeping of contracts (cf. Smith, 1762–1763/1978, i.9ff.; Nutzinger, 1991, pp. 93ff.). One century later, Mill argued in view of the social problems of expanding capitalism that a market economy also had to meet certain demands of distributive justice in order to achieve general utility (see Mill, 1861/1962, pp. 296ff.; 1879/1987, S. 65ff.).

Obviously, contractual justice is also not a major issue in contemporary economics. Yet, two highly influential economic theorists paid some attention to it, namely Léon Walras und Friedrich A. Hayek. Walras repeatedly underlined that a market would meet the demands of justice, if and only if all participants had, first, equal access to it (which, he thought, would exclude private property on natural resources, including land) and, secondly, complete freedom to pursue their interests without interference (cf. Walras, 1896/1990; de Gijsel, 1984, pp. 18ff.). For one condition of justice would be that "there is only one definite price on the market," and another that "the exchange process favours neither buyers nor sellers at the disadvantage of others" (quote following de Gijsel, 1984, p. 19, my translation). Although Hayek was much less demanding as to the initial conditions of a just market, he joined the view "that prices determined by just conduct of the parties in the market, i.e. competitive prices arrived at without fraud, monopoly and violence, was all that justice required," so that "only that 'natural' price could be regarded as just which would be arrived at in a competitive market" (Hayek, 1976, pp. 73, 75).

I deem this conception of contractual justice, according to which a well-functioning market facilitates just exchange relationships, highly plausible under one precondition, which, I guess, is also implicitly presupposed by the authors mentioned, except Hayek. This precondition is to the effect that a just exchange does also require an acceptable *initial distribution* of relevant individual assets, such as legal competences, individual liberties and property rights, a distribution which, at least, is not extremely unjust in light of reasonable and widely accepted demands of justice, including distributive justice. On the basis of these results, I would like to try to present a more refined statement of the *fair market conception of contractual justice* in two steps: I will begin with a highly demanding idealized version and then proceed to a more moderate practical understanding.

The *idealized version*, which may be understood as a regulative idea of market justice, runs as follows: *Contractual exchange transactions are just, if they take place under conditions of a perfect or well-functioning market that is based on an acceptable initial distribution of relevant assets*. As to contractual transactions which do not occur in the context of a market, one could add that such transactions are just if they come about through a procedure that mimics those conditions. In short, contractual justice ideally requires both a well-functioning market and an acceptable initial distribution. If both requirements are met, it can reasonably be assumed that the respective transactions are in the best interest of all parties involved, so that none of them has reason to complain about the outcomes of the exchange process. These requirements, however, are extremely demanding, for the ideal of a perfect or well-functioning market includes a number of highly idealized assumptions which are virtually never satisfied in reality, such as complete rationality and information on the part of all persons involved, the absence of negative external effects of their transactions and the irrelevance of inequalities of their power. Furthermore, the precondition of an acceptable initial distribution relies on standards of distributive justice which, in most cases, are greatly contested. Consequently, the idealized version of contractual justice has to be weakened in order to arrive at a more moderate practical understanding that can be applied to social reality.

A *practical understanding* of the fair market conception should reduce the requirements mentioned to a level which, on the one hand, allows a sufficient scope for binding contractual transactions, and, on the other, secures their voluntary nature and mutual utility by preventing obvious forms of fraud and exploitation. It is not easy to determine this level in detail. It appears clear, however, that an ethical conception of contractual justice, in order to get some bite, may raise stronger demands on a market order than a positive legal order that plausibly ties the validity of contracts to very weak minimum conditions only. Otherwise, it were impossible to regard a legally valid contract as unjust, as we sometimes actually do. So I want to suggest that a market order, in general, qualifies as just, if it by and large meets the conditions of a fair market by making sure that contractual transactions (1) are concluded freely among sufficiently rational and informed people, (2) have no significant negative effects on others, (3) are not distorted by pertinent inequalities of power and, last but not least, (4) take place on the basis of an acceptable, at least not greatly unjust initial or previous situation.

I would like to conclude my paper with some hints on the consequences of the proposed conception of contractual justice in comparison with current neoliberal thinking.

5. Concluding Notes

Most advocates of neoliberalism assume, following Robert Nozick (1974, pp. 149ff.), that a market whose transactions start from an acceptable initial distribution of resources and conform to contractual justice would automatically generate just outcomes, even though they may disagree about the standards as to the acceptability of the initial distribution. So they believe that just market transactions under acceptable initial conditions grant just results. This belief, however, cannot be true, if the previously sketched conception of contractual justice, particularly its practical understanding, is by and large plausible. On the basis of this understanding, whose requirements on just transactions are pretty weaker than the conditions of a perfect market, it may easily occur that a sequence of just market transactions eventually results in a state of affairs which appears clearly unjust, be it because it distorts the market process in a way that undermines the contractual justice of further transactions or because it violates widely accepted demands of distributive justice, provided that such demands, contrary to Hayek's view, do apply to the economic order, as I think (cf. Koller, 2006). Thus, not even a fair market whose transactions are by and large just can guarantee the justice of its outcomes. And the less can real markets do that, as most of them suffer from significant distortions, such as differentials of power, insufficiencies of rationality, lack of information and, not to forget, occurrences of fraud and cheating that remain uncorrected.

Many adherents of neoliberalism regard the market as a 'morally free zone' in the sense that, in a well-ordered market, individuals are free to pursue their own interests without being constrained by any requirements of morality and justice, since the framing conditions of such a market would guarantee the efficiency or general utility of individual activities (cf. Gauthier, 1986, pp. 83ff.). Apart from the fact that this thesis is somewhat misleading, it is also built on sandy ground. It is misleading, since it does not really mean what it says. What it means is not that, in the market, people have no moral duties, but rather that they are not subject to moral duties which exceed the requirements that define a well-functioning market. But even in this weaker reading, the thesis is without foundation. As it only applies to a well-functioning market in which, by definition, all transactions accord to a set of pretty demanding requirements, it is clearly not true of real markets which more or less deviate from the ideal. If one assumes that, in these markets, people are not completely free to pursue their own self-interests at all costs, but required to act in a way that brings their transactions closer to contractual justice, then they are subject to various moral demands, such as the requirement that their transactions ought to be not distorted by power differentials (cf. Rothschild, 1971). So one could rather state that a proper market is, or ought to be, a *power free zone*. Furthermore, the view that a well-functioning market is a morally free zone confounds morality with mere efficiency, because it forbears from any reference to a morally acceptable initial distribution. So it must rely on the respective *status quo*, which, however, may appear completely unacceptable from the viewpoint of general morality or distributive justice. Yet, it is certainly not plausible to regard the results of a market process as morally legitimate, when this process starts from a state of affairs which appears to be morally unacceptable. This leads me to my final point which concerns the relationship between market efficiency and contractual justice.

Market efficiency and contractual justice have much in common, but they differ in an important point: Efficiency assesses the market's operation and results in relation to the respective *status quo*, irrespective of how it may be judged from the moral viewpoint, whereas contractual justice evaluates market processes against the background of a morally legitimate starting point and, therefore, does also require a generally acceptable initial distribution of individual rights and assets. Consequently, market efficiency and contractual justice will coincide, as long as a market operates on the basis of an acceptable initial distribution, and they will more or less conflict, when the market transactions under consideration are based on a more or less questionable structure of individual rights and assets. In the latter case, contractual justice will require a correction of the prevailing state of affairs, a requirement which, in principle, should overrule opposite considerations of efficiency. It is true, however, that, in general, there is little agreement on the standards of morality and justice that could serve as a basis for judging distributive structures of individual resources. This may be one of the reasons why neoliberals are inclined to qualify the prevailing distribution of property and economic resources as acceptable, whether they deem it to be defensible in a way (Nozick, 1974; Buchanan, 1975, pp. 74ff.), or whether they think that no standards of morality and justice apply to the distribution of economic assets in large and advanced societies (Hayek, 1976). Yet, the matter is not so easy.

Admittedly, it does not make much sense to scrutinize the historical origins of a prevailing social state of affairs in light of current standards of morality and justice, since, when those origins appear morally contestable or even clearly unjust, there is usually no way to undo past wrongs long ago. In my view, however, contractual justice does by no means rely on the condition that the market process started from an acceptable initial distribution in a far past, on the basis of which a well-functioning market would always have lead to just outcomes that again would have provided an acceptable initial distribution for the subsequent market process, and so on. As a market system is an ongoing process in which each stage represents the initial situation of its subsequent transactions, contractual justice calls for enduring efforts to bring the distributive effects of the market process in accord with widely acknowledged and plausible standards of morality and justice, including distributive justice. Even though it is certainly true that these standards are susceptible to social change and subject to significant disagreement in detail, there is, at least in well-functioning social orders, usually a certain consensus on some fundamental standards of morality and justice on the basis of which it is possible to identify gross moral defects and injustices of a prevailing social situation that clearly contradict contractual justice. In present Western societies, for instance, it is widely consented that, in principle, a just social order in general and a fair market in particular demand some equality of opportunity, especially in the intergenerational context. And it is pretty obvious that this demand, however it may be interpreted in detail, cannot be satisfied by the market, not even a perfect one.

And there are many other features of market systems that tend to undermine contractual justice. Consequently, contractual justice not only requires appropriate framing conditions which facilitate a well-functioning market, but also a regulation of the ongoing market process that seeks to guarantee a generally acceptable distribution of economic assets as good as possible.

References

Aristotle (1954). *Nicomachian Ethics* (D. Ross, Ed.). Oxford University Press, London.
Arndt, H. (1979). *Irrwege der Politischen Ökonomie*. C. H. Beck, München.
Baumol, W. J. (2002). *The Free-Market Innovation Machine: Analyzing the Growth Miracle of Capitalism*. Princeton University Press, Princeton, NJ.
Braithwaite, J. and Pettit, P. (1990). *Not Just Deserts. A Republican Theory of Criminal Justice*. Clarendon Press, Oxford.
Buchanan, A. (1985). *Ethics, Efficiency, and the Market*. Rowman & Allanheld, Totowa, NJ.
Buchanan, J. M. (1975). *The Limits of Liberty. Between Anarchy and Leviathan*. University of Chicago Press, Chicago and London.
Chapman, J. W. and Pennock, J. R. (Eds.). (1989). *Markets and Justice*. New York University Press, New York and London.
Coleman, J. L. (1988). *Markets, Morals and the Law*. Cambridge University Press, Cambridge.
Coleman, J. L. (1992). *Risks and Wrongs*. Oxford University Press, Oxford.
Cordato, R. E. (1980). The Austrian theory of efficiency and the role of government. *The Journal of Libertarian Studies*, 4, 393–403.
Dworkin, R. (2000). *Sovereign Virtue. The Theory and Practice of Equality*. Harvard University Press, Cambridge, MA.
Feinberg, J. (1973). *Social Philosophy*. Prentice-Hall, Englewood Cliffs, NJ.
Fritsch, M., Wein, T., and Ewers, H.-J. (2001). *Marktversagen und Wirtschaftspolitik* (4th ed.). Vahlen, München.
Gauthier, D. (1986). *Morals by Agreement*. Clarendon Press, Oxford.
de Gijsel, P. (1984). Individuum und Gerechtigkeit in ökonomischen Verteilungstheorien. In *Wohlfahrt und Gerechtigkeit*, volume 2 of *Ökonomie und Gesellschaft. Jahrbuch* (14–66). Campus, Frankfurt and New York.
Graf, G. (1997). *Vertrag und Vernunft. Eine Untersuchung zum Modellcharakter des vernünftigen Vertrages*. Springer, Vienna and New York.
Grampp, W. D. (2000). What did Adam Smith mean by the invisible hand? *Journal of Political Economy*, 108, 441–465.
Haslinger, F. and Schneider, J. (1983). Die Relevanz der Gleichgewichtstheorie. In *Die Neoklassik und ihre Herausforderungen*, volume 1 of *Ökonomie und Gesellschaft. Jahrbuch* (1–55). Campus, Frankfurt and New York.
Hayek, F. A. (1945). The use of knowledge in society. *American Economic Review*, 35, 519–530.
Hayek, F. A. (1976). *Law, Legislation and Liberty. Vol. 2: The Mirage of Social Justice*. Routledge, London.
Hayek, F. A. (1978). Competition as a Discovery Procedure. In *New Studies in Philosophy, Politics, Economics and the History of Ideas* (179–190). Routledge, London. (Original work published 1968).

Hayek, F. A. (1979). *Law, Legislation and Liberty. Vol. 3: The Political Order of a Free People*. Routledge, London.
Höffe, O. (1987). *Politische Gerechtigkeit*. Suhrkamp, Frankfurt a.M.
Höffner, J. (1953). Der Wettbewerb in der Scholastik. *Ordo*, 5, 181–202.
Hunt, E. K. and Sherman, H. J. (1990). *Economics: An Introduction to Traditional and Radical Views* (6th ed.). Harper & Row, New York.
Kant, I. (1968). Grundlegung zur Metaphysik der sitten. In W. Weischedel (Ed.), *Kant-Werkausgabe in zwölf Bänden*, volume 7. Suhrkamp, Frankfurt a.M. (Original work published 1785).
Kaulla, R. (1904). Die Lehre vom gerechten Preis in der Scholastik. *Zeitschrift für die gesamte Staatswissenschaft*, 60, 579–602.
Kirzner, I. M. (1979). Hayek, Knowledge, and Market Processes. In *Perception, Opportunity, and Profit* (13–33). University of Chicago Press, Chicago.
Koller, P. (1995). Soziale Gleichheit und Gerechtigkeit. In H. P. Müller and B. Wegener (Eds.), *Soziale Ungleichheit und soziale Gerechtigkeit* (53–79). Leske & Budrich, Opladen.
Koller, P. (2000). Individualismus und Liberalismus bei Hayek und Nozick. In K. Seelmann (Ed.), *Wirtschaftsethik und Recht (Archiv für Rechts- und Sozialphilosophie, Beiheft 81)* (39–57). Franz Steiner, Stuttgart.
Koller, P. (2001). Zur Semantik der Gerechtigkeit. In *Gerechtigkeit im politischen Diskurs der Gegenwart* (19–46). Passagen Verlag, Vienna.
Koller, P. (2002). Das Konzept des Gemeinwohls. Versuch einer Begriffsexplikation. In W. Brugger, S. Kiste, and M. Anderheiden (Eds.), *Gemeinwohl in Deutschland, Europa und der Welt* (41–70). Nomos, Baden-Baden.
Koller, P. (2003). Soziale Gerechtigkeit—Begrifff und Begründung. *Erwägen Wissen Ethik*, 14(2), 237–250.
Koller, P. (2006). Ökonomische Verteilungsgerechtigkeit. In C. Langbehn (Ed.), *Recht, Gerechtigkeit und Freiheit. Festschrift für Wolfgang Kersting* (79–110). Mentis, Paderborn.
Koller, P. (2009). International law and global justice. In L. H. Meyer (Ed.), *Legitimacy, Justice and Public International Law* (186–206). Cambridge University Press, Cambridge.
Koslowski, P. (1988). *Prinzipien der Ethischen Ökonomie*. J.C.B. Mohr, Tübingen.
Langholm, O. (1998). *The Legacy of Scholasticism in Economic Thought. Antecedents of Choice and Power*. Cambridge University Press, Cambridge.
Lucas, J. R. (1980). *On Justice*. Clarendon Press, Oxford.
McCraw, T. K. (2007). *Prophet of Innovation. Joseph Schumpeter and Creative Destruction*. Belknap Press, Cambridge, MA and London.
Mill, J. S. (1962). *Utilitarianism. On Liberty. Essay on Bentham*, chapter Utilitarianism. Collins/Fontana, London and Glasgow. (Original work published 1861).
Mill, J. S. (1987). *On Socialism*. Prometheus Books, Buffalo, NY. (Original work published 1879).
Miller, D. (1976). *Social Justice*. Oxford University Press, Oxford.
Miller, D. (1999). *Principles of Social Justice*. Harvard University Press, Cambridge, MA and London.
Minowitz, P. (2004). Adam Smith's invisible hands. *Econ Journal Watch*, 1, 381–412.
Nozick, R. (1974). *Anarchy, State, and Utopia*. Basic Books, New York.
Nutzinger, H. G. (1991). Das System der natürlichen Freiheit bei Adam Smith und seine ethischen Grundlagen. In *Adam Smiths Beitrag zur Gesellschaftswissenschaft*, volume 9 of *Ökonomie und Gesellschaft. Jahrbuch* (79–100). Campus, Frankfurt and New York.
Rawls, J. (1971). *A Theory of Justice*. Harvard University Press, Cambridge, MA.

de Roover, R. (1958). The concept of the just price: Theory and economic policy. *The Journal of Economic History*, 18, 418–434.
Rothschild, K. W. (Ed.). (1971). *Power in Economics. Selected Readings*. Penguin Books, Harmondsworth.
Rothschild, K. W. (1980). Kritik marktwirtschaftlicher Ordnungen als Realtypus. In E. Streissler and C. Watrin (Eds.), *Zur Theorie marktwirtschaftlicher Ordnungen*, (13–37). J.C.B. Mohr, Tübingen.
Samuelson, P. A. and Nordhaus, W. D. (1998). *Economics* (16th ed.). Irwin/McGraw-Hill, Boston et al.
Schumpeter, J. A. (1961). *The Theory of Economic Development*. Oxford University Press, New York. (Original work published 1934).
Schumpeter, J. A. (1976). *Capitalism, Socialism and Democracy* (5th ed.). George Allen & Unwin, London. (Original work published 1942).
Sherman, H. (1972). *Radical Political Economy*. Basic Books, New York.
Smith, A. (1976a). *An Inquiry into the Nature and Causes of the Wealth of Nations*. Oxford University Press, Oxford, volume 2 of Glasgow edition). (Original work published 1776).
Smith, A. (1976b). *The Theory of Moral Sentiments*. Oxford University Press, Oxford, volume 1 of Glasgow edition). (Original work published 1759).
Smith, A. (1978). *Lectures on Jurisprudence*. Oxford University Press, Oxford, volume 5 of Glasgow edition). (Original work published 1762–1763).
Sterba, J. P. (1980). *The Demands of Justice*. University of Notre Dame Press, Notre Dame and London.
Stiglitz, J. E. (1997). *Economics* (2nd ed.). Norton, New York and London.
Streissler, E. (1980). Kritik des neoklassischen Gleichgewichtsansatzes als Rechtfertigung marktwirtschaftlicher Ordnungen. In E. Streissler and C. Watrin (Eds.), *Zur Theorie marktwirtschaftlicher Ordnungen*, (38–69). J. C. B. Mohr, Tübingen.
Streissler, E. (1988). The intellectual and political impact of the Austrian school of economics. In *History of European Ideas*, volume 9(2) (191–204). Elsevier, Pergamon Imprint, Oxford.
Sturn, R. (2002). Die Grenzen des Marktes—eine Typologie des Marktversagens. In *Alles käuflich*, volume 18 of *Ökonomie und Gesellschaft. Jahrbuch* (39–72). Metropolis, Marburg.
Sturn, R. (2007). Gerechter Preis und Marktpreis: Zur Interdependenz von Religion, Ökonomie und Sozialtheorie. In M. Held, G. Kubon-Gilke, and R. Sturn (Eds.), *Ökonomie und Religion*, volume 6 of *Normative und institutionelle Grundfragen der Ökonomik. Jahrbuch* (89–111). Metropolis, Marburg.
Sunstein, C. R. (1997). *Free Markets and Social Justice*. Oxford University Press, New York and Oxford.
Trusen, W. (1997). Äquivalenzprinzip und gerechter Preis im Spätmittelalter. In *Gelehrtes Recht im Mittelalter und in der frühen Neuzeit* (531–547). Keip Verlag, Goldbach. (Original work published 1967).
Walras, L. (1990). *Études d'économie sociale*, volume 9 of *Oeuvres économiques complètes*. Économica, Paris. (Original work published 1896).
Wood, D. (2002). *Medieval Economic Thought*. Cambridge University Press, Cambridge.

Part III

Workshops

Formal Methods in Philosophy

Shooting Right Without Collateral Damage

Pascal Engel

ABSTRACT

The problem with formal methods in philosophy is not whether to apply them or not. For most analytic philosophers they are basic and unquestionably legitimate and fruitful. The problem is their adequacy. Sometimes there is too much use of formal methods in a field, and their richness and the diversity of the formalisms overkills the part of the field and the questions that they are supposed to illuminate. At other moments, they miss the target and fail to capture what is essential. So the important question is when to use them so that they can shoot correctly at their target without too much collateral damage.

Introduction: No Drama

Once upon a time, the conflict between those among analytic philosophers who were friends of formal methods and those who were hostile to them raged. But the days of logical positivism have gone, as well as those of the opposition between Carnapian methods and ordinary language philosophy. Today the atmosphere is more peaceful. Almost nobody, I gather, believes that all philosophical problems can lend themselves to a formal treatment, and that formalisms are all encompassing. But there is still an opposition between those who think that one can go a long way with them, and those who are sceptical (very often for quite opposite reasons: for instance Wittgensteinians dislike them because they believe that there is a conceptual core in philosophy which cannot be represented by any formalism, and 'experimental philosophers' dislike them because they take them to be typical of the armchair method).

The question is: how far can we go, and what are the limits of scepticism about the use of formal methods? The problem is not to throw away formalism. The formalisms of logic and probability theory are part of the language and the natural equipment of today's philosophers: quantified modal logic, probability theory are the common language (they are almost Kuhnian paradigms). The problem is to assess their scope. I would like here to suggest that one of the main difficulties of the formalist method lies not with the use of formalism *per se*, but with the frequent lack of preliminary elucidation of the philosophical layout on which it is supposed to be applied.

1. Virtues and Vices of Formalisation

Formalism in philosophy has its virtues and its vices. According to its friends, formalism helps us to make philosophical problems and their possible answers precise: they improve its formulation, make more visible the possible assumptions, the possible solutions and their consequences, and help see clearly what the options are. In this sense they have a preparatory and heuristic role. They have also a negative and refutational role: if a thesis is not formalisable, it is very likely to be obscure and false, and we are led to remove ambiguities. But formalisms have a positive role as well: sometimes they can be used as heuristics. Formalism has also a number of vices. It has been objected that it misses the force of some philosophical problems by reducing them to their expression into a formalism, that it mistakes properties of the models for properties of the things to be modelled, that sometimes it is like using a sledgehammer to crack a walnut: as La Fontaine says in the fable *The bear and the amateur gardener*: "A foolish friend may cause more woe than could indeed the wisest foe."

In order to assess these criticisms, we need first to understand how formalisation works. Formalisation is not a mere regimentation of natural language into a symbolic language, such as predicate calculus, probabilistic models, or modal logic. It involves a preliminary regimentation of philosophical language into formal language, which has primitives, axioms, and metatheoretical properties, such as soundness and completeness—if possible. It involves a choice of formalisms and the possibility to compare them. It then involves the use of the properties of the formal language to clarify the assumptions of a fragment. Among the most familiar formalisms are quantified modal logic, model theory, probability theory, decision theory, utility theory and their many subbranches. As a number of proponents of the formalist approach have remarked,[1] these formalisms can be considered as models, in the same sense as the one with which we are familiar in the philosophy of science. Among their virtues are, in the first place, their heuristic importance: they suggest something that we might later be able to explain in a model-independent way. In the second place, models help us to explain features of science, i.e. to deal with more realistic (and, as it happens, more complicated) situations. As Stephan Hartmann says, "when different intuitions pull in different directions the philosophical model tells us which intuition wins" in which part of the parameter space. The virtues of models are thus described by Sven Ove Hansson:

> First, formalization incites definitional and deductive economy. It brings forth questions about the interdefinability of concepts and about minimizing the set of primitive principles of inference... Secondly, formalization serves to make implicit assumptions visible. Thirdly, formal theories can support delicate structures that would be much more difficult to uphold and handle in the less unambiguous setting of an informal language. Symbolic treatment has made it possible to penetrate some philosophical issues more deeply than what would otherwise have been possible. Fourthly, formalization stimulates strivings for completeness. The rigorousness of a formal language is, for instance, necessary to make it meaningful to search for a complete list of valid principles of inference. Often enough, this search may uncover previously unnoticed philosophical problems. (Hansson, 2000, p. 166–167)[2]

At the same time, Hansson is very lucid about the possible drawbacks of this method. They are the risks of oversimplification (reduction of primitive notions to a minimum), of false unification of concepts (for instance in deontic logic putting together all predicates of prescription (*should*, *must*, *ought*) under the same operator O may be a good idea in some contexts, not in others), false conceptual primitivity (selecting one concept because of its simplicity and elegance), invitation to ad hoc constructions, which are mere artefacts of the model, formalisation involves implicit ontological assumptions that are not innocuous, enigmatic style where important philosophical choices are made without explanation. Igor Douven and Leon Horsten (2008) are also critical when they say:

> If formal methods function, in some sense, as paradigms in the philosophy of science, then it should not come as a surprise that for every formal method there comes a point of diminishing returns. When a formal method

---

[1] I rely here on the very useful treatments of these issues by Douven and Horsten (2008), Hansson (2000), Hartmann (2008).

[2] See also the preface of his book (Hansson, 2001).

has been applied in one area of the philosophy of science, it is very natural to try to apply the same technique to other branches of the philosophy of science. But at some point the new applications begin to look forced and somehow unnatural: the formal method does not succeed in shedding (new) light on the conceptual problems at hand.

I quite agree, and just want here to confirm these diagnoses, without ignoring the fact that they come from researchers who illustrate the virtues of the method of formalisation.

2. Good Shots and Bad Shots of Formalisation

Everyone agrees that often formalisation hits its target, and that it is successful. Thus philosophy has benefited from the following examples of successful formalisation:

- Anselm's ontological proof (Gödel, Plantinga, Lewis, Oppy)
- Truth in L by Tarski
- Lewis's trivialisation result for conditionals
- Lewis on belief as desire
- Skyrms' formalisation of the social contract
- Gärdenfors impossibility result about the Ramsey test
- Bayesianism and Hume's argument on miracles (Salmon, Earman)

Why are such results successful? Because, it seems to me, they all combine a careful formulation of a philosophical problem in the first place, then its formulation in formal terms, and then a result which shows that one thesis is wrong or flawed or that a certain approach is fruitful. Of course, none of these formalisation solves the problem at issue, but they all give us clearer insight into the nature of the problem at hand, and help us in testing the coherence of a given thesis. Take for instance Lewis (1988) treatment of desire-as-belief. Lewis considers the view in moral psychology, defended by cognitivists against Humeans, that some normative beliefs—and not desires as the Humean has it—might motivate us to act, which we might call 'besires' since they would both be beliefs and states capable of having a motivating role. To that effect Lewis formulates in decision theoretic terms the notion of a desire-as-belief or 'besire,' desire-like states that are reducible to belief-like states: the desirability of X is the probability that X is good— and shows that if there were such states, decision theory would be crippled. This is not a knockdown argument against cognitivism about moral motivation, but it sets limits to it. It shows that if one wants to defend cognitivism, decision theory has to be revised. It also clarifies its link with the Humean theory of motivation. Like Lewis' trivialisation result for conditionals, it is a very interesting formal results which tells where a philosophical thesis leads, and where it cannot lead.

But not all formalisations are so successful. The problems raised by other philosophical claims based on formalisation is that either they draw too bold consequences from a formal result, or that they misconceive the preliminary interpretative step, which consists in setting clearly what the philosophical problem is, and what one can expect from a formal treatment of it.

As an example of the former difficulty, overgeneralisation, we can consider Hans Rott's comparison between belief change theory and decision theory. Hans Rott, in his impressive book, *Change, choice and inference* (2001) shows important representation theorems, and in particular that all operations of belief change that are generated by rational choice functions, with the choices satisfying certain coherence constraints, satisfy corresponding rationality postulates for belief change. He shows that conversely all operations of belief change that satisfy certain rationality postulates can be represented as operations that are generated by rational choice functions, with the choices satisfying corresponding coherence constraints. What does it show?

The formalisms were devised independently from each other, with different objectives. The correspondence results show that theoretical and practical reason obey the same structure, and since the logic belief change is a special case of the logic of rational choice, the latter has primacy over the former. To simplify a lot, take the familiar AGM postulates:

($*1$) $K^*\alpha = Cn(K^*\alpha)$;
($*2$) $\alpha \in K^*\alpha$;
($*3$) $K^*\alpha \subseteq Cn(K \cup \{\alpha\})$
($*4$) *if* $\neg\alpha/K$, *then* $Cn(K \cup \{\alpha\}) \subseteq K^*\alpha$
($*5$) *if* $Cn(\alpha) = L$, *then* $K^*\alpha = L$
($*6$) *if* $Cn(\alpha) = Cn(\beta)$, *then* $K^*\alpha = K^*\beta$.
($*7$) $K^*(\alpha \wedge \beta) \subseteq Cn(K^*\alpha \cup \{\beta\})$;
($*8$) *if* $\neg\beta \notin K^*\alpha$, *then* $K^*\alpha \subseteq K^*(\alpha \wedge \beta)$.

+ *Non monotonic principles (where inf (α) stands from the set of sentences which can be inferred non monotonically from α)*

(Or) $\text{Inf}(\alpha) \cap \text{Inf}(\beta) \subseteq \text{Inf}(\alpha \vee \beta)$

(RMon) if $\beta \notin \text{Inf}(\alpha)$, then $\text{Inf}(\alpha) \subseteq \text{Inf}(\alpha \wedge \beta)$.

Consider now the rational choice postulates:

(I) If $S \subseteq S'$, then $S \cap \delta(S') \subseteq \delta(S)$; (Property α)
(II) $\delta(S) \cap \delta(S') \subseteq \delta(S \cup S')$; (Property σ)
(III) if $S \subseteq S'$ and $\delta(S') \subseteq S$, then $\delta(S) \subseteq \delta(S')$; (Aizerman's Axiom)
(IV) if $S \subseteq S'$ and $\delta(S') \cap S \neq \emptyset$, then $\delta(S) \subseteq \delta(S')$. (Property $\beta+$)

Rott's results are that if we are given a choice function on sets of sentences, we can, for define a revision operation by letting $\beta \notin K^*\alpha$ hold if and only if $\alpha \to \beta \notin \delta(Cn(\neg\alpha))$, and similarly for contraction and inference. Given such conceptual bridges there is a 1–1 correspondence between postulates for belief change and inference, on the one hand, and postulates for rational choice, on the other hand. Rott interprets his results thus:

> The central and most surprising result . . . is that all postulates of theoretical reason are derivable from more general, practical principles of rational choice. (Rott, 2001, p. 5)
>
> . . . Philosophically, I take this to be a strong indication of the unity of practical and theoretical reason. (Rott, 2001, p. 214)

The problem with this interpretation of the results, as it was first noted by Hansson, is that Gärdenfors' original axioms of belief change were modelled after Lewis' models for counterfactuals, and as Olsson (2003) has remarked "the revision postulates were, in effect, partly motivated by the fact that in the context of belief models they jointly generate the same logic of conditionals as do prominent choice principles in the context of possible worlds models." If this is the case, there is less than it meets the eye in these results, which show partial overlaps between principles of rational choice and principles of belief change. They do not show that the *whole* structure of theoretical reason is dependent upon that of practical reason. It does not follow that there no such parallels and that Rott's work has not illuminated them. But the parallels have to be motivated by a certain theory of belief change or of inquiry. Isaac Levi has long insisted that we should conceive of inquiry in a decision theoretic way. A large body of work in philosophy is devoted today to bringing together epistemology and the theory of practical rationality, but it is not clear that we can derive the unity of practical and theoretical reason from the structures of their formalisms.

For an example of the second difficulty, insufficient elucidation of the initial formulation of a philosophical problem and overhasty formalisation, let us consider Heinrich Wansing's (2006) recent work on doxastic voluntarism. In a very interesting article, Wansing considers how a formalisation within deontic logic can vindicate a certain version of doxastic voluntarism. The paper, however, starts with an unpromising note:

> In this paper, I will not discuss what is maybe the most fundamental question concerning doxastic voluntarism, namely: What exactly does the doxastic voluntarist claim? Doxastic voluntarism has been characterized by philosophers in many ways, not all being equivalent and some being quite nonspecific or even unclear. .... In terms of belief formation (understood as an action of belief acquisition), the thesis would be that a doxastic subject may sometimes form a belief (as a result of deciding to form it). (Wansing, 2006, p. 202)

Wansing is right that the claim is, in much of the recent philosophical literature, unclear. His objective is to show that one can give to the claim a more precise formulation. He starts from the formulation of a kind of deontologism about belief:

α is justified in believing that p if and only if α is permitted to form (or voluntarily acquire) the belief that p.

He then proposes to use the '*Dstit*' modal logic of agency, which is based on the operator:

Dstit p = deliberately sees to it that p

and claims that the *Dstit* language allows us to evaluate arguments against voluntarism, in particular the well known 'anti-voluntarist' formulated by Feldman (2000):

(1) if deontological judgements about belief are true, then belief is under the control of the will

(2) but belief is not under the control of the will

(3) therefore deontological judgements about beliefs are false.

Wansing, however, does not deal with this argument, but notes that there can be unintentional actions, hence that we do not need to consider the concept of intention in deal with agency. According to him the *Dstit* formalisation meets this requirement. After having consider briefly the phenomenological argument against doxastic voluntarism (it is phenomenologically implausible that we could acquire beliefs at will) and the possibility of indirect doxastic voluntarism (one can bring oneself to believe that p by indirect means), Wansing proceeds to formulate a semantics of belief formation ascription, by which he combines the semantics of the *dstit*-operator and the semantics of standard doxastic logic. Wansing then uses the Prior-Thomason branching time structures.

> Each doxastic subject is supposed to be an agent who by her or his actions can influence the future course of the world. In *stit*-theory this idea is accounted for by assuming that for every individual agent, the histories passing through a moment are partitioned into sets of histories choice-equivalent for the agent. If two histories h and h' are choice-equivalent for an agent α at moment m, then α cannot discriminate by her or his actions

at m between h or h'. The sets of histories choice-equivalent for an agent at a moment m represent the 'choice-cells' of the agent at m. (Wansing, 2006, p. 214)

So we are given the following definitions

Definition 1
α deliberately sees to it that A
$[\alpha \text{ dstit}: A]$ is true in $\langle T, \leq, Agent, Choice, v \rangle$ at (m,h)
iff
(i) $\forall h' \in Choice_\alpha^m(h)$ A is true at (m,h'), and
(ii) $\exists h' \in H_m$ such that A is not true at (m,h').

Then we are given an interpretation of "agent α voluntarily acquires the (implicit) belief (knowledge) that A" (or "α forms the (implicit) belief (knowledge) that A") as $[\alpha \text{ dstit}: B_\alpha A]$ ($[\alpha \text{ dstit}: K_\alpha A]$).

Wansing then claims that the *dstit* models must be augmented by a doxastic (epistemic) accessibility relation between moment/history-pairs.

At this stage we may ask: where does Wansing takes his notion of forming the knowledge that P *voluntarily*? How can he assimilate in his definitions the notion of an implicit belief with that of an implicit knowledge? This is puzzling, for the problem of doxastic voluntarism, at least as it has been examined since Williams' (1973) seminal paper, is the problem of whether it is possible *consciously* to decide to belief at will. The definition proposed here of a state of implicit belief seems to beg this question.

Wansing then proposes 'doxastic *dstit* models'

Definition 2
$[\alpha \text{ dstit}: B_\alpha A]$ ($[\alpha \text{ dstit}: K_\alpha A]$) is true in the doxastic (epistemic) *dstit* model $\langle T, \leq, Agent, Choice, R, v \rangle$ at (m,h)
iff
(i) $\forall h' \in Choice_\alpha^m(h)$ $\forall h'' \in H_m$,
if $(m,h')R_\alpha^m(m,h'')$ then A is true at (m,h''), and
(ii) $(\exists h', h'' \in H_m$ such that $(m,h')R_\alpha^m(m,h'')$
and A is not true at (m,h'').

We are told that his allows us to make sense of a notion of "deciding to know implicitly." But what is the meaning of this notion? The assimilation of the concept of belief to that of knowledge, which seems required to fit in the problem with epistemic logic, begs the question, at least because even though it might make sense to talk of voluntary belief (in a sense to be determined), it makes little sense to talk about *knowing* at will. Knowledge is factive: how could one bring about knowledge of truths at will? Wansing is conscious of the difficulty and says here:

> Initially, it might sound highly implausible that an agent may decide not only to believe implicitly but even to *know* implicitly. What could this possibly mean? If epistemic voluntarism is a species of doxastic voluntarism, since knowledge is a kind of belief, how should an agent decide to know something? (Wansing, 2006, p. 220)

Wansing's apparent answer to this question is that there can be voluntary knowledge *formation*:

> An agent α decides to know that p at a moment/history-pair (m, h) iff at ever moment/history pair (m, h') such that h' is choice-equivalent with h for α at m, α knows that p at (m, h'). It is clear that "α sees to it that α knows that p" is satisfiable in some epistemic *dstit* model if p is neither a tautology nor a contradiction. Thus, if it is true at (m, h) that α sees to it that she knows that p, by her actions at (m, h) the agent can make sure that the future course of events comprises only histories h' such that p is true at every moment/history-pair (m, h'') compatible with what α knows at (m, h'). It is quite conceivable that agent α has the latter capacity. ... There are unintentional actions. In an unintentional performance of a generic action, the agent need not be aware of performing that action. When voluntarily acquiring the implicit knowledge that p, the agent need not be aware of forming this implicit knowledge. (Wansing, 2006, p. 220)

Wansing also argues that "an agent α can see to it that another agent β forms a certain belief only if α is not independent of α." The model also allows us to make sense of the following possibility:

> Thus, if α and β are distinct and mutually independent agents, it is logically impossible that α sees to it that β sees to something. And therefore, under this assumption about α and β, α cannot see to it that β forms a belief. Whereas [α dstit: $B_\beta A$] is satisfiable, [α dstit: [β dstit: $B_\beta A$]] is not. It follows that α is never responsible for β's acts of belief formation, but may be responsible for β's beliefs. Agent α might, for example see to it that β believes something by applying hypnosis or, perhaps, brain surgery. (Wansing, 2006, p. 222)

If I understand correctly this means that α can make it the case that β believes that p, through indoctrination, hypnosis, or whatever. But this possibility of indirect belief formation, through influencing an other agent's beliefs is granted by all parties within the doxastic voluntarism debate. It was never in question. That the *dstit* model allows us to formulate it does not seem to me to count in favour of the model.

Wansing's conclusion is that the *dstit* model allows to make sense of, and in this sense of partially vindicating, a kind of doxastic voluntarism. But I do not see how this can be the case without begging a number of questions. In spite of its interest, this attempted formulation of the problem of doxastic voluntarism seems to me to beg too many questions. The very fact that a model allows the kind of "implicitly" knowing voluntarily that p involves a confusion between the indirect acquisition of a belief, which is indeed quite often voluntary, and its direct acquisition, which is the difficult issue. But this issue is never raised by Wansing's formalisation. It therefore seems to function like an idle wheel.

What has gone wrong here is not the kind of conclusions which one can draw from the model, which may be correct although it is not clear that they show anything about the possibility of believing at will, but the kind of preliminary layout needed to *formulate* the problem. In order to be able to formalise properly, it is necessary in the first place to be clear about the truth conditions of given claims such as: *the can be voluntary belief formation*. And this syndrome is quite widespread. I am not claiming that all formalisations fall into such difficulties, but that it is a permanent temptation of the method that formalism is applied without enough taking into account the nature of the problem at hand.[3]

Conclusion: Uses and Misuses of Formal Models

To conclude, I would like to suggest some differences between philosophy and the exercise of modelling in science, since the comparison is often made between the two. Philosophy can use models, but it is not modelling. Models, including formal models of various philosophical notions are enormously useful in philosophy. But philosophy is not modelling. In the first place models in philosophy are not like models in science, e.g. the Bohr model of the atom: they do not help visualising or give an intuitive summary of a theory which is otherwise developed fully: they *are* theories. For instance bayesianism is a theory of belief, not part of a larger and more comprehensive theory which could be formulated at a later stage. It involves strong assumptions which do not enter innocuously within the attempts, for instance, to model belief change. In the second place, models in philosophy cannot be tested against experiments. They can be tested only against intuitions (this claim is notoriously problematic, since it is often unclear what counts as an intuition and how they are tested, but in any sense of the notion of intuition here, we cannot equate it to experiments in science, in particular answers to questionnaires about a given notion, *pace* "experimental" philosophers, does not count as hypothesis testing in the scientific sense). In the third place, the activity of modelling is not neutral. Models

---

[3] As it was suggested to me by Igal Kvart, who has an unpublished paper on these issues.

modelise philosophical concepts and theses. Formal models in philosophy do not model uninterpreted concepts; they model already existing philosophical *theories*. But in general there is not only one theory in the offing; there are several rival ones. The model must be strong enough to compare several philosophical theories; but this is very rare. In the fourth place, as the example of doxastic voluntarism above shows, modelisation must not be question begging. This was Hansson's worry about the choice of primitives. A model must not beg the question against a certain philosophical theory in being used to model another one. Certainly, the formaliser is not alone in philosophy in being guilty of this: even an ordinary language philosopher, or a philosopher hostile to logical or mathematical modelling can beg questions! But the formalist is all the more in danger of doing so that his models have to be interpreted. Philosophy needs to identify the terms of a problem and to explore various ways to solve it. Among these ways, models can play a role. But only in so far as the specific problem is *identified*. If it is not identified, the modelling is idle or misleading.[4]

References

Douven, I. and Horsten, L. (2008). Formal models in the philosophy of science. *Studia Logica*, 89, 151–162.
Feldman, R. (2000). The ethics of belief. *Philosophy and Phenomenological Research*, 50, 667–695.
Hansson, S. O. (2000). Formalization in philosophy. *Bulletin of Symbolic Logic*, 6(2), 162–175.
Hansson, S. O. (2001). *The Structure of Values and Norms*. Cambridge University Press, Cambridge.
Hartmann, S. (1996). The world as a process: Simulations in the natural and social sciences. In R. Hegselmann, U. Mueller, and K. G. Troitzsch (Eds.), *Modelling and Simulation in the Social Sciences from the Philosophy of Science Point of View* (77–100). Kluwer Academic Publisher, Dordrecht.
Hartmann, S. (2008). Modeling in philosophy of science. In M. Frauchiger and W. Essler (Eds.), *Representation, Evidence, and Justification: Themes from Suppes*, volume 1 of *Lauener Library of Analytical Philosophy*. Ontos Verlag, Frankfurt.
Lewis, D. (1988). Desire as belief. *Mind*, 97, 323–332.

---

[4] I thank Hannes Leitgeb, Igor Douven, and Stephan Hartmann for having invited me at the workshop on formal models held in the ECAP meeting in Kraków, Wlodek Rabinowicz, Tomasz Placek, Leon Horsten, and the participants in the workshop for their reactions, and Katarzyna Kijania-Placek for her editorial work and her patience.

Olsson, E. (2003). Belief revision, rational choice and the unity of reason. *Studia Logica*, 73, 219–240.
Rott, H. (2001). *Change, Choice and Inference*. Oxford University Press, Oxford.
Wansing, H. (2006). Doxastic decisions, epistemic justification and the logic of agency. *Philosophical Studies*, 128, 201–227.
Williams, B. (1973). Deciding to believe. In *Problems of the self* (136–151). Cambridge University Press, Cambridge.

Formal Methods In Philosophy of Science

Thomas Müller

Philosophy of science in the tradition of logical empiricism was characterised by an almost exclusive reliance on formal methods. The historicist turn in philosophy of science in the 1960s initiated a mostly critical attitude towards such methods. After sketching this development in section 1, I will argue that the availability of new formal methods and an increased sensitivity for the uses and limitations of formal approaches makes possible a fresh case for the usefulness of formal methods in philosophy of science. There are success stories to be told about the application of new formal methods—I will list some of these developments in section 2. In the final section 3 I will then speculate about a possible further success story: the benefits of a formal approach for getting clear on the status of modality in philosophy of science.

1. Historical Background

Philosophy of science is an old subject and a new subject, depending on how one looks at it. There are good reasons for viewing philosophy of science as a historically unified enterprise with roots as far back, perhaps, as the 13th century (not to mention Aristotle); this historical lineage is the subject of the flourishing field of history of philosophy of science. On the other hand, the current academic sub-discipline of philosophy of science is a development of the late 19th and the early 20th centuries—Ernst Mach in 1895 was the first person to hold a chair in philosophy of science at Vienna, and the *Verein Ernst Mach*, subsequently the *Vienna Circle*, together with the *Berlin Circle* in the 1920s and 1930s were the birthplace of logical empiricism, which played a key role in forming and establishing the discipline of philosophy of science. When it comes to the role of formal methods in philosophy of science, this more recent historical lineage is crucially important.

Logical empiricism was, broadly speaking, an attempt at turning philosophy into a respectable scientific discipline. In the eyes of the propounders of this doctrine this meant to abolish metaphysics, where no clear scientific standards were discernible, and instead to embrace the strict standards of reasoning made possible by formal logic. Carnap in his programmatic paper on "the old and the new logic" (Carnap, 1930, p. 26) put the matter thus:[1]

> To pursue philosophy means nothing but: clarifying the concepts and sentences of science by logical analysis.

This idea of a formal study of science can be linked to the widespread formal self-understanding of science. The idea that proper science needs to be mathematical had been strong since the 17th century—witness Galileo's image of the book of nature being written in the language of mathematics, or Kant's later pronouncement that a purported science was a science only insofar as it was mathematical (cf. Galilei, 1623 and Kant, 1786). Two of the main new ideas of logical empiricism were a demand for the unity of science, seen as an ideal for various reasons, and the idea that a philosophical approach to this unified science had to proceed by means of formal analysis. The messy details of actual science notwithstanding, the unified science was to be rationally reconstructed using the formal methods of logic—which of course meant: of the logic of the time.[2]

Philosophy of science developed as a subject proper mainly in the U.S., following the emigration of many of the leading logical empiricists due to the rise of Nazism.[3] In the 1950s, the field consolidated around a positivist orthodoxy, leading to compendia such as Nagel's *The Structure of Science* (1961). Formal accounts of explanation, confirmation, theory reduction, laws of nature, and other key concepts had been worked out by then. The

---

[1] German original: "Philosophie betreiben bedeutet nichts Anderes als: die Begriffe und Sätze der Wissenschaft durch logische Analyse klären."

[2] It should not be forgotten that the "left wing" Vienna Circle besides Carnap also included philosophers like Neurath, who proposed a pragmatic approach to the philosophy of science including psychological and sociological studies, cf. Uebel (2001) on Neurath (1932). This idea however had little impact on the development of the subject of philosophy of science in the years after the Second World War.

[3] The historical context of logical empiricism is described in detail in the essays of Stadler, Hoffmann and Reisch in Richardson and Uebel (2007).

cracks were however already beginning to show: the adequacy of those formal accounts appeared doubtful.

Initially, logical empiricism could respond to criticisms about the descriptive adequacy of proposed accounts by pointing to their status as first steps in a research program. When the account of scientific concepts remained questionable vis-à-vis actual practice over decades, however, it appeared that the research program had failed to deliver. Historical and sociological studies of actual science such as Kuhn's (1962) *Structure of Scientific Revolutions* (published in the logical empiricists' own book series) were seen as more important than logical constructions that increasingly seemed to be built of thin air.

I have sketched this historical background—very roughly, to be sure—because it may help to explain the generally critical attitude towards formal methods that is, or at least was, prevalent among many philosophers of science.[4] Defending formal methods in philosophy of science nowadays means to be aware of this historical baggage, and to take up the challenge of showing how the criticism leveled against logical empiricism's deployment of formal methods can be met.

2. Formal Methods in Philosophy of Science: Some Success Stories

To be sure, formal methods never vanished from philosophy of science. Many of the early formal-logical accounts—e.g., the deductive-nomological account of explanation—have always remained important for the field, not at least in teaching the subject, and not just because of their historical significance, but also because they remain systematically significant due to their clarity and exactness.

The examples of the employment of formal methods in philosophy of science that I wish to summarize here are however different: they are *new* success stories, made possible at least in part by the advent of new formal methods. Such methods are nowadays not limited to the traditional field of formal logic—which by itself has expanded vastly, providing for modal, temporal and other logics and giving much formal insight into the important notion of a model, or a structure. The methods also include a significant amount of probability theory and aspects of game theory, graph theory, computer simulations and other techniques of formal modelling. It should also

---

[4] For a more detailed overview, cf., e.g., Richardson (2007).

be emphasized that in this development, philosophy of science does not play the merely passive role of employing off-the-shelf techniques developed in other disciplines, but has also led to the development of new techniques.[5]

In the following short descriptions of formal success stories, the contrast is always between the way matters were seen within the original paradigm of logical empiricism focusing on inferential relations among sentences, and new approaches based on an extended array of formal methods.

I do not claim any originality for the accounts of the employment of formal methods given in this section. These accounts are rather meant to illustrate my main point, which is that we are witnessing a return of the fruitful employment of formal methods in philosophy of science. Consequently the following sketches of confirmation (section 2.1), reduction (section 2.2), social aspects of science (section 2.3), quantum logic (section 2.4), and determinism (section 2.5) will be rather brief.

2.1. Confirmation

One of the key questions for philosophy of science is how scientific knowledge comes about, and how scientific theories can be confirmed by empirical data. If a scientific theory is viewed as a collection of universally quantified statements, as logical reconstruction suggests, then the matter seems to be quite simple: Any empirical statement that instantiates one of those quantified statements would seem to confirm the theory, while any statement contradicting the theory apparently disconfirms it. Confirmation thus seems to boil down to logical consistency. This view however has many counterintuitive consequences, among them the "raven paradox": the general statement that all ravens are black is confirmed, on this view, by the observation of a yellow lemon, which after all instantiates the logically equivalent general statement that all non-black things are non-ravens.

The radical move of denying the significance of confirmation flies in the face of scientific practice. So a better account of confirmation is needed. Historical and sociological studies can elucidate the use of confirmation in scientific practice for sure—but is there a formal account that helps, too? Early probabilistic models were beset with problems. Nowadays, however, it seems that Bayesian confirmation theory has gone a long way towards resolving the old puzzles, even though open issues remain (Vranas, 2004; Maher, 2004; Fitelson and Hawthorne, 2005; Huber, 2005). The advent of

---

[5] Cf., e.g., Leitgeb (2009), who also echoes the earlier programmatic paper of van Benthem (1982). Cf. also Horsten and Douven (2008) for a state-of-the-art survey.

new formal techniques—in this case, Bayesian confirmation theory—has brought an old dispute closer to resolution.

2.2. Reduction vs. Intertheoretic Relations

What is the relation between a scientific theory and the theory that historically takes its place—like, e.g., the Newtonian theory of universal gravitation superseding Galileo's law of falling bodies? The new theory should at least account for the same empirical facts as the old one. Thus, within the logical empiricist paradigm of theories as collections of general statements, it seemed that some relation of logical derivability or reduction would be appropriate: the new theory should allow one to derive all empirical statements of the old one, plus some more. It is easy to see that this idea breaks down even in the case of the example of Galileo vs. Newton (ironically used as an illustration by Nagel (1961)): In the earth's non-uniform gravitational field, the Galilean law is only an approximation to what Newton's theory predicts.

Again, the move to present-day probabilistic methods seems promising. Rather than focus on the "reduction" of one theory by another, a wider picture of intertheoretic relations that also includes the data the theories account for remains much closer to actual scientific practice (Batterman, 2008; Hartmann, 2008).

2.3. Social Aspects of Science

In the paradigm of rational reconstruction, social aspects of science were not deemed to be relevant. Epistemic subjects and their interaction remained unaccounted for—there was no place for them in a picture of theories as sets of sentences. The historicist turn of the 1960s forcefully pointed out this blind spot of philosophy of science—and surely the influence of social factors on actual scientific practice cannot be denied. How could formal modelling hope to give an account of this?

In fact, the employment of formal models in the study of social interaction within science is a young and vivid field of research. Formal techniques developed, e.g., in economics or in the study of voting systems (Hartmann and Bovens, 2008) can be applied to a number of problems in the philosophy of science. One prominent object of study recently is the question of how individuals pool together their beliefs if there are logical relations among them—the so-called problem of judgement aggregation (Dietrich, 2006; Pigozzi, 2006). This question is crucial for an understanding of science as a social practice. So even though the formal study of social interaction is still a rather new addition to the toolbox of philosophy of science and will have to be developed further, one can already see that a supposed non-issue for formal approaches has moved into the reach of such methods.

2.4. Quantum Logic: Old and New

The quantum logic of Birkhoff and von Neumann (1936) was an attempt at reading off a "new logic" from the mathematical structure of quantum mechanics. Initially the idea was to find an interpretation of propositional connectives like conjunction and negation that would be a formal counterpart to operations on the set of subspaces of a Hilbert space that constitutes the state space of a quantum system. A fascinating possibility was that the "true" logic could turn out to be different from classical propositional logic—and for empirical reasons.

Present-day logic paints a different picture. Quantum logic never came to replace classical logic—but the logic community has also become much more open towards the idea that there could be different logics, each suited to a specific domain. Furthermore, there are new tools within logic that can be fruitfully employed in a study of quantum mechanics. In fact dynamic logics seem to be very well suited for a description of quantum operations studied in quantum information theory (Baltag and Smets, 2008). Thus, advanced formal methods allow one to leave old normative questions (about "the" logic) behind and work towards a better understanding of science as actually practised.

2.5. Determinism and Indeterminism of Theories

The question of whether a given scientific theory is deterministic or not, was approached mostly informally before Montague (1962) introduced a model theoretic approach. In this field many advanced methods of mathematical physics have been employed, and the formal technical level of discussion is very high (witness Earman, 2007). In fact here the deployment of formal methods has significantly advanced other discussions, too, in that the importance of precise definitions of, e.g., the notion of state has been recognised. Questions of theory determinism or indeterminism are furthermore relevant not just for philosophy of science, but also for science itself.

3. On the Way Towards Another Success Story? Modality in Philosophy of Science

The last of the examples given above directly leads to the topic of this section, which is modality. Determinism after all is a modal notion: it signifies the absence of open possibilities. Modality arguably plays a role in many other concepts of science, too: laws of nature, essences and natural kinds, causation and intervention, and probability. I will sketch a few aspects of a possible success story about modality in philosophy of science in the same way as in the examples above, i.e., starting from the state of affairs at the time of logical empiricism.

From the point of view of logical empiricism, there were two problems about modality in science. Firstly, modality was interpreted as *logical modality*, where logical possibility just means the absence of formal contradiction—but this is not the notion of modality that is needed to analyse the mentioned scientific concepts. The notion of logical possibility is too broad: many things that are physically impossible are still logically possible (think, e.g., of going faster than the speed of light). Secondly, modality apparently has poor empiricist credentials. This continues to stand in the way of a fruitful employment of modal notions in philosophy of science. After all, mere possibilities—possibilities that are not actualised—are empirically inaccessible because they are unreal, so how could they be important for empirical science?

My claim is that the advent of new techniques of formal modal logic and a balanced use of formal and informal methods leads the way towards a fruitful integration of modality into philosophy of science. The results thus attainable can also be useful if one remains agnostic or even negative with respect to the metaphysical status of unrealized possibilities.

The first important step towards an employment of modality in philosophy of science is to take a lead from the discussion about different modalities. This discussion developed out of formal research into the semantics of modal logic since the 1950s. Initially one may view this semantic enterprise as a quest for a formal representation of *the* meaning of "possibly" and "necessarily." The semantics that was established, the so-called Kripke semantics that spells out the modalities in terms of relations among possible worlds, showed however that there is much leeway in specifying different modal logics with different semantics. The initial assessment of this fact was rather critical: among all those options, it seemed that one still had to find the right one to specify what "possibly" and "necessarily" *really* meant. This assessment has changed in the meantime, and the many options for a semantics of modality are now seen as a good thing. It has become common to acknowledge a number of different kinds of modality: there isn't just logical modality, but there are various other kinds of modality that may have different formal properties and a different metaphysical status. It will be best to explain some of these differences in terms of possibility; the consequences for the dual modality of necessity follow immediately.[6]

As mentioned, there is logical possibility: the absence of formal contradiction. This notion is rather broad. Famously Ramsey pointed out to Wittgenstein that his *Tractatus* theory, which relied on logical possibility in postulating the independence of elementary propositions, was flawed because it could not, e.g., account for the rather straightforward impossibility of the same patch's being both red and green—no formal contradiction is involved here, since "red" and "green" just figure as two different predicates, and it is logically possible for one and the same thing to fall under any number of different predicates. The colour overlap in question is however clearly impossible in another sense. It has become common to speak of *metaphysical* possibility here, and to base philosophical arguments on metaphysical rather than logical possibility. For philosophy of science, however, a notion of *physical* possibility seems to play an even more important role. Physical possibility is often taken to be what laws of nature express, and insofar as science is a quest for the laws of nature, science is really about physical possibility.

Questions about the interrelation of various kinds of modality are important, but also difficult to resolve. There are arguments in favour of modal monism (the claim that there is one single fundamental modality, to which all other modal notions can be reduced), but also in favour of modal pluralism (the claim that there are different irreducible modalities). Thus, the question of whether physical possibility is just a restricted version of logical or metaphysical possibility has been debated: e.g., Fine (2005) argues convincingly that physical and metaphysical modality are independent and indeed believes that they are both fundamental, thus providing an argument in favour of modal pluralism.

My conviction is that physical possibility is not fundamental, and that a fruitful explanation of the use of possibility in philosophy of science needs to refer to a different notion of possibility: *real possibility*, also known as historical possibility because of its link with temporality.[7] The peculiarities

---

[6] Possibility and necessity are dual in the following sense: It is necessary that p if and only if it is not possible that non-p.

[7] Fine, in the mentioned work, explicitly excludes real ("historical") modality from his discussion, but gives no reason for this (cf. Fine, 2005, 237, n. 4). This strikes me as odd,

of that notion of possibility are best explained via some of its specific formal properties, thus making good on what I said about the importance of formal methods especially in this field (section 3.1). In section 3.2 I will then argue that physical modality is best viewed as derived from an underlying notion of real possibility.[8] I will finish by sketching some consequences of this result for the discussion about *ceteris paribus* laws (section 3.3).

3.1. The Formalities of Real Possibility

The formalities of real possibility have been worked out since the 1950s. Prior's *Time and Modality* (1957) set the agenda for research into the interrelation between modality and tense, whose formal similarities as sentence-modifying operators had by then just been recognized. Prior (1967) and subsequently Thomason (1970) developed models for so-called "branching time" in which the tempo-modal notion of an open future serves as the basis for a semantics of both the tenses and the modalities of real possibility and real necessity. In a model of branching time, possible courses of events, also called *histories*, are maximal linear subsets of a branching tree of open possibilities. A modern description of the branching time framework is given by Belnap, Perloff and Xu (2001, Chap. 6–8).

In terms of formal properties, real possibility is special because of its interaction with the tense operators. We will employ the standard formalisations of "F" for the future operator "it will be the case that" (the past tense "it was the case that" is accordingly symbolized as "P"), and "$\Diamond$" and "$\Box$" for the modal operators "possibly" and "necessarily," respectively. A specific aspect of real possibility is the satisfiability of the formulae

$$\Diamond p \, \& \, F\neg\Diamond p \tag{F1}$$

and

$$\Diamond p \, \& \, \neg F\Diamond p, \tag{F2}$$

which express the temporality of real possibility. (F1) says that some p that is now possible, will at some future point in time not be possible any more— a fact that we know all too well, as witnessed by the fact that we sometimes complain about missed opportunities. (F2) is even stronger, saying that p,

---

since he himself has contributed to the development of the formalities of real possibility; cf. Prior and Fine (1977).

[8] Systematically speaking I believe that real possibility can also serve as a basis for explaining most if not all other modal notions, but my task in this paper is restricted to the case of physical possibility.

which is now possible, will cease to be possible immediately in the future—it's now or never, so to speak. Instances of this are also well known.

These formulae are *not* satisfiable if "◊" is read as logical or as metaphysical possibility; those modal notions are abstract, without any link with the passage of time. What is logically possible now will remain so forever, and has in fact always been logically possible—if those temporal determinations make any sense at all.[9] For further formal properties of real possibilities based on branching time, cf. again Belnap et al. (2001).

The mentioned formal framework of branching time has been extended in order to overcome one of its major shortcomings: While real possibility is possibility in a concrete and thus concretely localised situation, branching time does not capture that *spatial* aspect. In the extended formal framework of *branching space-times* (BST; Belnap, 1992) this aspect is explicitly recognised, as histories (possible courses of events) in that framework do not have the form of a single temporal chain of events, but of a single space-time. In BST it is therefore possible to express the fact that something that is possible here now, is not possible now somewhere else.[10] Belnap's BST is the most advanced formal framework for studying real possibility available to date, and it has been used in a number of applications to problems of metaphysics, philosophy of language, and philosophy of physics.[11]

3.2. Physical Possibility Based on Real Possibility

Physical possibility, the modal notion that determines the laws of nature, belongs to the same group of abstract, a-temporal modalities as logical and metaphysical possibility: what is physically possible now, has always been

---

[9] This question is mirrored in the case of mathematics, where there are different opinions as to whether "It is now the case that $2 + 2 = 4$" makes any sense at all.—Do not be misled by the fact that, e.g., a logical possibility may be *instantiated* as a real possibility, which then *is* temporal. E.g., it is logically possible that crows fly, and it may be really possible that a certain concrete crow that is now before you should fly within the next five minutes. This, however, is not the same as the mentioned abstract logical possibility, but also depends on many local and temporal factors, e.g., the state of the crow's feathers and the air pressure.

[10] In view of BST's compatibility with relativity theory, the "now" of course has to be taken with a grain of salt. Technically, possibilities are linked to space-time locations in BST, in a manner that is fully compatible with the absence of a notion of absolute simultaneity in special relativity theory.

[11] Cf., e.g., Belnap (2005) for causation, Weiner and Belnap (2006) and Müller (2005) for objective single-case probabilities, Müller, Belnap and Kishida (2008) for modal correlations, Placek and Müller (2007) for counterfactuals, and Müller and Placek (2001) as well as Placek (this volume) for Bell-type correlations.

physically possible and will remain so forever.[12] Real possibility, on the other hand, is possibility in a concrete, indexically specifiable situation: it is right there before us. The main question about the interrelation of real vs. physical possibility is how scientific practice, which is based on real, concrete experiments and observations, can help us gain access to abstract physical possibility. This question is similar to the question about the interrelation of theory and observation in the sciences, but phrasing it in terms of possibilities gives it an importantly different twist.

Real possibilities rule in the lab and in scientific work generally: Every concrete run of an experiment reveals one of the outcomes that are *really* possible in the given, concrete situation—including, in almost all cases, the real possibility that the experiment may fail due to some sort of interference. Even though experiments thus primarily reveal something about real possibilities, they can sensibly be seen as probes of physical possibility, too. At least that is what experiments are designed for: Generally speaking, in an experiment one wants to find out not about the really, but about the physically possible outcomes, together with their probabilities, of an experimental set-up with given, experimenter-controlled initial conditions. One will therefore disregard certain runs as not pertinent to the question about physical possibility (e.g., because somebody kicked the apparatus), even though the pertinence of these runs for the issue of real possibility cannot be questioned. One will also smooth out the observed distribution of results in various ways. Details vary by case—here a connection with Bogen and Woodward's (1988) data/phenomena distinction suggests itself: physical possibilities appear as phenomena distilled from real possibility figuring as data, with all the well-known idiosyncrasies of that step. It is generally acknowledged that there is no formal way of inferring phenomena from data.

Physical possibilities as summed up in laws of nature and physical theories are thus determined via the notion of real possibility that has primacy in scientific practice. In concrete runs of experiments, real possibilities are actualized. Both the concrete initial situation of the respective runs and the concrete outcomes are then described via a number of variables, giving rise to stable, repeatable phenomena. The aim of the experimenter in such a description is to record all salient variables, not everything at all. Physical possibilities (which in a given case may be physical necessities) are then arrived at from real possibilities: so-called laws of nature are established as generalisations covering many experiments, and considerations of saliency

---

[12] At least this is so if one disregards scenarios in which the laws of nature change over time. I will ignore such scenarios in what follows. The point about abstractness would remain in any case.

again play a crucial role here, as in any case in which phenomena are inferred from data.

Statements about laws of nature on this account have an unquestionable modal content: they simply report what is physically possible or necessary. Their genesis via statements about real possibilities also shows a way towards resolving some of the puzzles about *ceteris paribus* clauses.

3.3. Consequences for the *Ceteris Paribus* Discussion

It has often been observed that laws of nature appear to be open to counterinstances. Does this mean that the physically impossible can happen? Consider a specific case: Yes, it is a law of nature that unsupported bodies fall to the ground in the gravitational field of the earth, but no, this does not rule out a concrete case in which a seagull catches a falling breadcrumb in mid-flight (cf. Keil, 2005). A typical first reaction to such cases is the addition of specific exception clauses to the initial formulation of the law in question that are meant to immunize them against the counterexample in question. This strategy, however, only leads to the formulation of new counterexamples, and there is no hope that the race will ever end. This consideration has led many philosophers of science to postulate an unspecific "*ceteris paribus*" clause to be appended to any purported law statement, which would act as a sort of safety net and neutralize all exceptions. There are substantial problems with this approach, many of which Keil (ibid.) refers to. From a formal point of view, the matter can be stated as follows:

Laws of nature state abstract physical possibilities or necessities. What is possible in a concrete situation, on the other hand, depends on all concrete aspects of that situation—not just the ones that are addressed in the physical laws. The picture behind the problem of exceptions is that what holds in a concrete situation is assumed to be just be a formal instantiation of what holds generally. Logically speaking, this is still the story of logical empiricism, where laws are treated as general statements of the form

$$\forall x (Fx \to Gx) \qquad \text{(LAW)}$$

(" all F's are G's"), and an instance of such a law accordingly has the form

$$Fa \to Ga. \qquad \text{(INST)}$$

Taking the modal nature of laws and concrete instances seriously, however, one arrives at a different picture. Let us consider a law that states a physical necessity (with the subscript P at the box signalling this physical type of modality),

$$\Box_P \forall x (Fx \to Gx). \qquad \text{(LAW')}$$

Here, the predicates (or more generally, open formulae) Fx and Gx can carry no indexical reference to the current time or place, since laws are abstract.[13] A statement about a specific situation, on the other hand, has to involve the notion of real possibility or necessity, and has to include indexical reference to the current time and place. Such a statement may be expressed in the form

$$\Box_R \phi, \qquad \text{(INST')}$$

where the R subscript indicates real necessity and the different, indexical content is signified by the use of a Greek letter.

From this formalization it is clear that the 'instantiation' of a physical as a real necessity is not a straightforward formal task. It rather involves the mirror image of the process of arriving at physical possibilities from real possibilities, which I had linked to the inference of phenomena from data. The move from data to phenomena always means that not all details present in the data are retained, and conversely, the move from physical possibility back to real possibilities means that local, situation-dependent details have to be filled in. If all goes well, i.e., if a law of nature is formulated such that most salient features of actual situations are mentioned, then counterinstances may be rare, and the instantiation of a law may look superficially like the move from (LAW) to (INST). Concrete counterexamples to purported laws of nature however remind us that the matter is indeed more intricate than that.

A look at the formal details of the different types of modality involved in laws of nature and their instances thus points to the *informal* nature of that step. Purported counterinstances of physical possibilities are concrete situations that would not have supported the move from real to physical possibility inherent in the formulation of laws of nature—but they are real, and really possible, situations nonetheless.

Acknowledgements

I would like to thank the audience and my co-symposiasts at the *Workshop on Formal Methods in Philosophy*, Kraków, 24 August 2008, and at the ESF Conference *The Present Situation in the Philosophy of Science*, Vienna, 18 December 2008, for helpful discussions. Support by the *Deutsche Forschungsgemeinschaft* is gratefully acknowledged.

---

[13] Even in the case of laws of nature changing over time, what would be temporalized would be the necessity operator $\Box_P$, not these predicates.

References

Baltag, A. and Smets, S. (2008). A dynamic-logical perspective on quantum behavior. *Studia Logica*, 89(2), 187–211.
Batterman, R. (2008). Intertheory relations in physics. In E. N. Zalta (Ed.), *The Stanford Encyclopedia of Philosophy*.
URL = http://plato.stanford.edu/archives/fall2008/entries/physics-interrelate/.
Belnap, N. (1992). Branching space-time. *Synthese*, 92, 385–434.
Belnap, N. (2005). A theory of causation: *Causae causantes* (originating causes) as inus conditions in branching space-times. *The British Journal for the Philosophy of Science*, 56, 221–253.
Belnap, N., Perloff, M., and Xu, M. (2001). *Facing the Future. Agents and Choices in Our Indeterminist World*. Oxford University Press, Oxford.
Birkhoff, G. and von Neumann, J. (1936). The logic of quantum mechanics. *Annals of Mathematics*, 37, 823–843.
Bogen, J. and Woodward, J. (1988). Saving the phenomena. *Philosophical Review*, 97(3), 303–352.
Carnap, R. (1930). Die alte und die neue Logik. *Erkenntnis*, 1, 12–26.
Dietrich, F. (2006). Judgment aggregation: (im)possibility theorems. *Journal of Economic Theory*, 126(1), 286–298.
Earman, J. (2007). Aspects of determinism in modern physics. In J. Butterfield and J. Earman (Eds.), *Handbook of the Philosophy of Physics*, (1369–1434). Elsevier, Amsterdam.
Fine, K. (2005). *Modality and Tense*. Oxford University Press, Oxford.
Fitelson, B. and Hawthorne, J. (2005). How Bayesian confirmation theory handles the paradox of the ravens. In E. Eells and J. Fetzer (Eds.), *The Place of Probability in Science*. Open Court, Chicago.
Galilei, G. (1623). *Il Saggiatore*. Giacomo Mascardi, Rome. English translation as "The Assayer". (1960). In S. Drake and C. D. O'Malley (Eds.), *The Controversy on the Comets of 1618*. University of Pennsylvania Press, Philadelphia.
Hartmann, S. (2008). Between unity and disunity: A Bayesian account of intertheoretic relations. (Forthcoming).
Hartmann, S. and Bovens, L. (2008). Welfare, voting and the constitution of a federal assembly. In M. Galavotti, R. Scazzieri, and P. Suppes (Eds.), *Reasoning, Rationality and Probability* (61–76). CSLI Publications, Stanford.
Horsten, L. and Douven, I. (2008). Formal Methods in the Philosophy of Science. *Studia Logica*, 89(2), 151–162.
Huber, F. (2005). What is the point of confirmation? *Philosophy of Science*, 72, 1146–1159.
Kant, I. (1786). *Metaphysische Anfangsgründe der Naturwissenschaft*. Hartknoch, Riga. English translation by M. Friedman (2004): *Metaphysical Foundations of Natural Science*. Cambridge University Press, Cambridge.
Keil, G. (2005). How the ceteris paribus laws of nature lie. In J. Faye (Ed.), *Nature's Principles* (167–200). Springer, Heidelberg.
Kuhn, T. S. (1962). *The Structure of Scientific Revolutions*. International Encyclopedia of unified Science. University of Chicago Press, Chicago.
Leitgeb, H. (2009). Logic in general philosophy of science: Old things and new things. In V. Hendricks (Ed.), *PHIBOOK, Yearbook of Philosophical Logic*. Automatic Press, Copenhagen.
Maher, P. (2004). Probability captures the logic of scientific confirmation. In C. Hitchcock (Ed.), *Contemporary Debates in Philosophy of Science*. Blackwell, Oxford.

Montague, R. (1962). Deterministic theories. In D. Willner (Ed.), *Decisions, Values and Groups* (325–370). Pergamon Press, Oxford. Reprinted In R. H. Thomason (Ed.). (1974). *Formal Philosophy* (303–359). Yale University Press, New Haven, CT.

Müller, T. (2005). Probability theory and causation: A branching space-times analysis. *The British Journal for the Philosophy of Science*, 56, 487–520.

Müller, T., Belnap, N., and Kishida, K. (2008). Funny business in branching space-times: infinite modal correlations. *Synthese*, 164, 141–159.

Müller, T. and Placek, T. (2001). Against a minimalist reading of Bell's theorem: lessons from Fine. *Synthese*, 128, 343–379.

Nagel, E. (1961). *The Structure of Science*. Hartcourt, Brace and World, New York.

Neurath, O. (1932). Soziologie im Physikalismus. *Erkenntnis*, 2, 393–431.

Pigozzi, G. (2006). Belief merging and the discursive dilemma: an argument-based account to paradoxes of judgment aggregation. *Synthese*, 152(2), 285–298.

Placek, T. and Müller, T. (2007). Counterfactuals and historical possibility. *Synthese*, 154, 173–197.

Prior, A. N. (1957). *Time and Modality*. Oxford University Press, Oxford.

Prior, A. N. (1967). *Past, Present and Future*. Oxford University Press, Oxford.

Prior, A. N. and Fine, K. (1977). *Worlds, Times and Selves*. Duckworth, London.

Richardson, A. (2007). Thomas Kuhn and the decline of logical empiricism. In Richardson and Uebel (2007), pp. 346–369.

Richardson, A. and Uebel, T. E. (Eds.). (2007). *The Cambridge Companion to Logical Empiricism*. Cambridge University Press, Cambridge.

Thomason, R. H. (1970). Indeterminist time and truth-value gaps. *Theoria (Lund)*, 36, 264–281.

Uebel, T. (2001). Carnap and Neurath in exile: Can their disputes be resolved? *International Studies in the Philosophy of Science*, 15, 211–220.

van Benthem, J. (1982). The logical study of science. *Synthese*, 51, 431–472.

Vranas, P. B. M. (2004). Hempel's raven paradox: A lacuna in the standard Bayesian solution. *The British Journal for the Philosophy of Science*, 55(3), 545–560.

Weiner, M. and Belnap, N. (2006). How causal probabilities might fit into our objectively indeterministic world. *Synthese*, 149, 1–36.

If in Doubt, Treat'Em Equally
A Case Study in the Application of Formal Methods to Ethics

Wlodek Rabinowicz

ABSTRACT

The so-called 'presumption of equality' requires that individuals be treated equally in the absence of relevant information that discriminates between them. Our objective is to make this principle more precise, viewed as a principle of fairness, and to determine why and under what conditions it should be obeyed.

Presumption norms are procedural constraints, but their justification can be sought in the possible or expected outcomes of the procedures they regulate. This is the avenue pursued here. The suggestion is that in the absence of information that would discriminate between the individuals, equal treatment minimizes the expected unfairness of the outcome. Another suggestion is that equal treatment under these circumstances also minimizes the maximal possible unfairness of the outcome. Whether these suggestions are correct or not depends on the properties of the underlying unfairness measure.

1. Introduction

This paper examines the so-called Presumption of Equality, which enjoins us to treat different individuals equally, if we can't discriminate between them, in relevant respects, on the basis of the available information. I will view this principle as a requirement of *fairness*—more precisely, as a procedural principle whose goal is to promote fair outcomes. I would like to make Presumption of Equality more precise and determine why and under what conditions it should be obeyed. The paper can be viewed as a case study the purpose of which is to show how problems in ethics can be amenable to formal methods.

Why, then, should Presumption of Equality be obeyed? A rather obvious answer is that, in the absence of relevant discriminating information, unequal treatment is bound to be *arbitrary* and thus could invite a suspicion of partiality. I don't think, though, that this is all there is to it. While arbitrariness considerations are important, they are not decisive. In some cases in which the discriminating information is missing, equal treatment might be wrong, and unequal treatment might be right, despite the arbitrariness of the latter. To give an example, suppose individuals i and j compete for two scholarships, the more attractive scholarship A and the less attractive B. Your task is to make the decision. While it is given that both applicants are qualified, your information doesn't discriminate between their merits: While you might know that one of the candidates is more deserving, you have no clue which candidate it is and there is no time to gather further information. Since the scholarships aren't equally attractive, to give one to i and the other to j would be to treat the applicants unequally and to that extent arbitrarily. At the same time, you *could* treat them equally by withholding the scholarships from both. However, such 'levelling down' would be even more unsatisfactory given that they both deserve a scholarship. Avoidance of arbitrariness is not everything that counts. To justify equal treatment, in those cases in which such treatment *can* be justified, we need to rely on other considerations.

At this point, I expect the objection: Why not simply decide the scholarship assignment by a toss of a coin? This would give each of the applicants a fifty-fifty chance of getting the more attractive scholarship, but even in case of a loss the applicant will not come away empty-handed: She will receive the less attractive scholarship instead. Such a lottery, it seems, is itself a form of equal treatment. By tossing a coin, we would avoid arbitrariness but at the same time see to it that both applicants do get scholarships, which they deserve.

It is true that, in the case at hand, drawing lots or tossing a coin is the obvious thing to do. Avoidance of arbitrariness *is* important. I would deny, however, that an equal lottery on unequal treatments itself amounts to equal treatment. On an outcome-oriented approach to fairness, this is not so. While an equal lottery on unequal outcomes gives different individuals equal chances, the *outcome* of the lottery will still be unequal. And inequality in outcome may well matter from the point of view of fairness.

Here is the suggestion I would instead want to examine: Principles of fairness can be constraints on procedures or constraints on outcomes. Presumption norms constrain procedures, but procedural constraints could be justified in terms of the expected outcomes of the procedures they regulate. This avenue is pursued in my paper. In particular, equal treatment should be chosen because and to the extent that it *minimizes the expected unfairness*. When the available information does not discriminate between the individuals concerned, it will typically be the case, I would like to suggest, that expected unfairness will be at its lowest if individuals are treated equally. Not always, though. As will be seen, the scholarship example provides an exception to this rule.

It has been suggested that all presumption norms are grounded, to some extent at least, in 'the differential acceptability of the relevant sorts of expected errors.'[1] Thus, for example, presumption of innocence in criminal law is justified by the asymmetry between potential moral costs of punishing an innocent and those of letting a guilty person go free. My approach to Presumption of Equality is different. As will be seen, for this principle to hold, the moral cost of treating equals unequally need not be greater than that of the equal treatment of unequals.

Some presumptions might have more to do with the differential probability of errors than with their differential costs. In such cases, what is being presumed is deemed to be so probable as to be used as a default assumption. But my argument does not assume that equality in deserts is probabilistically privileged in this way. Indeed, it may be very *im*probable that the individuals in the case at hand are equally deserving.

While the justification I offer does not appeal to the differential costs of errors or to the differences in error probabilities, it does appeal to the differences in the *expected* costs of errors. In the absence of discriminating information, the suggestion is that we should treat individuals as if they were equal because such treatment minimizes the expected moral cost of error: the expected unfairness in treatment. To get the main idea, consider an analogy. Suppose that, in wartime, you are being sent to make a rendezvous with a group of partisans behind the enemy lines. You will be parachuted somewhere in the area where the partisans operate. Only afterwards you will be notified by radio of their exact location. Where in the area should you make your landing in order to minimize the expected distance to the rendezvous point? Barring special considerations (road conditions, an irregular shape of the area, etc.), the answer is obvious: You should position yourself right in the center. If the probability of partisan location is uniformly distributed over the area, or at least if it is distributed symmetrically with respect to the different geographical directions, the expected distance to the target will be minimal in the center.

---

[1] See Ullmann-Margalit, E. (1983). On Presumption. *The Journal of Philosophy*, 80, 143–163.

Now, unfairness might also be seen as a kind of distance between the way individuals are treated and the way they deserve to be treated. A treatment is more unfair, the farther away it is from the (perfectly) fair treatment. Its expected unfairness can therefore be seen as its expected distance from the fair treatment. Now, the conjecture is that treating people equally is like positioning yourself in the center of a designated area. Just as choosing the center-point minimizes expected distance to the target in the absence of information that discriminates between the geographical directions, equal treatment is conjectured to minimize expected distance from the fair treatment, i.e. to minimize expected unfairness, in the absence of information that discriminates between the individuals.

I want to examine under what conditions equal treatment will in fact have this feature. What conditions on the unfairness measure do we need to impose to guarantee this result? I will also inquire what happens if we instead choose a 'minimax' approach, i.e., if we opt for a treatment which is least unfair if 'worst comes to worst.' In other words, I will also examine under what conditions equal treatment minimizes maximal possible unfairness.

2. Individuals and Treatments

In my discussion, I will make use of an abstract model that allows of different interpretations. Clearly, there are dangers in abstraction: Unless we are cautious, we might inadvertently leave out important considerations. But, on the positive side, our conclusions gain in generality in this way and we avoid unnecessary complications.

Two main components of the model are a non-empty finite set $\mathbf{I} = \{i_1,\ldots,i_n\}$ of *individuals* who are to be subjected to treatment and a non-empty set $\mathbf{T} = \{a,b,c,\ldots\}$ of possible alternative *treatments*. (Other components will be introduced later.) Every treatment a in $\mathbf{T}$ is a vector $(a_1,\ldots,a_n)$, where $a_k (1 \leq k \leq n)$ is the way individual i_k is treated in a. a is equal iff for all individuals i_k and i_m, $a_k = a_m$. We shall sometimes use the notation $a(i_k)$ for a_k.

The model allows of different interpretations. The interpretations I will suggest below are themselves relatively abstract. Each may in turn be exemplified in many different ways.

Interpretation 1: Cake-divisions

A 'cake' is a homogeneous object that is to be divided among the individuals in $\mathbf{I}$. A treatment a is a particular division of the cake and thus may be seen as a vector of real numbers, $(a_1,\ldots,a_n)$, with each a_k being the share of the cake assigned by a to i_k. Each such share is non-negative and together they sum up to one. **T** is the set of all possible vectors of this kind. The equal treatment divides the cake equally among the members of **I**: $(1/n, \ldots, 1/n)$.

Representing cake-divisions in this schematic way means that we view them as *types* rather than tokens. Thus, if two cake pieces are of the same size, this interpretation does not distinguish between giving one piece to i and the other to j or assigning these pieces the other way round. Since the cake is homogenous, the question who gets which equal-sized piece is irrelevant from the point of view of fairness. There is therefore no reason to make this distinction in the model. This is a general feature of our approach. Treatments are interpreted as types that incorporate the relevant characteristics of the treatment tokens. As a result, any two treatments in the model are supposed to be relevantly different from each other.

Interpretation 2: Rankings

On this interpretation, a treatment is a ranking of the individuals in **I**: **T** is the set of all such possible rankings. That a treatment a ranks i at least as highly as it ranks j means that i is treated in a at least as well as j. Such an interpretation is appropriate when the *ordinal* differences between the individuals are all that matters from the point of view of fairness, i.e., when fairness only requires that the more deserving individuals should be better treated and that the equally deserving individuals should be treated equally well.

Formally, a ranking may be represented as an assignment of ordinal numbers, $1, 2, 3 \ldots$, to individuals, with 1 being the highest level in the ranking, 2 the second highest level, and so on. The assignment of levels starts from the highest one and continues downwards. Thus, the equal treatment is a ranking in which every individual is assigned the same highest level: $(1, \ldots, 1)$. (A different, but equivalent representation of rankings will be made use of below, in section 3.)

Interpretation 3: Indivisible goods

Suppose that G is a set of indivisible objects that are to be distributed, with or without remainder, among the individuals in **I**. Each treatment a in **T** is an assignment of disjoint subsets of G to individuals in **I**. Thus, $a(i)$ is the subset of G that treatment a assigns to an individual i. For some i, $a(i)$ may be empty, and for distinct i and j, $a(i) \cap a(j) = \emptyset$. But it is not required that $\bigcup a(i)_{i \in \mathbf{I}} = G$: Some objects in G may be withheld from the distribution. The scholarship case provides an example. There, G consists of the two scholarships, A and B, and the possible treatments amount to different possible partial or total distributions of G among the two individuals involved.

What would the equal treatment consist in, according to this interpretation? If the objects in G are relevantly different from each other, as in the scholarship case, then the equal treatment must be $(\emptyset,\ldots,\emptyset)$, i.e. the distribution in which every i is assigned the empty subset of G. But if G contains sufficiently many relevantly similar objects, say, $A_1,\ldots,A_n$, the number of equal treatments would rapidly rise: One such treatment, α, would assign A_1 to i_1, A_2 to i_2, ..., and A_n to i_n; another, β, would assign A_2 to i_1, A_3 to i_2, ..., and A_1 to i_n; etc. These treatments, however, would be relevantly similar to each other, thereby violating the 'relevance' restriction we have imposed on the model. To satisfy this restriction, we would need to re-interpret the notion of a treatment. Thus, we could partition G into equivalence classes with respect to the relevant similarity relation and then let a treatment be a function that specifies, for every individual i, how many objects i is to receive from each equivalence class. Given this re-interpretation, we would not have to make irrelevant distinctions. Thus, for example, if the equivalence class C contains the objects $A_1\ldots,A_n$, then the two equal treatments α and β, described above, would not be distinguished. Both would be tokens of the same treatment (type) that consists in each individual being assigned one object in class C.

However, such an interpretation, while satisfying the relevance restriction, would still violate another constraint I want to impose: For simplicity, I *exclude* decision problems in which there is more than one equal treatment available. And the case described above would be an example of such a problem: The individuals could be here equally treated either by not being given anything at all or, say, by being given exactly one C-object each.

The same problem would arise, by the way, in Cake-Division, if we allowed divisions in which part of the cake is withheld from the distribution. The number of equal treatments would then increase from one to infinity: The set of real-numbered vectors $(a_1,\ldots,a_n)$ such that $a_1 = \ldots = a_n \geq 0$ and $a_1 + \cdots + a_n \leq 1$ is non-denumerable.

Here, I want to exclude such interpretations. To keep the model relatively simple, I will be assuming that there is a unique equal treatment in **T**. In addition, I will assume that **T** is closed under permutations on individuals. Thus, we impose two conditions on the set of treatments:

A1. For every permutation f on **I** and every a in **T**, **T** contains some b such that for every i in **I**, $b(f(i)) = a(i)$.[2]

---

[2] A *permutation* on a set X is any one-one function from X onto that set itself, i.e. any as-

A2. There is a unique element of **T**, call it **e**, such that **e** is an equal treatment.

We have seen that A2 is a controversial claim. So is A1, which is a kind of completeness requirement on **T**. If **T** is the set of actually *available* treatments, this set might in some cases be too small for A1 to be satisfied. Thus, suppose the agent can give the whole cake to i_1, but, for some reason, he cannot give it to i_2. Then the set of available cake-divisions contains $(1, 0, 0)$ but not $(0, 1, 0)$. Under these circumstances, **T** wouldn't be closed under permutations on individuals that assign i_2 to i_1: There is no available treatment that gives to i_2 what $(1, 0, 0)$ gives to i_1. Here, however, I want to ignore this difficulty. I shall assume that **T** is sufficiently 'roomy' for A1 to be satisfied.[3]

If A1 holds, every permutation f on **I** induces a permutation f_T on **T** such that for every treatment a and every individual i, $f_T(a)$ treats $f(i)$ in the same way as a treats i: $f_T(a)(f(i)) = a(i)$. A1 implies the existence of such a permutation on **T**. And it is easily seen that this permutation must be unique. Had there been two of them, f_T and f_T', then there would exist some a in **T** such that $f_T(a) \neq f_T'(a)$. This would require the existence of some i in **I** for which $f_T(a)(f(i)) \neq f_T'(a)(f(i))$. Which contradicts the assumption that, for every i, both f_T and f_T' treat $f(i)$ in the same way as a treats i, i.e., that they both assign $a(i)$ to $f(i)$.

Now, suppose that f is a permutation on **I** and let f_T be the permutation on **T** induced by f. I will refer to the union of f and f_T as an *automorphism* and use symbols p, p', etc., to stand for different automorphisms on **I**∪**T**. Intuitively, an automorphism is a simultaneous permutation of individuals into individuals and of treatments into treatments, in which the former permutation induces the latter:

Automorphism: An automorphism, p, is a simultaneous permutation of **I** and of **T** such that for all i in **I** and all a in **T**, $p(a)(p(i)) = a(i)$.

Observation: Every permutation on **I** is included in exactly one automorphism.[4]

---

signment of elements of X to the elements of X that (i) assigns different elements to different elements, and (ii) assigns each element to some element. (If X is finite, but not otherwise, (ii) follows from (i).)

[3] This assumption is made for the sake of simplicity. We are interested in whether equal treatment minimizes expected unfairness (and/or maximal possible unfairness), as compared with other available treatments, in the absence of information that discriminates between the individuals. Restricting the set of available treatments would obviously not threaten this conjecture.

[4] The analogue of Observation obviously does not hold for permutations on **T**. Some

This notion of an automorphism will come in handy below. Here follow some examples. Suppose that $\mathbf{I} = \{i_1, i_2, i_3\}$ and let $\mathbf{T}$ be the set of cake-divisions among the members of $\mathbf{I}$. One automorphism would then permute i_1 into i_2, i_2 into i_3 and i_3 into i_1. This would effect a corresponding permutation on cake-divisions. For example, $(0, 2/3, 1/3)$ would be transformed into $(1/3, 0, 2/3)$. Analogously, if $\mathbf{T}$ is the set of rankings of $\mathbf{I} = \{i_1, i_2, i_3\}$, the automorphism that permutes i_1 into i_2, i_2 into i_3 and i_3 into i_1 effects a corresponding permutation on rankings. For example, it transforms the ranking with i_1 on top, followed by i_2 and i_3, in that order, into a ranking with i_2 on top, followed by i_3 and i_1. In general, it is easy to see that only the equal treatment $\mathbf{e}$ stays invariant under all automorphisms: For all automorphisms p, $p(\mathbf{e}) = \mathbf{e}$, and for all $a \in \mathbf{T}$, if $a \neq \mathbf{e}$, then for some automorphism p, $p(a) \neq a$.

We now define an equivalence relation that's going to play an important role in the discussion below—the relation of structural identity between treatments.

Structural Identity: A treatment a is *structurally identical* to a treatment b iff there exists some automorphism p such that $p(a) = b$.

Intuitively, this relation obtains between two treatments if we can get one from the other just by moving individuals around, so to speak. In the case of Cake-Divisions, this means that the cake in both treatments is cut in the same way and the only difference consists in to whom the shares are distributed. In the case of Rankings, we have structural identity if two rankings would look exactly the same if one in each of them were to replace individuals by individual variables. Finally, in the case of Indivisible Goods, two assignments of goods are structurally identical if they both involve the same collection of individual 'baskets' (subsets of G). The difference between structurally identical goods-assignments only appears when it comes to the question which basket goes to whom.

Structural identity is an equivalence relation, i.e., it is reflexive, symmetric, and transitive.[5] Therefore, we can partition $\mathbf{T}$ into *structures*, which are equivalence classes of treatments with respect to the relation of structural identity. I shall refer to different structures as S, S', S'', etc. As an

---

such permutations might not be included in any automorphism. (Think for example of a permutation on cake-divisions that assigns $(1, 0, 0)$ to $(1/2, 1/2, 0)$.) And if $\mathbf{T}$ is limited in scope, some permutations on $\mathbf{T}$ might be included in several automorphisms. (Thus, in the extreme case in which $\mathbf{T}$ just consists of the equal treatment, the identity permutation on $\mathbf{T}$ is included in every automorphism.)

[5] This follows because the set of automorphisms contains the identity automorphism and is closed under inverses and relative products.

example, suppose that **T** is the set of cake-divisions among the individuals in $\mathbf{I} = \{i_1, i_2, i_3\}$. Consider a cake-division $a = (1, 0, 0)$. Its structure consists of three treatments:

$$(1, 0, 0), (0, 1, 0) \text{ and } (0, 0, 1).$$

On the other hand, the structure of $b = (1/2, 1/3, 1/6)$ consists of six treatments. In b, each individual gets a different share and there are six ways in which we can assign three different shares to three different individuals. As for the equal division, $\mathbf{e} = (1/3, 1/3, 1/3)$, its structure contains nothing but **e** itself.

While the number of treatments in a structure may vary, this number is always finite given that **I** is finite: If **I** consists of n individuals, the number of treatments in a structure is at most equal to $n!$, which is the number of possible permutations on **I**. Different automorphisms correspond to different permutations on **I** and the number of treatments in a structure cannot be greater than the number of automorphisms.

For any treatment a in **T**, let S_a be the structure of a. As we have seen, the number of treatments in a structure may vary. In some cases, several permutations on **I** induce automorphisms transforming a into the same treatment, which decreases the size of S_a. This will be the case whenever there are two or more individuals that in a are treated equally. Thus, for example, there are two permutations on individuals that give rise to automorphisms transforming $(1, 0, 0)$ into $(0, 1, 0)$. Each assigns i_2 to i_1, while they differ in their assignments to i_2 and i_3. However, since i_2 and i_3 are treated equally in $(1, 0, 0)$, in both cases $(1, 0, 0)$ is transformed into the same treatment.

At the one extreme, all the individuals are treated in the same way in **e**. Therefore, **e**'s structure contains only **e** itself. At the other extreme, if all the individuals are treated differently in a, the number of treatments in a's structure will be equal to the number of automorphisms. Any two distinct automorphisms will then transform a into different treatments.

3. Unfairness Measure

The last component of the model is an unfairness measure. Before I introduce it, let me first impose a new simplifying assumption: In the situations to be considered, it will be assumed that, according to the agent's information, there is exactly one (perfectly) fair treatment in **T**, i.e. exactly one treatment that fully satisfies the demands of fairness.

This 'uniqueness of fairness'-assumption would be unacceptable in the absence of the relevance constraint introduced in section 2. If a treatment could have variants in **T** that differ from each other only in irrelevant respects, then **T** could obviously contain any number of (perfectly) fair treatments. But even in the presence of the relevance constraint, the 'uniqueness of fairness' is a non-trivial constraint on the model.

The unfairness measure **d** is a function from pairs of treatments to real numbers. $\mathbf{d}(a,b)$ measures how unfair treatment a would be *if b were the unique (perfectly) fair treatment*. The unfairness of a on the hypothetical assumption that b is the fair treatment can be understood as a kind of *distance* between a and b: If a is farther away from b, its unfairness is greater. Given this interpretation, **d** can be assumed to satisfy the standard conditions on a distance measure:

(D0) $\mathbf{d}(a,b) \geq 0$;   *(Non-negativity)*
(D1) $\mathbf{d}(a,b) = 0$ iff $a = b$;   *(Minimality)*
(D2) $\mathbf{d}(a,b) = \mathbf{d}(b,a)$ *(Symmetry)*
(D3) $\mathbf{d}(a,b) + \mathbf{d}(b,c) \geq \mathbf{d}(a,c)$ *(Triangle Inequality)*

Interpreting **d** as distance gives our enterprise a geometrical flavour. The pair (**T**, **d**) is what the mathematicians call a *metric space*: a set of points with a distance measure defined on point-pairs. The set **I** of individuals might then be seen as the set of *dimensions* of that space: A point $a = (a_1, \ldots, a_n)$ is defined by its coordinates on the different dimensions i_k in **I**.

As will be shown below, in the section on expected unfairness, interpreting unfairness as distance goes beyond what is needed for our purposes. Thus, for example, it needn't be assumed that the unfairness measure is symmetric or that it satisfies triangle inequality.[6] Nevertheless, this geometrical interpretation is plausible and independently attractive: Our model becomes more intuitive and easier to grasp.

D0–D3 are not the only conditions that might be imposed on **d**. One such further condition is Impartiality, which requires **d** to be invariant under automorphisms:

---

[6] As Richard Bradley and Mark Schroeder have pointed out to me, we do not even need to assume that there is a single unfairness metric on **T**. We could allow that every b in **T** imposes its own fairness standards, if we view it as *the* fair treatment, and that it thereby induces its own unfairness measure on **T**. As an analogy, we can think of David Lewis' semantics for counterfactuals, in which each possible world induces its own similarity ordering on the set of possible worlds and the orderings induced by different worlds are potentially independent of each other. Given this way of looking at the problem, neither symmetry nor triangle inequality need to be generally valid.

Impartiality Condition: For all automorphisms p and all a, b in **T**,
$$\mathbf{d}(p(a), p(b)) = \mathbf{d}(a,b).$$

According to Impartiality, if one permutes the individuals in two treatments in the same way, the distance between the so permuted treatments doesn't change. In other words, the unfairness measure pays no attention to personal identities. Thus, for example, giving all of the cake to the individual who only deserves a small share is equally unfair independently of who it is who gets this unfair advantage.

How is **d** to be understood on the different interpretations of our model? Consider Cake-Divisions first. It seems plausible that the distance between two cake-divisions should be an increasing function of the differences between the shares of the cake that go to each individual in these treatments, i.e., an increasing function of the absolute differences $|a_1 - b_1|, \ldots, |a_n - b_n|$. The simplest function of this kind is the sum:

$$\mathbf{d}(a,b) = \sum_{i=1}^{n} |a_i - b_i|$$

This kind of measure is sometimes called *city-block* distance. If we instead measure distance by the square root of the sum of the squared differences between vector components, we get the Euclidean distance measure. There are, of course, many other possible interpretations of **d** within the framework of Cake-Divisions. Thus, city-block and the Euclidean measure are only two instances of a much broader class of *Minkowski-distance* functions. **d** belongs to this class iff, for some $k \geq 1$, and for all a, b in **T**,

$$\mathbf{d}(a,b) = \sqrt[k]{\sum_{i=1}^{n} |a_i - b_i|^k}$$

If $k = 1$, **d** is the city-block distance; if $k = 2$, the formula above gives us the Euclidean distance. The higher k is, the greater is the weight carried by the larger differences between corresponding vector components, as compared with the smaller differences. Only when k equals 1, as in the city block, all the differences between the components are given equal weight, independently of size. But already for $k = 2$, as in the Euclidean measure, the larger absolute differences between vector components are given a disproportionate influence, by exponentiation.

How is the distance to be understood for Rankings? Again, there are several possibilities, but one suggestion seems especially plausible. As a preliminary, note that a ranking might be represented as a set of ordered pairs of individuals. On this representation, a pair (i, j) belongs to a ranking a iff *a* ranks *i* at least as highly as it ranks *j*. More precisely, a set of ordered pairs is a ranking iff it is transitive and complete. It is *transitive* iff it contains (i,k) whenever it contains (i,j) and (j,k), and *complete* iff it contains (i,j) or (j,i) for all i and j in **I**. The equal ranking, **e**, ranks each i at least as highly as it ranks each j. Thus, **e** is the universal relation on **I**: the set of all pairs (i,j).

The distance between two rankings can be measured by the number of ordered pairs with respect to which the rankings differ from each other. Thus, on the representation of rankings as sets of ordered pairs, the distance between two rankings equals the cardinality of their symmetrical difference:

$$\mathbf{d}(a,b) = \text{card}(a\text{-}b \cap b\text{-}a).^7$$

It is easy to show that this definition satisfies our conditions D0–D3.

The case of Indivisible Goods is more difficult: Here we lack a plausible definition of distance that is *generally* applicable. Different situations that exemplify the general structure of Indivisible Goods invite different specifications of the unfairness measure. Therefore, instead of discussing the general case, let us focus on the scholarship example we have started with. The example involves two individuals, i and j, and two scholarships, one more attractive, A, and the other, B, less so. According to our information, both individuals deserve a scholarship, though we don't know their relative merits. But we do know that one of the candidates (though we don't know which one) is more deserving than the other.[8]

There is no need to fully specify a suitable unfairness measure **d**. I shall only assume that the following holds: the distance between alternative treatments in which each applicant gets a scholarship, i.e., the distance between $a = (\{A\},\{B\})$ and $a' = (\{B\},\{A\})$, is shorter than the distance to each of them from the equal treatment $\mathbf{e} = (\{\emptyset\},\{\emptyset\})$, in which none of the appli-

---

[7] I.e., $\mathbf{d}(a,b)$ = the number of pairs (i,j) such that (i,j) belongs either to a or to b but not to both. This is the so-called "Kemeny-Snell distance measure." Its axiomatic characterization was given in Kemeny, J. G. and Snell, J. L. (1962). *Mathematical Models in the Social Sciences*. MIT Press, Cambridge, chapter *Preference Rankings: An Axiomatic Approach*. This measure is the only distance function on the set of rankings that satisfies the following intuitive condition: If ranking b is between rankings a and c, then $\mathbf{d}(a,b) + \mathbf{d}(b,c) = \mathbf{d}(a,c)$. Kemeny and Snell obtain their uniqueness result by using this (possibly somewhat controversial) definition of betweenness: b is *between a* and *c* iff b includes the intersection of a and c and is included in their union.

[8] Without this last assumption of unequal deserts, a perfectly fair treatment would not necessarily be available, given that the two scholarships are of unequal value. In the fair distribution, greater deserts are more highly rewarded and equal deserts are rewarded equally.

cants gets anything. To put it formally,

$$\mathbf{d}(\mathbf{e},a) > \mathbf{d}(a,a') < \mathbf{d}(\mathbf{e},a').$$

Here is the intuitive motivation of this assumption. $\mathbf{d}(a,a')$ measures the unfairness of a under the hypothesis that a' is the fair treatment. If a' is fair, both individuals deserve a scholarship. But then withholding the scholarships from both is even more unfair than giving the somewhat better scholarship to the less deserving individual. Thus, $\mathbf{d}(a,a') < \mathbf{d}(\mathbf{e},a')$. Analogously, $\mathbf{d}(a',a) < \mathbf{d}(\mathbf{e},a)$. Since $\mathbf{d}$ is symmetric, $\mathbf{d}(a',a) = \mathbf{d}(a,a')$. Therefore, $\mathbf{d}(a,a') < \mathbf{d}(\mathbf{e},a)$.

4. Information Measure and Expected Unfairness

I, **T**, **e**, and **d** are the fixed components in our model. In addition, the model contains a variable component: the agent's probability distribution P on **T**. (Fixed components are marked in bold, while the italics are used for the variable component.) P reflects the agent's information about the case at hand.[9] For any a in **T**, $P(a)$ stands for the agent's probability for a being the (perfectly) fair treatment.

Since, as we have assumed, the agent is certain that there is exactly one fair treatment in **T**, we take it that the P-values for different treatments sum up to one. There is a difficulty here, though. On some interpretations, such as Cake-Divisions, the number of treatments in **T** is infinite. In such cases, even though the agent is certain that one of the treatments in **T** is fair, the sum of P-values for different treatments might be lower than one. In fact, it might be zero if the probability is distributed uniformly over **T**. Then each treatment gets the probability zero (unless we allow infinitesimals as probability values) and the sum of zeros is zero. This problem can be dealt with by replacing summation with integration, but since I want to avoid mathematical complications, I shall simplify the model by imposing a restriction on P: I shall assume that there exists a finite subset of **T** such that the agent is certain that the fair treatment belongs to that subset. This will keep the calculations at the elementary level: P-values of different treatments will sum up to one, and, for any subset Y of **T**, the probability that Y contains

---

[9] We take P be a variable (rather than fixed) component of the model, since our purpose is to examine whether equal treatment ought to be chosen independently of the particular contents of our information, as long as that information does not distinguish between the individuals.

the fair treatment will be the sum of P-values assigned to the members of Y:

$$P(Y) = \Sigma_{a \in Y} P(a).$$

How does P have to look like if the agent lacks relevant information that discriminates between individuals? As we know, two treatments are structurally identical iff one can be transformed into the other by an appropriate permutation on **I**. But then, to say that the available information does not discriminate between the individuals in **I** must mean that structurally identical treatments are assigned the same P-values. Thus, we are led to the following definition:

> *P does not discriminate between the individuals* iff for all structurally identical a and b in **T**, $P(a) = P(b)$.[10]

To put it differently but equivalently, P does not discriminate between the individuals iff it is *invariant under automorphisms*. That is, iff for every automorphism p and every treatment a, $P(a) = P(p(a))$. As we remember, two treatments are structurally identical iff there is an automorphism that transforms one into the other.

It is easy to define the *expected unfairness* of a treatment a with respect to a given probability function P:

Expected unfairness: $\mathrm{EI}_P(a) = \Sigma_{b \in \mathbf{T}} P(b) \mathbf{d}(a, b)$

(I don't use 'EU' for expected unfairness, to avoid conflation with expected utility. 'EI' is short for 'expected injustice.') Thus, the expected unfairness of a treatment is a weighted sum of its distances to different possible treatments. For each treatment b, the distance to b is weighted with the probability of b being *the* fair treatment.

For the expected value to be a meaningful notion, it is enough if the underlying value function is unique up to positive affine transformations, i.e., up to the choice of unit and zero. Representing unfairness as a *distance* function means that the only arbitrary decision concerns the choice of the unit. (To give an analogy, it does not matter whether we measure spatial distance in meters or yards.) The zero-point for distance is non-arbitrary: That each point's distance to itself, and only to itself, equals zero is a defining feature of a distance measure. Thus, letting the unfairness measure **d** be a distance function is more than we need in order to give meaning to the

---

[10] The finiteness constraint on P formulated above does not create any problems here; it does not hinder P from being indiscriminative between the individuals. This is because, as we have seen, every structure is finite given that **I** is finite.

notion of expected unfairness. Neither the assumption of a fixed zero-point nor the requirements of symmetry or triangle inequality are needed for this purpose. But, as I have suggested above, treating unfairness as distance has its advantages: a geometric model is easier to grasp. At the same time, these extra assumptions about the unfairness measure are independently plausible.

5. Expected Unfairness and Equal Treatment

Consider the following hypothesis:

> *EI-minimization*: For every P that does not discriminate between the individuals, the expected unfairness is minimized by the equal treatment $\mathbf{e}$. That is, for every treatment a in $\mathbf{T}$, $\mathrm{EI}_P(\mathbf{e}) \leq \mathrm{EI}_P(a)$.

In other words, on this hypothesis, equal treatment minimizes expected unfairness in the absence of discriminating information. We want to know under what conditions EI-minimization is going to hold.

If Y is a finite set of treatments, let $\bar{\mathbf{d}}(a, Y)$ stand for a's *average distance* to the treatments in Y:

$$\bar{\mathbf{d}}(a, Y) = \Sigma_{b \in Y} \mathbf{d}(a, b) / \mathrm{card}(Y).$$

As we shall see, the following condition on $\mathbf{d}$ is both *necessary and sufficient* for EI-minimization:

> *Structure Condition*: For every structure $S \subseteq \mathbf{T}$ and every $a \in \mathbf{T}$,
> $$\bar{\mathbf{d}}(\mathbf{e}, S) \leq \bar{\mathbf{d}}(a, S).$$

In other words, for every structure, equal treatment has a minimal average distance to that structure, as compared with other available treatments.

> Sufficiency: Structure Condition $\Rightarrow$ EI-minimization

Proof.

> *Claim*: If P does not discriminate between the individuals, then for every $a \in \mathbf{T}$, $\mathrm{EI}_P(a) = \Sigma_{S \subseteq \mathbf{T}} P(S) \bar{\mathbf{d}}(a, S)$.

I.e., in the absence of discriminating information, a's expected unfairness is a weighted sum of its average distances to different structures, with weights being the probabilities of the structures in question. Here's the proof of the Claim:

$$\begin{aligned}
\mathrm{EI}_P(a) &= \Sigma_{b \in \mathbf{T}} P(b) \mathbf{d}(a,b) && \text{[by the definition of EI]} \\
&= \Sigma_{S \subseteq \mathbf{T}} \Sigma_{b \in S} P(b) \mathbf{d}(a,b) && \text{[since } \mathbf{T} \text{ can be partitioned into structures]} \\
&= \Sigma_{S \subseteq \mathbf{T}} \Sigma_{b \in S} (P(S)/\mathrm{card}(S)) \mathbf{d}(a,b) && \text{[since } P \text{ is indiscriminative]} \\
&= \Sigma_{S \subseteq \mathbf{T}} P(S) (\Sigma_{b \in S} \mathbf{d}(a,b)/\mathrm{card}(S)) && \text{[by algebra]} \\
&= \Sigma_{S \subseteq \mathbf{T}} P(S) \bar{\mathbf{d}}(a,S) && \text{[by the definition of } \bar{\mathbf{d}}\text{]}
\end{aligned}$$

Given Structure Condition, the average distance from $\mathbf{e}$ to any structure S never exceeds the corresponding distance from any a to S. Consequently, Claim implies that $\mathrm{EI}_P(\mathbf{e}) \leq \mathrm{EI}_P(a)$. □

We now want to prove that Structure Condition is *necessary* for EI-minimization:

Necessity: Structure Condition ⇐ EI-minimization

Proof. We want to show that if Structure Condition is violated by our model, i.e. if for some structure S° and treatment a^*, $\bar{\mathbf{d}}(a^*, S^\circ) < \bar{\mathbf{d}}(\mathbf{e}, S^\circ)$, then EI-minimization is violated as well, i.e., there exists a probability function P that does not discriminate between the individuals and some treatment a in $\mathbf{T}$ such that, with respect to P, a's expected unfairness is lower than $\mathbf{e}$'s. To construct such a P, we simply let it be the uniform probability distribution on S°. I.e., for every b in S°, $P(b) = 1/\mathrm{card}(S^\circ)$. Clearly, P assigns equal values to structurally identical treatments, which means that it doesn't discriminate between individuals. At the same time, given this specification of P, the expected unfairness of every treatment equals its average distance to S°. In particular, then, $\mathrm{EI}_P(a^*) = \bar{\mathbf{d}}(a^*, S^\circ)$ and $\mathrm{EI}_P(\mathbf{e}) = \bar{\mathbf{d}}(\mathbf{e}, S^\circ)$. Since, by hypothesis, $\bar{\mathbf{d}}(a^*, S^\circ) < \bar{\mathbf{d}}(\mathbf{e}, S^\circ)$, it follows that $\mathrm{EI}_P(a^*) < \mathrm{EI}_P(\mathbf{e})$. □

To avoid possible misunderstandings, it should be pointed out that for a *particular P* that does not discriminate between the individuals, $\mathbf{e}$ might still minimize expected unfairness with respect to that P even if the underlying unfairness measure $\mathbf{d}$ happens to violate Structure Condition. However, Structure Condition is necessary if $\mathbf{e}$ is to minimize expected unfairness for *all* possible P that do not discriminate between the individuals, i.e. independently of how such a non-discriminative P otherwise looks like.

Is Structure Condition satisfied by the different interpretations of our model? I think it is fair to say that this condition *usually* holds. Thus, it can be shown to hold for all Minkowski-distance measures on cake-divisions.[11] It can also be shown to hold for the interpretation of treatments

---

[11] For the proof, see Rabinowicz, W. Presumption of Equality. (2008). In M. L. Jönsson (Ed.), *Proceedings of the 2008 Lund-Rutgers Conference* (109–155), Appendix A. Lund

as rankings.[12] On the other hand, this condition is violated in the scholarship example. There, as we remember, the distance between treatments $a = (\{A\},\{B\})$ and $a' = (\{B\},\{A\})$ is shorter than the distance to each of them from the equal treatment $\mathbf{e} = (\{\emptyset\},\{\emptyset\})$. Since the set $\{a,a'\}$ is a structure, it immediately follows that there exists a structure S such that the average distance from a to S is shorter (in fact, more than twice as short) than the corresponding average distance from $\mathbf{e}$ to the structure in question:

$$\bar{\mathbf{d}}(a,S) = (\mathbf{d}(a,a) + \mathbf{d}(a,a'))/2 = \mathbf{d}(a,a')/2 < (\mathbf{d}(\mathbf{e},a) + \mathbf{d}(\mathbf{e},a'))/2 = \bar{\mathbf{d}}(\mathbf{e},S).$$

Since Structure Condition is violated in this case, it follows that there exists a probability function P that does not discriminate between the individuals and with respect to which $\mathbf{e}$'s expected unfairness exceeds the expected unfairness of a: One such P is the uniform probability distribution on $\{a,a'\}$. If we are certain that the fair treatment is either a or a', with each of these treatments being an equally likely candidate to the title, treating the individuals equally by withholding the scholarships from both will not minimize expected unfairness.

According to Structure Condition, the average distance from $\mathbf{e}$ to any structure is minimal. It might be thought that in those cases in which Structure Condition does hold, this condition could be strengthened to strict inequality:

(∗) For any structure $S \subseteq \mathbf{T}$ and any treatment $a \in \mathbf{T}$, if $a \neq \mathbf{e}$,
$$\bar{\mathbf{d}}(\mathbf{e},S) < \bar{\mathbf{d}}(a,S).$$

If (∗) were to hold, then, in the absence of discriminating information, $\mathbf{e}$'s expected unfairness would not only be minimal but also uniquely minimal:

Philosophy Reports 2008. In that appendix, I consider a more general interpretation according to which $\mathbf{T}$ consists of all possible real-number assignments to individuals, i.e., not only those in which the assigned numbers are non-negative and add up to 1 (as in Cake-Divisions). On this interpretation, there are non-denumerably many equal treatments in $\mathbf{T}$. I show that every Minkowki-distance measure on such a set $\mathbf{T}$ satisfies the following condition:

Generalized Structure Condition: For every $a \in \mathbf{T}$, there exists an equal treatment $\mathbf{e}_a \in \mathbf{T}$ such that for every structure $S \subseteq \mathbf{T}$, $\bar{\mathbf{d}}(\mathbf{e}_a,S) \leq \bar{\mathbf{d}}(a,S)$.

For Minkowski spaces, the equal treatment $\mathbf{e}_a$ that satisfies Generalized Structure Condition with respect to a treatment a is the one that we obtain from a by averaging:

For every individual i, $\mathbf{e}_a(i) = (a_1 + \cdots + a_n)/n$.

Clearly, when $\mathbf{T}$ is restricted to the set of Cake-Divisions, in which the values assigned to the different individuals are non-negative and always add up to 1, $\mathbf{e}_a$ will coincide with $\mathbf{e}$, for every cake-division a. Therefore, for this restricted set of treatments, the simple Structure Condition will hold.

[12] For the proof, see Rabinowicz, W. Presumption of Equality. (2008). Appendix B.

It would be lower than the expected unfairness of any other treatment. However, (∗) cannot be accepted. Thus consider the Euclidean version of Cake-Division and assume that $\mathbf{I} = \{i_1, i_2\}$. The set of cake-divisions between two individuals forms a line within a two-dimensional space. It corresponds to the line aa' in the diagram:

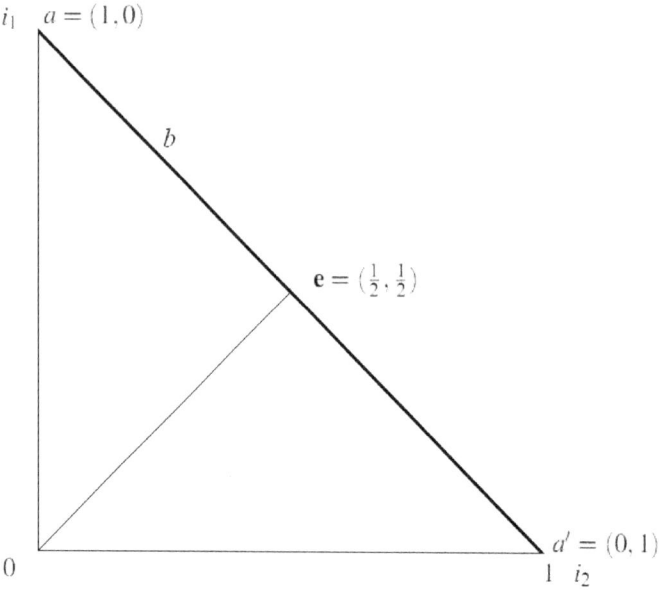

Now, it is easy to see that for *every* cake-division b, its average Euclidean distance to the structure $\{a, a'\}$ equals one half of the Euclidean distance between a and a':

$$\bar{\mathbf{d}}(b, \{a, a'\}) = (\mathbf{d}(b, a) + \mathbf{d}(b, a'))/2 = \mathbf{d}(a, a')/2.$$

Since $\{a, a'\}$ is a structure, (∗) is violated.

But what if we restrict (∗) to cases in which $\mathbf{I}$ contains at least *three* individuals? Then it can be shown that (∗) does hold for the Euclidean interpretation. In a Euclidean space, the points (= cake-divisions) that form a structure (any structure) are vertices of a regular figure, with $\mathbf{e}$ situated right in the center of the figure in question. The figure is located within an n-dimensional space if $\text{card}(\mathbf{I}) = n$, but the figure itself only has $n - 1$ dimensions. Thus, it is a line segment within a two-dimensional space when $\text{card}(\mathbf{I}) = 2$. In the three-dimensional case, i.e. when $\text{card}(\mathbf{I}) = 3$, the set of cake-divisions forms an equi-lateral triangle within a three-dimensional space, with (1, 0, 0), (0, 1, 0) and (0, 0, 1) as vertices. The figure determined by any structure is a regular polygon within that triangle: either a single point, if all three individuals receive equal shares (the structure of $\mathbf{e}$), or an equilateral triangle, if two of the three individuals receive equal shares (cf. as in the structure of $(1, 0, 0)$ or in the structure of, say, $(1/4, 1/4, 1/2)$) or a regular hexagon, if each of the individuals gets a different share (as in the structure of $(5/8, 3/8, 0)$). In the four-dimensional case the set of cake-divisions forms a regular tetrahedron: The figure determined by a structure is a single point in the case of equal treatment, a regular tetrahedron, if three of the four individuals receive equal shares, or a regular polyhedron with six, twelve, or twenty-four vertices, in the remaining cases. And so on, for spaces of higher dimensions. The point **e** is situated in the center of each such regular figure. Consequently, it is the unique point that minimizes the average Euclidean distance to the vertices.

But even with the restriction to cases involving at least three individuals, $(*)$ is satisfied neither by the city-block distance nor by the ranking interpretation. For the city-block, it holds for *every* $n \geq 1$ that if $\mathrm{card}(\mathbf{I}) = n$ and a is a cake-division that assigns the whole of the cake to one of the individuals in **I**, then the average city-block distances from a to S_a and from **e** to S_a both equal $2(n-1)/n$.[13] Analogously, in the case of rankings, for *every* $n \geq 1$, if $\mathrm{card}(\mathbf{I}) = n$ and a is a ranking that linearly orders the individuals in I, then the average distances from a to S_a and from **e** to S_a both equal $n(n-1)/2$.[14]

Condition $(*)$ is therefore unacceptable: Its applicability is too limited. However, already given Structure Condition, **e**'s expected unfairness will be lower than that of any other treatment, if P not only does not discriminate between the individuals but also assigns some positive probability, however small, to **e** itself. To be more precise, Structure Condition implies:

---

[13] As we remember, S_a is the structure of treatment a. In the case of $\mathbf{e} = (1/n, \ldots, 1/n)$ and $a = (1, 0, \ldots, 0)$, **e**'s city-block distance to every treatment in S_a, and thus its average city-block distance to S_a, equals $(1-1/n) + ((n-1)(1/n - 0)) = ((n-1)/n) + (1/n(n-1)) = 2(n-1)/n$. At the same time, a's city-block distance to itself equals 0, while its corresponding distance to every other treatment in S_a (there are $n-1$ such treatments) equals 2. Thus, a's average distance to S_a is also $2(n-1)/n$.

[14] The equal ranking, **e**, ranks each i at least as highly as each j. Thus, **e** is representable as the set of all pairs (i, j) such that $i, j \in \mathbf{I}$. Therefore, if a linearly orders the individuals in I, **e**'s distance to each ranking b in S_a, and thus its average ranking distance to S_a, equals the number of pairs (i, j) such that b does not rank i at least as highly as j. For a linear ranking b, there are $n(n-1)/2$ such pairs.

Proof. For the highest ranked individual in b, j_1, the number of such pairs (i, j_1) is $n-1$. For the next highest one, the corresponding number is $n-2$, and so on until we reach $n-n$ for the last individual. Now, $(n-1) + (n-2) + \cdots + (n-n) = n^2 - (1 + 2 + \cdots + n) = n^2 - (n^2 + n)/2 = (2n^2 - n^2 - n)/2 = n(n-1)/2$. □

The proof of the claim that the average distance of a linear ranking a to the rankings in S_a also equals $n(n-1)/2$ is left to the reader.

Unique EI-minimization: For every P that does not discriminate between the individuals, if $P(\mathbf{e}) > 0$, then $\mathbf{e}$ is the unique treatment that minimizes expected unfairness. That is, for any a in $\mathbf{T}$, if $a \neq \mathbf{e}$, $\mathrm{EI}_P(\mathbf{e}) < \mathrm{EI}_P(a)$.

Observation: Structure Condition $\Rightarrow$ Unique EI-minimization.

Proof. As we have seen above, if P not discriminate between the individuals, the expected unfairness of any treatment a is the sum of the products $P(S)\bar{\mathbf{d}}(a,S)$, for all structures S. By Structure Condition, (i) for all structures S, $P(S)\bar{\mathbf{d}}(a,S) \geq P(S)\bar{\mathbf{d}}(\mathbf{e},S)$. Furthermore, if $P(\mathbf{e}) > 0$, condition D1 on $\bar{\mathbf{d}}$ implies that (ii) in one case at least, when $S = S_\mathbf{e}(= \{\mathbf{e}\})$, $P(S)\bar{\mathbf{d}}(a,S) > P(S)\bar{\mathbf{d}}(\mathbf{e},S)$ for every $a \neq \mathbf{e}$. (i) and (ii) together imply that if $a \neq \mathbf{e}$, $\mathrm{EI}_P(a) > \mathrm{EI}_P(\mathbf{e})$. $\square$

As we have seen in this section, Structure Condition is both sufficient and necessary if equal treatment is to minimize expected unfairness. At the same time, this condition is not especially transparent or intuitive. It has a feel of a constraint that itself should be derivable from some more basic and simple conditions. What these conditions might be is not clear to me, though. One of them would probably be Impartiality mentioned in section 3, which implies that for each structure, all its elements are equidistant from the equal treatment. But we obviously need other conditions as well. These would guarantee that equal treatment lies in the center of each structure. Together with Impartiality, they would therefore give us Structure Condition. Finding these underlying conditions is, as I see it, a major outstanding problem that is left for further inquiry.

6. Minimax

Given Structure Condition, equal treatment minimizes expected unfairness in the absence of discriminating information. However, for some of us, this might not be a decisive consideration. The proper course of action might not be to minimize expected disvalue but rather to minimize the maximal potential disvalue, i.e. to minimize unfairness in *the worst possible outcome*, with a possible outcome being understood as one that has some positive probability, however small. What is the position of equal treatment from this 'minimax' perspective?

For all a in $\mathbf{T}$ and all probability distributions P on $\mathbf{T}$, a's *maximal possible unfairness* with respect to P can be defined as the maximal distance from a to a positively P-valued treatment:

Maximal possible unfairness: $\mathrm{MI}_P(a) = \max\{\mathbf{d}(a,b) : b \in \mathbf{T}\,\&\,P(b) > 0\}$.[15]

With this notion in hand, we can consider the following hypothesis:

MI-minimization: For all P that do not discriminate between the individuals, and all a in $\mathbf{T}$, $\mathrm{MI}_P(\mathbf{e}) \leq \mathrm{MI}_P(a)$.

It is easy to show that the following condition on $\mathbf{d}$ is both necessary and sufficient for the validity of MI-minimization:

Minimax Condition: For every a in $\mathbf{T}$ and every structure $S \subseteq \mathbf{T}$,
$$\max\{\mathbf{d}(\mathbf{e},b) : b \in S\} \leq \max\{\mathbf{d}(a,b) : b \in S\}.$$

According to Minimax Condition, equal treatment minimizes maximal distance to every structure, as compared with other treatments in $\mathbf{T}$.

Sufficiency: Minimax Condition $\Rightarrow$ MI-minimization

Proof. Let c be any treatment such that $P(c) > 0$ and $\mathrm{MI}_P(\mathbf{e}) = \mathbf{d}(\mathbf{e},c)$. Let S_c be the structure of c. Consider any a in $\mathbf{T}$. By Minimax Condition,
$$\max\{\mathbf{d}(\mathbf{e},b) : b \in S_c\} \leq \max\{\mathbf{d}(a,b) : b \in S_c\}.$$
But then, for some $c' \in S_c$, $\mathbf{d}(a,c') = \max\{\mathbf{d}(a,b) : b \in S_c\} \geq \max\{\mathbf{d}(\mathbf{e},b) : b \in S_c\} = \mathbf{d}(\mathbf{e},c)$. Since $P(c) > 0$ and P does not discriminate between the individuals, $P(c') = P(c) > 0$. Consequently, $\mathrm{MI}_P(a) \geq \mathbf{d}(a,c')$. Since $\mathbf{d}(a,c') \geq \mathbf{d}(\mathbf{e},c) = \mathrm{MI}_P(\mathbf{e})$, it follows that $\mathrm{MI}_P(a) \geq \mathrm{MI}_P(\mathbf{e})$. □

Necessity: Minimax Condition $\Leftarrow$ MI-minimization

Proof. Suppose that for some a and S, Minimax Condition is violated:
$$\max\{\mathbf{d}(\mathbf{e},b) : b \in S\} > \max\{\mathbf{d}(a,b) : b \in S\}.$$
Let P be the uniform probability distribution on S. Then P does not discriminate between the individuals, but $\mathrm{MI}_P(\mathbf{e}) = \max\{\mathbf{d}(\mathbf{e},b) : b \in S\} > \max\{\mathbf{d}(a,b) : b \in S\} = \mathrm{MI}_P(a)$. Which means that MI-minimization is violated as well. □

If the distance measure satisfies Structure Condition and Impartiality (i.e. invariance under automorphisms), there is no need to impose Minimax Condition as an independent constraint. It can be shown that:

---

[15] If the set $\{\mathbf{d}(a,b) : b \in \mathbf{T}\,\&\,P(b) > 0\}$ is infinitely large, it might not have a maximum. However, this problem won't arise as long as we hold on to our simplifying assumption that the number of treatments in $\mathbf{T}$ with positive P-values is finite.

Structure Condition & Impartiality ⇒ The Minimax Condition.

Proof. Impartiality implies that for every a in **T** and every automorphism p, $\mathbf{d}(\mathbf{e},a) = \mathbf{d}(p(\mathbf{e}),p(a)) = \mathbf{d}(\mathbf{e},p(a))$. Since a and b are structurally identical iff $b = p(a)$ for some automorphism p, and since **e** is invariant under automorphisms, the following must hold given Impartiality:

For all structurally identical treatments a, b in **T**, $\mathbf{d}(\mathbf{e},a) = \mathbf{d}(\mathbf{e},b)$.

Thus, for every structure S, **e** has the same distance to every treatment in S. Now, consider any treatment a in **T**. By Structure Condition, $\bar{\mathbf{d}}(\mathbf{e},S) \leq \bar{\mathbf{d}}(a,S)$. Therefore, since **e**'s distance to different treatments in S is constant, the maximal distance from **e** to the elements of S cannot exceed the maximal distance from a to the elements of S. Minimax Condition follows. □

Thus, given impartiality of the distance measure, equal treatment will automatically minimize not only expected unfairness but also maximal possible unfairness, if that measure satisfies Structure Condition.[16] There is therefore no need for an independent worry about minimax considerations. However, an interesting question is whether there are any plausible interpretations of our model in which the distance measure satisfies Minimax but violates Structure Condition. On such interpretations, it will be possible to have probability distributions that do not discriminate between individuals and with respect to which **e** does not minimize expected unfairness, even though on all non-discriminative probability distributions **e** will minimize maximal possible unfairness. The issue whether to opt for equal treatment will on such interpretations sometimes depend on whether we adhere to the minimization of expected disvalue or to minimaxing. I don't know, however, of any plausible interpretation of this kind.

7. Extensions

The model we have presented rests on a series of simplifying assumptions. While it always is a good idea to start out with a simple formal framework, this makes our approach unrealistic in several respects. It is therefore natural to consider possible extensions of the modeling. Here are some rather obvious questions we might ask:

---

[16] This means, in particular, that equal treatment minimizes maximal possible unfairness in Rankings and in Cake-Divisions with Minkowski distance, but does *not* minimize it in the scholarship example. If all probability is equally distributed between treatments $(\{A\},\{B\})$ and $(\{B\},\{A\})$, the equal treatment $(\emptyset,\emptyset)$ will be a bad choice for a minimaxer.

(i) What if the set of treatments contains *more than one equal treatment*? In some such cases, I suppose, there need not exist any equal treatment that minimizes expected unfairness or maximal possible unfairness as compared with all other treatments in **T**. What one might hope, though, under such conditions is at least that for every treatment a in **T** there is always *some* equal treatment that is at least as preferable as a in terms of minimization of expected unfairness and/or minimization of maximal possible unfairness. More precisely, if we just focus on minimization of expected unfairness, what one might hope for is that the following generalization of EI-minimization holds:

Generalized EI-minimization: For every treatment a in **T**, **T** contains *some* equal treatment $\mathbf{e}_a$ such that for every P that does not discriminate between individuals, $\mathrm{EI}_P(\mathbf{e}_a) \leq \mathrm{EI}_P(a)$.

It can easily be proved that the condition on the distance measure that is both necessary and sufficient for Generalized EI-minimization is a generalized version of Structure Condition:

Generalized Structure Condition: For every $a \in \mathbf{T}$, there exists an equal treatment $\mathbf{e}_a \in \mathbf{T}$ such that for every structure $S \subseteq \mathbf{T}$, $\bar{\mathbf{d}}(\mathbf{e}_a, S) \leq \bar{\mathbf{d}}(a, S)$.

In other words, according to this condition, for each treatment a there exists some equal treatment $\mathbf{e}_a$ whose average distance to every structure never exceeds the average distance from a itself to the structure in question.[17] A natural question is whether Generalized Structure Condition can be derived from some set of more intuitive and basic assumptions about the distance measure.

(ii) What if the available information does not discriminate between the individuals *within a limited subgroup $X \subseteq \mathbf{I}$*, but not necessarily outside that subgroup? Let's introduce some definitions:

If $X \subseteq \mathbf{I}$, p is an X-automorphism iff p permutes X onto X.

A probability distribution P on **T** does not discriminate between the individuals in $X \subseteq \mathbf{I}$ iff P is invariant under X-automorphisms, i.e., iff for all X-automorphisms p and all $a \in \mathbf{T}$, $P(p(a)) = P(a)$.

If $X \subseteq \mathbf{I}$ and $a, b \in \mathbf{T}$, a is X-structurally identical with b iff there is some X-automorphism p such that $p(a) = b$.

---

[17] For a discussion of that condition, see footnote 9 above.

If we just focus on the minimization of expected unfairness, we might be interested in the following hypothesis:

Subgroup-Generalized EI-minimization: For every $X \subseteq \mathbf{I}$ and every $a \in \mathbf{T}$, $\mathbf{T}$ contains *some* equal treatment $\mathbf{e}_{a,X}$ such that for every P that does not discriminate between individuals in X, $\text{EI}_P(\mathbf{e}_{a,X}) \leq \text{EI}_P(a)$.

Since X-structural identity is an equivalence relation, just like the ordinary structural identity, $\mathbf{T}$ can be partitioned into X-structures—equivalence classes with respect to X-structural identity. The condition on the distance measure that is necessary and sufficient for Subgroup-Generalized EI-minimization is a further generalization of Generalized Stucture Condition:

Subgroup-Generalized Structure Condition: For every $X \subseteq \mathbf{I}$ and every $a \in \mathbf{T}$, there exists an equal treatment $\mathbf{e}_{a,X} \in \mathbf{T}$ such that for every X-structure $S \subseteq \mathbf{T}$, $\bar{\mathbf{d}}(\mathbf{e}_{a,X}, S) \leq \bar{\mathbf{d}}(a, S)$.[18]

It is easy to see that this condition entails Generalized Structure Condition as a special case (for $X = \mathbf{I}$), but it would be interesting to know how to derive this condition from some more basic conditions on a distance measure.

(iii) What if there might be *several (perfectly) fair treatments*, and not just one? We would then need to work with a modified unfairness measure:

$\mathbf{d}(a,Y)$—the unfairness of a on the hypothetical assumption that Y is the set of all fair treatments in $\mathbf{T}$.

We take it that $\mathbf{d}(a,Y)$ is defined only if Y is non-empty and finite. (Finiteness is assumed mainly for the sake of simplicity.) An obvious requirement on this measure is that $\mathbf{d}(a,Y) = 0$ iff $a \in Y$. In other words, the unfairness of a is zero if a is one of the (perfectly) fair treatments. In fact, $\mathbf{d}(a,Y)$ could simply be defined as the minimal distance from a to the elements of Y.[19]

We would also need a modified measure of information:

$P(Y)$—the probability that Y is the set of all fair treatments in $\mathbf{T}$.

---

[18] In the case of Cake-Divisions, we can conjecture that $\mathbf{e}_{a,X}$ is obtainable from a as follows: (i) for every i in X, $\mathbf{e}_{a,X}$ assigns to i the average of the values assigned by a to the members of X; while (ii) for every i outside X, $\mathbf{e}_{a,X}(i) = a(i)$. It is less clear how to construct $\mathbf{e}_{a,X}$ in the case of Rankings.

[19] This definition reduces distance to sets to distances to singleton-sets. $\mathbf{d}(a,b)$ has now to be read as a short for $\mathbf{d}(a,\{b\})$, i.e. as the unfairness of a on the assumption that $\{b\}$ is the set of fair treatments. Note also that if we allowed Y to be infinite, $\mathbf{d}(a,Y)$ could be defined as the lower bound of $\{\mathbf{d}(a,b): b \in Y\}$, but only provided that the latter set does have the lower bound. Otherwise, $\mathbf{d}(a,Y)$ would have to be left undefined, which would create problems for this proposal.

For such a probability measure P, the notion of non-discrimination would have to be appropriately re-defined:

> P does not discriminate among the individuals iff for all automorphisms p, $P(p(Y)) = P(Y)$, where $p(Y) = \{b : \exists a \in Y \; p(a) = b\}$.

And we would need to correspondingly re-define the notions of expected unfairness and maximal expected unfairness:

$\mathrm{EI}_P(a) : \Sigma_{Y \subseteq \mathbf{T}} P(Y) \mathbf{d}(a, Y)$.
$\mathrm{MI}_P(a) = \max\{\mathbf{d}(a, Y) : Y \subseteq \mathbf{T} \;\&\; P(Y) > 0\}$

The obvious question is: What condition on the unfairness measure that is modified in this way in order to allow for a larger number of fair treatments will then guarantee that, in the absence of discriminating information, equal treatment will minimize expected unfairness and/or maximal expected unfairness? Will some condition analogous to our Structure Condition do the job?[20]

These are just some of the follow-up questions that could be raised. But their examination has to await another occasion.[21]

---

[20] The condition I have in mind states that for every family Ω of subsets of $\mathbf{T}$ such that, for some finite $Y \subseteq \mathbf{T}$, $\Omega = \{Z \subseteq \mathbf{T} : \text{for some automorphism } p, \; p(Y) = Z\}$, $\bar{\mathbf{d}}(\mathbf{e}, \Omega) \leq \bar{\mathbf{d}}(a, \Omega)$ for all $a \in \mathbf{T}$. If some automorphism permutes Y into Z, then sets Y and Z can be said to be structurally identical. If $Y \subseteq \mathbf{T}$, $\Omega = \{Z \subseteq \mathbf{T} : \text{for some automorphism } p, p(Y) = Z\}$ can therefore be seen as the structure of the set Y. Thus, the condition I have just suggested says the $\mathbf{e}$ minimizes the average distance to the structure of every finite set of treatments.

[21] The main ideas of this paper were presented at a workshop on impartiality and partiality in Reading 2007, at a meeting of the Choice Group at LSE 2008, at the Lund-Rutgers philosophy conference in Lund 2008, at a symposium on equality in Jerusalem 2008 and at the workshop on formal methods organized in connection with the European Congress of Analytic Philosophy in Cracow in the same year. I am much indebted to the participants of these various events for their comments. Especially, I want to thank Richard Bradley, Hannes Leitgeb, Christian List, and Mark Schroeder.

Extending the Hegselmann-Krause Model II

Alexander Riegler and Igor Douven

ABSTRACT

Hegselmann and Krause have developed a computational model for studying the dynamics of belief formation in a population of epistemically interacting agents who try to determine, and also get evidence concerning, the value of some unspecified parameter. In a previous paper, we extended the Hegselmann-Krause (HK) model in various ways, using the extensions to investigate whether, in situations in which random noise affects the evidence the agents receive, certain forms of epistemic interaction can help the agents to approach the true value of the relevant parameter. This paper presents an arguably more radical extension of the HK model. Whereas in the original HK model each agent is solely characterized by its belief, in the model described in the current paper, the agents also have a location in a discrete two-dimensional space in which they are able to move and to meet with other agents; their epistemic interactions depend in part on who they happen to meet. We again focus on situations in which the evidence is noisy. The results obtained in the new model will be seen to agree qualitatively with the results obtained in our previous extensions of the HK model.

Modern science is largely a community enterprise; scientists working in relative isolation from their colleagues, in the way Kepler or Newton did, are exceptionally rare nowadays. Interactions between working scientists are multiple and multifarious, ranging from jointly carrying out experiments and coauthoring papers to discussing half-baked ideas during coffee breaks. As a result of this, or perhaps as a direct effect of a separate form of epistemic interaction, a scientist's belief on a given matter will often be influenced to at least some extent by his or her colleagues' beliefs on the same matter. But—one may ask—is this for better or for worse? It is not entirely inconceivable that in general scientists would do best, in terms of achieving whatever epistemic goals their research is meant to serve, by going purely by the data, and not allowing their beliefs to be affected by those of any of their colleagues. At a minimum, the said form of interaction might be inessential from a strictly scientific perspective.

In previous work, we have begun studying this matter from a truth-tracking perspective, that is, with an eye towards answering the question whether various forms of adjusting one's belief in response to learning the beliefs of others can help one to achieve the scientific goal of approximating the truth (see Douven, 2010 and Douven and Riegler, 2009). In that work, we could build on pioneering research carried out by Rainer Hegselmann and Ulrich Krause, who in a series of papers have developed a model for investigating the dynamics of belief formation in societies of truth-seeking agents.[1] While the merits of their work are beyond doubt, it has also been observed that their model has severe limitations. Especially if one wants to use it to study the aforementioned questions, many of the assumptions of the Hegselmann-Krause (HK) model cannot but strike one as being too idealized. In our own earlier work, we have sought to take some first steps towards "concretizing" the HK model (in the sense of Nowak 1980), that is, we have proposed extensions of the model that do away with some of the idealizations inherent in it. Most importantly perhaps, we gave up the assumption that the agents in the model receive "noise-free" experimental data. The resulting model enabled us to argue that particular ways of responding to learning the beliefs of certain colleagues are, from a truth-tracking viewpoint, clearly preferable as an epistemic strategy to ignoring those beliefs and going purely by the data, at least in environments in which the data are noisy—as they tend to be in typical empirical settings.

Still, many more limitations of the HK model remain to be addressed if it is to inform us about the value (or otherwise) of the said form of epistemic interaction as it is to be observed in the practice of science. One of the major impediments in this respect is that, in the HK model, all agents are supposed to be privy to the beliefs of all the other agents at any time. Due to this, the model would seem to apply to only a very limited range of actual situations. One may claim that the members of a research group are mostly aware of each other's beliefs (insofar as these are relevant to their ongoing research). But epistemic interaction is not restricted to members of one and the same research group, and it would certainly be false to suppose

---

[1] See Hegselmann and Krause (2002, 2005, 2006).

that scientists are generally aware of the beliefs of all others—members of different research groups included—working in their field. To overcome this problem, below we present a further extension of the HK model in which agents are only aware of the beliefs of the subgroup of agents who they happen to meet at a given time. Specifically, and in contrast to the HK model, we endow agents with a location in a discrete two-dimensional space in which they are able to move about and to meet with other agents. These meetings determine, at least partly, with whom they epistemically interact. We examine various aspects of this new model. Again, we also focus on situations in which the agents receive noisy data. It will be interesting to see that the results obtained in the new model agree, at least qualitatively, with the results obtained in our previous extensions of the HK model.

1. The Hegselmann–Krause Model and Beyond

The HK model was devised to investigate the dynamics of belief formation in populations of epistemically interacting truth-seeking agents. In the model, a population consists of n agents, where each agent i ($1 \leqslant i \leqslant n$) holds a belief $x_i(t)$ at any time step t. The agents try to ascertain the value τ of a given parameter (which could be the mass of some particle, the viscosity of a fluid, the probability of life on Mars, etc.). It is assumed that $\tau \in [0,1]$, and that the agents know this antecedently; thus $x_i(t) \in [0,1]$ for all i and t. At discrete time steps, the agents simultaneously update their beliefs, where the following rule gives the belief of agent i at time step $t+1$ as a function of the agents' beliefs at t and the true value of the parameter:

$$x_i(t+1) = \alpha \frac{\sum_{j \in X_i(t)} x_j(t)}{|X_i(t)|} + (1-\alpha)\tau. \qquad (1.1)$$

Here, $|X_i(t)|$ is the cardinality of $X_i(t)$, which is defined to be the set of agents whose beliefs are "close enough" to the belief of agent i at t, or, in Hegselmann and Krause's terminology, that are within i's *bounded confidence interval* (BCI) at t. More exactly, $X_i(t) = \{j : |x_i(t) - x_j(t)| \leqslant \varepsilon\}$, for some real-valued ε. If $j \in X_i(t)$, we shall say that j is an *epistemic neighbor* of i at t; note that, trivially, every agent is its own epistemic neighbor at any time. Furthermore, $\alpha \in [0,1]$ is a weighting factor that determines the degree to which the agents' beliefs depend on those of their epistemic neighbors. In short, at each time step the new belief of an agent is calculated as the weighted average of the average of the beliefs of all its epistemic neighbors and the true value of the parameter. We shall say that an agent *talks to its epistemic neighbors* iff both $\alpha > 0$ and $\varepsilon > 0$. We might thus say that, if in addition $\alpha < 1$, an agent's new belief is the outcome of the combination of talking to its epistemic neighbors and making experiments, where the results of the experiments point in the direction of τ. In their papers, Hegselmann and Krause present some analytic results about this model, but for the most part they explore its properties by means of computer simulations.[2] In particular, they study the relation between the various parameters of the model and the convergence of the agents' beliefs to τ.

In our earlier-cited papers, we noted that the assumption, inherent in the HK model, that the agents receive precise evidence pointing in the direction of τ is not particularly realistic; in reality, scientists have to live with measurement errors and other factors that make their data noisy. For this reason, the assumption was dropped in the extension of the HK model presented in Douven (2010); this model explicitly allows data to be noisy. A further assumption of the original HK model that seemed questionable to us was that the agents' beliefs all weigh equally heavily in the updating process; the model is thereby unable to reflect the fact that in scientific practice the beliefs of some count for more than those of others. These considerations led us in Douven and Riegler (2009) to propose a further extension of the HK model that replaces (1.1) by

$$x_i(t+1) = \alpha \frac{\sum_{j \in X_i(t)} x_j(t) w_j}{\sum_{j \in X_i(t)} w_j} + (1-\alpha)(\tau + \mathrm{rnd}(\zeta)). \qquad (1.2)$$

In this equation, w_j denotes the fixed reputation of agent j, and $\mathrm{rnd}(\zeta)$ is a function returning a unique uniformly distributed random real number in the interval $[-\zeta, +\zeta]$ each time it is invoked ($\zeta \in \mathbb{R}$).

The main result of the computer simulations that we used to investigate this model can be put thus: in situations in which evidence obtained in experiments is noisy, populations of agents that attribute more weight to talking to each other (i.e., that have a higher value for α) converge to τ more accurately, albeit more slowly, than do populations that give less weight to talking and rely more on the evidence. See Douven (2010) and Douven and Riegler (2009) for the details. Reputation, on the other hand, appeared to have little to no effect on the outcome of the simulations. See Douven and Riegler (2009) for the details.

While we thus did away with some limitations of the HK model, it still holds for the models described in the aforementioned papers that all agents

---

[2] For the many virtues of this approach for investigating belief dynamics in multi-agent systems, see Humphreys (1991), Epstein and Axtell (1996), Hartmann (1996), Gaylord and D'Andria (1998), and Winsberg (1999), as well as the papers cited in the previous note.

are supposed to know, at any time, the beliefs of all the other agents. As intimated, this reduces the scope of the model considerably; the number of colleagues with whom scientists collaborate, or whom they meet at conferences, is generally just a fraction of the number of their epistemic neighbors. In the following, we describe an extension of the HK model that deviates more drastically from the original and that eliminates the said unrealistic assumption by letting agents move about in a two-dimensional environment in which they encounter others and interact only with the epistemic neighbors they happen to meet.

2. Adding Spatial Dimensions

In order to accurately model the aspect of encountering epistemic neighbors, we extended the HK model by introducing *spatial dimensions*. More concretely, each agent (exclusively) populates a site in a discrete two-dimensional toroidal grid, that is, a grid whose outside borders wrap around to the opposite side in order to avoid edge effects. (Such effects would, for example, reduce the number of possible adjacent neighbors for any agent at an edge, or in a corner, of the grid.) Our usage of discrete two-dimensional environments is neither original nor arbitrary, as they have long proven to be valuable tools in disciplines such as artificial life and sociology (see, e.g., Epstein and Axtell, 1996 and Gaylord and D'Andria, 1998).

To be more specific, the environment consists of a two-dimensional grid of 25×25 sites each of which can either be empty or occupied by an agent. Agents face one of the four cardinal points of the compass. They move about by leaping to the adjacent free site they are facing. As noted above, unlike in the original HK model and our earlier extensions thereof, in which an agent at each time step interacts with *all* its epistemic neighbors, in the present model it interacts only with those of its epistemic neighbors that are to be found in its *spatial* neighborhood. That is to say, for all i, j, and t, agent i epistemically interacts with agent j at time t iff:

1. j is in the *epistemic* neighborhood of i, that is, $|x_i(t) - x_j(t)| \leqslant \varepsilon$, and
2. j is in the *spatial* neighborhood of i.

The notion of spatial neighborhood can be—and has been—defined in various ways. In our simulations, we made use of the three neighborhood structures depicted in Figure 1, which are relatively common in the literature (see, e.g., Gaylord and D'Andria, 1998). Following a suggestion of Gaylord and D'Andria (1998), we also considered a variation of the von Neumann neighborhood in which only those agents are considered that face one's own po-

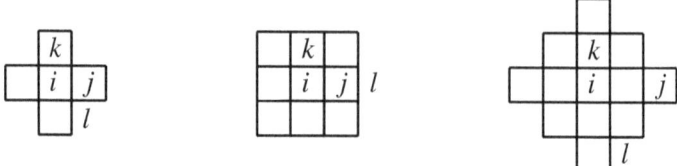

Figure 1: Three neighborhood structures: von Neumann (left); Moore (center); Gaylord-Nishidate (right). In each case, j and k are spatial neighbors of i, whereas l is not

sition (for the other two neighborhoods this extra "facing" condition makes no sense).

In accordance with the distinction between epistemic and spatial neighborhoods, the development of agents through time is characterized by both a belief-update rule and a migration rule. As in the HK model, the update rule takes care of the belief transitions an agent can undergo from one time step to another. The migration rule determines how agents move about in their environment. The former looks thus:

$$x_i(t+1) = \begin{cases} \alpha \frac{\sum_{j \in X_i(t)} x_j(t)}{|X_i(t)|} + (1-\alpha)(\tau + \text{rnd}(\zeta)) & \text{if } |X_i(t)| > 1, \\ x_i(t) & \text{otherwise,} \end{cases} \quad (2.1)$$

where $X_i(t)$ now designates the set of i's epistemic neighbors at t that are also within its spatial neighborhood at that time. Note that the upper clause of equation (2.1) corresponds to equation (1.2) with reputation $w_j = 1$ for all j. This clause is invoked whenever there is at least one epistemic neighbor present in the spatial neighborhood besides the agent itself. If $|X_i(t)| = 1$, agent i's belief remains unchanged from t to $t+1$. The migration rule simply says that after an agent has updated its belief, it moves one step to the adjacent site it faces if that is free and not faced by at least one other agent, or else it randomly changes its orientation to any of the four cardinal directions (see Gaylord and D'Andria, 1998 for details).

3. Exploring the Model

In this section, we present the results of computer simulations we conducted in order to explore systematically the properties of our two-dimensional model. Like Hegselmann and Krause in the investigations of their model, we were particularly interested in questions concerning the relation between the values for the parameters α and ε and a population's ability to track the

3.1. Different Spatial Neighborhoods

Given the different definitions of the notion of spatial neighborhood, one may ask which of these yields the best results in terms of accuracy and speed of convergence of the agents' beliefs to the value of τ. We put the four types described in the previous section to test by comparing two populations of $n = 100$ agents that attribute different weights to talking ($\alpha = .1$ and $\alpha = .9$, respectively). For both these populations, it holds that $\zeta = .1$ and $\varepsilon = .1$. We ran 100 simulations for each of the two populations and each of the four spatial neighborhoods, yielding eight trials of 100 simulations in total. In these, as in all other simulations we performed for the present paper, we set $\tau = .75$. For each trial, we calculated, at each time step, the average over the 100 simulations of the average distance from the truth of the beliefs of all the agents in the population, where the distance from the truth of agent i's belief at t was simply taken to be $|x_i(t) - \tau|$.

The results are shown in Figure 2. The most accurate and also fastest convergence is achieved by the Gaylord-Nishidate neighborhood (solid line), followed by Moore (dashed), von Neumann (dotted), and von Neumann facing (dash-dotted). This is so regardless of whether the agents give much weight to the evidence (low value for α) or rather to talking to others (high value for α), although in the former case the differences between the various neighborhoods are less pronounced. We opted for the Gaylord-Nishidate neighborhood in all further experiments. That the

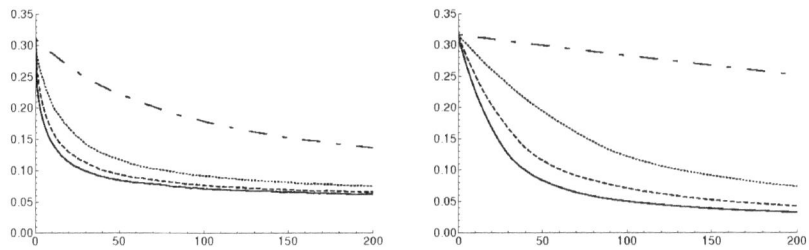

Figure 2: Convergence in different neighborhoods. The abscissa represents time steps, the ordinate is the average distance from τ. Left: $\alpha = .1$; right: $\alpha = .9$

Gaylord-Nishidate neighborhood, which encompasses more surrounding sites than any of the other neighborhoods, is superior in terms of convergence to the truth can already be considered as a first indication that, at least if the evidence the agents receive is noisy, the present model makes talking to others come out as a good epistemic strategy.

3.2. The Weight of Talking

The foregoing would clearly square with the earlier-cited conclusion of Douven and Riegler (2009) that, in situations in which evidence is noisy, giving much weight to talking to one's epistemic neighbors is an effective scientific strategy. But does this conclusion really carry over to the two-dimensional model? Given that in this model most of one's epistemic neighbors may be unavailable for talking at any given time step, the answer is not straightforward.

Talking is not a yes/no affair in the original HK model, in our previous extensions of that, or in the two-dimensional model. "How much" talking is going on in a population depends in all these models on the values of α and ε. To investigate the impact of varying these values in the two-dimensional model, we first systematically increased the value of ε between 0 and .3, in steps of .05, in two populations, one with $\alpha = .1$, the other with $\alpha = .9$. We ran 100 simulations with a population of $n = 100$ agents for each combination of α and ε. Interpolated results for the entire range of ε-values are shown in Figure 3.

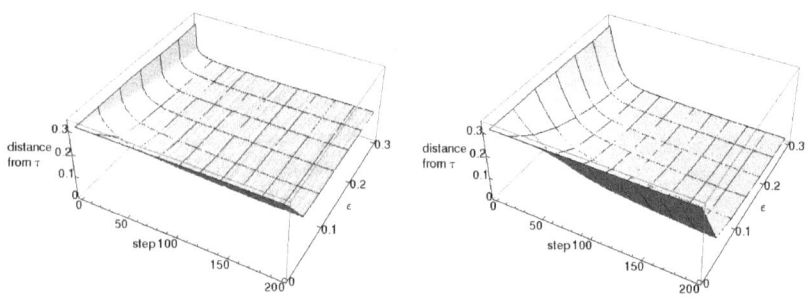

Figure 3: Varying ε for $\zeta = .3$. Left: $\alpha = .1$; right: $\alpha = .9$

The results suggest that, at least for the given level of noise, the larger ε the more accurate the convergence. Still, the value of α is seen to matter as well. For populations of agents that assign a value of only .1 to α, convergence to τ is far less accurate than for populations of agents that assign a value of .9 to α.

For additional clarity, Figure 4 shows the convergence to τ for only four selected combinations of values for α and ε. It indicates clearly the advantage of assigning higher values to both α and ε for approaching τ. (We also

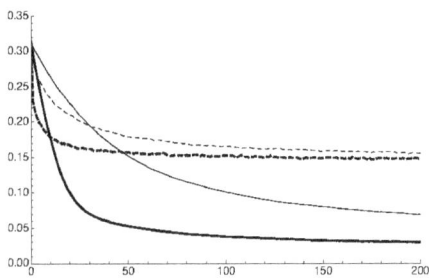

Figure 4: Dashed lines: $\alpha = .1$; solid lines: $\alpha = .9$. Thin lines: $\varepsilon = 0$; thick lines: $\varepsilon = .3$

ran simulations for $\zeta = .1$ and $\zeta = .2$, but this did not lead to a qualitative difference in the outcome. Quantitatively, the differences between the different parameter settings are a bit more pronounced with $\zeta = .3$, which is why we used this value for producing the graphs.)

In further computer experiments, we systematically increased the value of α between 0 and 1, again in steps of .05, keeping ε fixed at .2.[3] In these experiments, it appeared that the closest approximation to the truth was reached if α was given a value close to 1, as can be gleaned from Figure 5. While a high value for α delays the convergence, convergence is ultimately more accurate, that is, it leads to a smaller average distance of the agents' beliefs from τ. Based on these results, it would seem that scientists interested in quick approximate solutions should give less weight to talking than to the evidence they obtain, whereas scientists who want to approach τ as closely as possible—which from a purely scientific perspective would seem to be the more desirable goal—are well advised to do the opposite and rather give less weight to the evidence and more to talking to others.

---

[3] Again we used various values for ζ, and again this did not significantly influence the results. We also again ran 100 simulations for each relevant combination of the parameters. All experiments to be reported below have the same set-up.

This is fully in line with the above-mentioned conclusion from Douven and Riegler (2009) about our simpler extensions of the HK model.

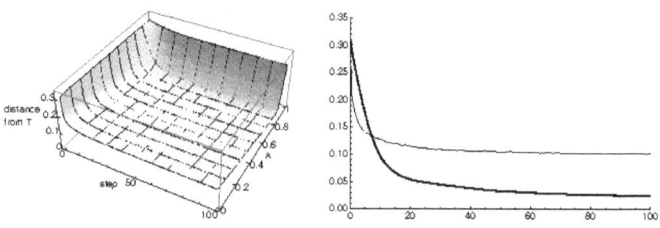

Figure 5: Varying α for $\zeta = .2$. Right: comparison of $\alpha = .1$ (thin) and $\alpha = .9$ (thick)

3.3. Elite Versus Normal Scientists

Following Douven and Riegler (2009), also in the two-dimensional model we performed experiments in which we tried to model a distinction between elite scientists, who are very skilled experimenters, able to obtain noise-free data, and normal scientists, who are not equally skilled and obtain noisy data. To that end, we introduced two classes of agents that differ with respect to their noise parameter: of the 100 agents in each population considered in the simulations, there are 75 normal agents with a noise parameter $\zeta_N = .2$ and 25 elite agents with $\zeta_E = 0$. For ε we chose a value of .1 and for α we experimented with values between 0 and 1. Figure 6 shows the results for selected values of $\alpha = .1$ and $\alpha = .9$.

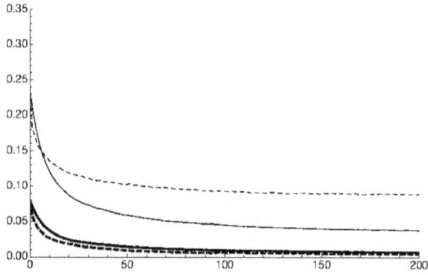

Figure 6: Thick lines: elite agents; thin lines: normal agents. Dashed lines: $\alpha = .1$; solid lines: $\alpha = .9$

These results exhibit that normal agents (thin lines) benefit greatly if all agents—elite and normal ones alike—give much weight to talking, as they (the normal ones) then approach τ much more closely than they do if all agents give only little weight to talking. On the other hand, elite agents (thick lines) are better off in the latter case. Nevertheless, since they do not do *much* worse in that case, a kind of epistemic utilitarianism would suggest that, provided the choice is between just these two options—either all agents assign a high value to α or all agents assign a low value to α—the former is preferable.

3.4. The Role of Reputation

Not only are some scientists better than others, some scientists are also more reputed than others. In many situations, the views of the highly reputed scientists weigh more heavily than the views of others. But *should* they do so? Might this aspect of scientific practice not bear *negatively* on the prospects for making progress? Perhaps scientists are generally bad at distinguishing the good from the not-so-good scientists, and tend to assign a higher reputation to the latter than to the former. This question can be investigated, to some extent, in our two-dimensional model by replacing the update rule (2.1) by

$$x_i(t+1) = \begin{cases} \alpha \frac{\sum_{j \in X_i(t)} x_j(t) w_j}{\sum_{j \in X_i(t)} w_j} + (1-\alpha)(\tau + \text{rnd}(\zeta)) & \text{if } |X_i(t)| > 1, \\ x_i(t) & \text{otherwise,} \end{cases} \quad (3.1)$$

where, as in equation (2.1), $X_i(t)$ designates the set of agents that are both i's epistemic and its spatial neighbors at t, and where, as in equation (1.2), w_j represents the reputation of agent j.

The specific question we wanted to address was whether, in order to improve the accuracy in truth approximation of a population in its entirety, elite scientists should be given a higher reputation than normal scientists, where the difference between these groups is understood in the way defined above. To investigate this question, we set up three scenarios. In the first, the agent's status, *qua* elite ($\zeta_E = 0$) or normal ($\zeta_N = .2$), corresponds to its reputation: elite agents are given a reputation w_j of 2, normal agents a reputation of 1. In the second scenario, this is reversed: elite agents are given a reputation of 1, normal agents a reputation of 2. And in the third scenario, both the elite and the normal agents have a reputation of 1.

In line with the findings of Douven and Riegler (2009), the results of various computer experiments (using various combinations for the values of the other parameters) suggest that there is no difference among the three settings, neither in terms of accuracy of convergence, nor in terms of speed of convergence. As also explained in the aforementioned paper, this should not be taken to support the cynical conclusion that, from a purely scientific perspective, it is immaterial whether we assign a high reputation to the best scientists or rather to the crackpots: reputation plays a role in science in many ways, most of which go unaccounted for in the present model. Nevertheless, it is noteworthy that in at least one way in which one would have guessed reputation to matter, it does *not* matter.

3.5. Choosing Between Talking and Experimenting

In the computer experiments described above, we have been assuming that whenever agents talk to other agents they also take into account evidence pointing in the direction of τ. Clearly, this is an idealization. Human scientists may perform an experiment one day and share their results only the next day. One can model this to some extent by letting each agent at each time decide which one of the following it wants to do: (i) gather evidence pointing in the direction of τ by performing an experiment (or making observations, or some such); (ii) talk to those of its epistemic neighbors that are in its spatial neighborhood at the given time (if any are). For this purpose, we once more changed the update rule, as follows:

$$x_i(t+1) = \begin{cases} \alpha x_i(t) + (1-\alpha)(\tau + \text{rnd}(\zeta)) & \text{if } |X_i(t)| > 1 \text{ and rnd}(1) > \theta, \\ \dfrac{\sum_{j \in X_i(t)} x_j(t)}{|X_i(t)|} & \text{if } |X_i(t)| > 1 \text{ and rnd}(1) \leqslant \theta, \\ x_i(t) & \text{if } |X_i(t)| = 1. \end{cases}$$

(3.2)

Here, $|X_i(t)|$ again designates the cardinality of the set of agents that are both within i's epistemic and within its spatial neighborhood at t; the function rnd(1) returns a uniformly distributed random real number in the interval $[0,1]$; and θ is a threshold influencing the agent's decision: a low value of this threshold corresponds to a greater likelihood of updating one's belief on the evidence and one's current belief only, a high value to a greater likelihood of updating by talking to others only.

Our results show a clear, although not very significant, difference in the ability to approach τ for different values of θ. As can be seen in Figure 7, convergence is more accurate, though slower, for higher values of the threshold, that is, for a greater likelihood of talking (the graph shows the results for $\varepsilon = .1$ and $\alpha = .1$; the results proved relatively robust for variations in these values).

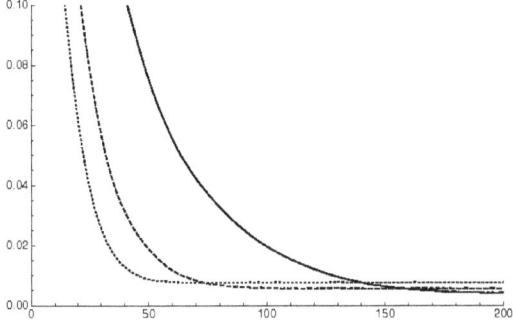

Figure 7: Dotted line: $\theta = .25$; dashed line: $\theta = .5$; solid line: $\theta = .75$

4. Conclusion

Hegselmann and Krause's model must be considered an important tool for studying the effects of certain types of epistemic interaction in populations of truth-seeking agents on these populations' abilities to track the truth. While their own model is highly idealized, it can easily be extended in ways that allow for more realistic interpretations. Above, we have presented an extension that introduces spatial neighborhoods, whereby one can drop the assumption of the original HK model that each agent knows at all times the beliefs of all other agents. The results about this extended model, which were obtained by means of computer simulations, reconfirm the main conclusion of Douven and Riegler (2009): assuming the evidence to be noisy, as in real life it tends to be, talking to others is an epistemically good strategy for a population of truth seeking agents in that it helps the agents to get, on average, closer to the truth, closer, at any rate, than if they disregard the beliefs of others and purely go by the data and their own current beliefs.

In closing, we note that the model presented above, while already much less idealized than the original HK model, is still rather limited in scope, as in it all agents are supposed to hold single beliefs only. Real scientists invariably have a great many beliefs that, moreover, tend to be logically interconnected. A model that equips the agents with propositional theories rather than a single numerical belief is described, and systematically explored, in Riegler and Douven (2009).[4]

---

[4] We are grateful to Christopher von Bülow for valuable comments on a draft version of this paper.

References

Douven, I. (2010). Simulating peer disagreements. *Studies in History and Philosophy of Science*. (Forthcoming).

Douven, I. and Riegler, A. (2009). Extending the Hegselmann-Krause model I. *The Logic Journal of the IGPL*. (In press).

Epstein, J. and Axtell, R. (1996). *Growing Artificial Societies*. MIT Press, Cambridge, MA.

Gaylord, R. J. and D'Andria, L. J. (1998). *Simulating Society*. Springer, New York.

Hartmann, S. (1996). The world as a process: Simulations in the natural and social sciences. In R. Hegselmann, U. Mueller, and K. G. Troitzsch (Eds.), *Modelling and Simulation in the Social Sciences from the Philosophy of Science Point of View* (77–100). Kluwer Academic Publisher, Dordrecht.

Hegselmann, R. (1996). Cellular automata in the social sciences. In R. Hegselmann, U. Mueller, and K. G. Troitzsch (Eds.), *Modelling and Simulation in the Social Sciences from the Philosophy of Science Point of View* (209–233). Kluwer Academic Publisher, Dordrecht.

Hegselmann, R. and Krause, U. (2002). Opinion dynamics and bounded confidence: Models, analysis, and simulations. *Journal of Artificial Societies and Social Simulation*, 5. available at URL = http://jasss.soc.surrey.ac.uk/5/3/2.html.

Hegselmann, R. and Krause, U. (2005). Opinion dynamics driven by various ways of averaging. *Computational Economics*, 25, 381–405.

Hegselmann, R. and Krause, U. (2006). Truth and cognitive division of labor: First steps towards a computer aided social epistemology. *Journal of Artificial Societies and Social Simulation*, 9. available at URL = http://jasss.soc.surrey.ac.uk/9/3/10.html.

Humphreys, P. (1991). Computer simulations. In A. Fine, M. Forbes, and L. Wessels (Eds.), *Proceedings of the PSA 1990*, volume 2 (497–505). East Lansing MI. Philosophy of Science Association.

Nowak, L. (1980). *The Structure of Idealization*. Reidel, Dordrecht.

Riegler, A. and Douven, I. (2009). Extending the Hegselmann-Krause model III: From single beliefs to theories. *Episteme*, 6, 145-163.

Winsberg, E. (1999). Sanctioning models: The epistemology of simulation. *Science in Context*, 12, 275–292.

Structured Meanings and Concepts

Concepts as Structured Meanings

Marie Duží and Pavel Materna

ABSTRACT

The paper is a summary of the approaches to the concept of structured meaning. Apart from recapitulating various semantic theories such as denotational, model-theoretic and inferentialist ones, we propose procedural semantics as a solution to the well-known problem of substitution within belief sentences. Using Transparent Intensional Logic (TIL), we demonstrate that furnishing expressions with TIL *constructions* as their senses meets the desirable semantic principles of compositionality, universal transparency and a near-match between syntax and semantic structures as encoded by (disambiguated) expressions. As a solution to the problem of the identity of the complements of beliefs we define *concepts* as closed constructions in normal form. Our main claims are that concepts are the structured meanings of expressions and that expressions are synonymous if they express one and the same concept.

1. Introduction

In this paper we argue in favour of construing concepts as structured meanings.[1] The fundamental question of semantics, "What is meaning?" receives many answers. Regardless of how meaning is conceived, assigning meanings to expressions ought to be governed by various constraints. Key constraints ought to include compositionality, referential transparency in all sorts of context (i.e. thoroughgoing anti-contextualism), and a near-match

---

[1] A version of this paper was read at the workshop *Structured Meanings and Concepts* as part of *ECAP 6*, August 21–26, 2008, Kraków.

between the syntax of (disambiguated) linguistic items and the semantic structures they encode. Thus meanings should be structured almost as finely as the syntactic structures they match.

The prevailing denotational approach to semantics conceives of the meaning of an expression E as the entity beyond language denoted (or referred to) by E. This approach has many advocates as well as opponents. Since the pioneering paper Frege (1892b) the advocates of denotational semantics like Carnap, Montague, and Cresswell have striven to define so-called structured meanings that would comply with the principles of compositionality and universal transparency. Various adjustments of Frege's semantic schema have been proposed, shifting the entity named by an expression from the level of atomic (physical/abstract) objects to the intensional level of abstract objects such as sets or functions/mappings. Yet it seems that natural language is rich enough to generate expressions so complicated that either compositionality or universal transparency (anti-contextualism) have to be abandoned. Frege, for example, adhered to compositionality and got stuck with contextualism.

Beginning with Quine, the opponents of the denotational approach have provided a different answer to the question what meaning is. Since natural language is so rich that, apparently, there are not enough entities beyond language apt to serve as meanings, there is no such thing as the meaning of an expression. Instead, the expression has the property of being *used* by some community in a certain way. This is the *pragmatic turn* propagated by, e.g., the later Wittgenstein or Brandom, Peregrin and other inferentialists.

Our position is a plea for a *realist procedural semantics*, which is at variance with set-theoretic semantics such as model theory and pragmatic semantics such as inferentialism. Language expressions represent, or rather encode, their structured meanings, which are abstract entities dwelling in a Platonic realm. The subject-matter of logical, *a priori* semantics is to study these entities independently of their encoding in a particular language. But the abstract entities that are assigned to expressions as their meanings are neither extensional atomic objects nor intensional set-theoretical mappings. Rather, they are hyperintensionally individuated, algorithmically structured *procedures* producing extensional/intensional entities (or lower-order procedures) as their products. This approach—which could be characterized as being informed by an *algorithmic* or *procedural* turn—has been advocated by, for instance, Moschovakis (1994). Yet much earlier, in the early 1970s, Tichý introduced his notion of *construction* and developed the system of Transparent Intensional Logic (TIL) (see Tichý, 1988, 2004). We base our procedural semantics on TIL, arguing for a robust concept of semantic structure as an extra-linguistic, abstract *procedure* (a generalized *algorithm* in the shape of a TIL construction), because procedures are inherently structured. They consist of one or multiple steps that have to be executed in order to arrive at the product (if any) produced by the respective procedure.

Yet to anticipate a common misapprehension, we wish to emphasize that procedures are not syntactic objects. Just like one and the same algorithm can be encoded by different programs possibly written in different programming languages, so one and the same procedure can be encoded by different pieces of syntactic items belonging to different languages. Moreover, the linguists' annotated trees depicting syntactic structure, though being illustrative diagrams, fall short of capturing semantic structure. These syntactic trees are set-theoretical, ordered n-tuples, and set-theoretic objects are not suitable candidates as structured meanings. Though having elements, an *instruction* to glue them together into a coherent structure is missing. Thus the procedural turn is not a syntactic turn.

The paper is organised as follows. Section 2 is a brief critical summary of particular approaches to logical semantics. In Section 3 we introduce the system of Transparent Intensional Logic and demonstrate its power by offering a solution to the problem of belief sentences. Section 4 deals with the category of concept. Section 5 sums up our approach and outlines further research.

2. Theories of Meaning

2.1. Frege's Schema

Frege's well-known semantic diagram is frequently accepted as one of the foundations of modern semantics. To explain why a true sentence of the form "$a = b$" can be informative, unlike a sentence of the form "$a = a$", Frege introduced an entity standing between an expression and the object denoted (*bezeichnet*) by the expression. He named this entity *Sinn* (sense) and explained the informative character of the true sentences of the form "$a = b$" by saying that 'a' and 'b' denote one and the same object but differ in expressing (*ausdrücken*) distinct senses. Thus Frege's semantic diagram can be visualized as in Figure 1.

There has been much dispute over Frege's schema. No wonder; Frege tackled a serious semantic problem and proposed a solution to it. Yet the proposal is controversial and not definite enough. Many objections have been raised against the schema, among which perhaps the most pressing are the infinite ambiguity argument (see Burge, 2005, pp. 198–204) and the circularity argument (see Tichý, 1986, pp. 252–256; Tichý, 2004, pp. 650–654), both directed against Frege's contextualism.

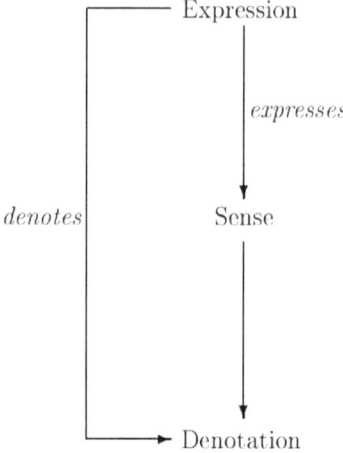

Figure 1: Frege's semantic diagram

On a rational reconstruction, Frege attempts first to apply universal transparency to his puzzles like 'The Morning Star' vs. 'The Evening Star', or 'The author of *Waverley*' vs. 'The author of *Ivanhoe*'. An agent *a* can easily believe that the author of *Waverley* is a poet without believing that the author of *Ivanhoe* is a poet, and without *a* being guilty of inconsistency or irrationality. Yet since both 'The author of *Waverley*' and 'The author of *Ivanhoe*' refer to Sir Walter Scott, believing the former without believing the latter does seem to amount to inconsistency and irrationality. Frege wanted to observe compositionality. Thus he opted for contextualism, making the semantics of an expression depend on the linguistic context in which it is embedded. In 'ordinary' extensional contexts 'The author of *Waverley*' and 'The author of *Ivanhoe*' both refer to Sir Walter Scott, but in 'oblique' contexts both descriptions refer to their senses. Compositionality is saved (the expressions possessing distinct senses); the price exacted is contextualism. The price is very high, indeed.

First, no expression can denote an object, unless a particular kind of context is provided. But expressions can embed within other expressions to various degrees. Compare

> Tom believes that the author of *Waverley* is a poet.

with

> Charles knows that Tom believes that the author of *Waverley* is a poet.

The expression 'The author of *Waverley*' should now refer to the 'normal' sense of the 'normal sense' of itself. Adding still further layers of embedding sets off an infinite hierarchy of senses, which is to say that 'The author of *Waverley*' has the potential of being infinitely ambiguous. This seems plain wrong.

Second, the problem with the distinction between two kinds of semantic context is that it is *circular*, as the following dialogue illuminates:

Q: When is a context extensional?
A: A context is extensional if and only if it validates the rules of
    (∗) substitution of co-referential singular terms and
    (∗∗) existential generalisation.
Q: And when are (∗), (∗∗) valid?
A: These rules are valid if and only if all the contexts they are applied to are extensional.

Third, it is insufficient to let 'The F' denote a *Sinn* in an oblique context. If a believes that the F is a G then 'The F' denotes a *Sinn*—but a does not believe that some *Sinn* is a G. The advocates of reference shift need to explain how, in an oblique context, the *Sinn* of a term is to descend to an entity capable of being a G.

The overall problem, though, is that Frege never defined *sense*. All he says is that it is a 'mode of presentation' (*Art des Gegebenseins*) of the denotation. The frequent interpretation of sense in contemporary semantics has it that sense is an *intension*, whereas the extension of an expression is the object or set of objects referred to, pointed to, or indicated by, the expression. Thus the extension of 'The Morning Star' is a certain planet, Venus. The extension of a *predicate* is the set of all objects to which the predicate truly applies. The extension of 'is red' is the set of all red things, etc. 'Intension' can be interpreted in various ways. In Montague or Carnap it is *the intension of* an expression. Thus the intension of a predicate is a property; the intension of a sentence is a proposition.

Carnap's method of intensions and extensions is a well-elaborated semantic theory based on Frege's schema (see Carnap, 1947). According to Carnap, an expression possesses an intension and an extension, the former corresponding to Frege's *Sinn*, the latter to Frege's *Bedeutung*.[2] Yet Carnap was aware of the problems connected with Frege's contextualism and rejects reference shift as well as what he calls Frege's 'naming theory'. In his view only the intension of an expression is semantically salient while its extension is a factual matter. Thus he defined the L-equivalence of expressions

---

[2] Church in (1954) has 'denotation'.

sharing the same intension as a stronger relation of semantic individuation than the merely factual equivalence of expressions sharing the same extension. He also defined rules of substitution *salva veritate* in what he called intensional and extensional contexts. But Carnap's method failed in the context of 'belief sentences'. The reason is this. Intensional semantics cannot distinguish between L-equivalent expressions which denote the same intension yet fail to be synonymous. Thus Carnap characterised the character of 'belief sentences' as being neither extensional nor intensional, and eventually admitted that his method failed to provide a satisfactory solution to the compositionality problem in belief sentences.

2.2. Truth-Conditional and Other Set-Theoretical Theories

Contemporary possible-world semantics (*PWS*) takes intensions to be functions whose domain is made up of possible worlds. On this approach the sense of a sentence consists in articulating particular truth-conditions. A widespread explication of *truth-condition* is as a *PWS proposition*. In general, the sense of an expression is a *PWS* intension (see, for instance, Kirkham, 1992/1997, p. 4 but also Montague, 1974). Since intensions are functions/mappings the identification of meanings with *PWS* intensions amounts to a model theory within the set-theoretical paradigm.

At the outset of this paper we mentioned the tenet that *meanings are structured*. However, *PWS* intensions, being sets, are simple objects. Thus *PWS* is a coarse-grained theory, and there are many problems connected with this conception. In what follows we list some of them.

a) Frege's first example where he encountered the problem of "$a = b$" being informative, unlike "$a = a$", concerned the medians of a triangle. The problem is this. The fact that the three medians of a triangle intersect at one point (the centroid of the triangle) is informative, unlike the fact that this point is self-identical. However, mathematical facts and functions are independent of possible worlds. Thus possible worlds are out of place in mathematics and the meanings of mathematical expressions cannot be intensions.

b) Empirical expressions may possess distinct meanings though being L-equivalent in Carnap's sense by denoting (having) the same *PWS* intension. For example, the sentences
"The Moon is smaller than the Earth" and
"The Earth is greater than the Moon"
have exactly the same truth-condition and thus denote the same *PWS* proposition. Moreover, all analytically true sentences denote the same proposition *TRUE*, taking the value True in all possible worlds. For instance, "Bachelors are men" and "Whales are mammals" certainly

do not have the same meaning. Yet they both have the same intension, *viz.* the proposition *TRUE*.

c) As the problem of belief sentences illustrates, some contexts are sensitive to the syntactic structure of an expression. Thus the structure of the meaning assigned to an expression should correspond to the syntactic structure of the expression. Intensions are simply sets (mappings) and thus do not posses any structure, let alone a sophisticated structure corresponding to the structure of the expression.

d) If meaning were an intension then understanding an empirical expression would amount to knowing the respective (uncountable) function with domain in possible worlds. Thus understanding a simple sentence like "The Moon is smaller than the Earth" would amount to knowing the list of all possible worlds together with the respective distributions of truth-values: something no human could possibly accomplish.

e) Denotation should be unambiguously determined by meaning. The interpretation of sense as *intension* and *denotation* as *extension* in the case of empirical expressions (like 'The Morning Star', 'The Evening Star') does not meet this requirement. Already Carnap, in (1947), realized that a logical analysis cannot provide the contingent values of intensions. If intensions are functions from possible worlds (and times, as in TIL) then we could logically determine the value of an intension in the actual world simply by knowing which of the possible worlds is the actual one. On any rational explication of the notion of possible world, this knowledge cannot be *a priori*; therefore, determining the value of an intension in the actual world must always be a matter of factual experience (rather than of logic).

At least since Frege's days logicians have striven to avail themselves of fine-grained and structured meanings. Bertrand Russell, to whom many theorists attribute the idea of *structured proposition*, held various views about the nature of propositions. The idea of structured proposition was defended by Russell in (1903). His structured propositions, unlike sets, consist of *parts*, and thus are more finely individuated than *PWS* propositions. Russell differed with Frege both over what the constituents of structured propositions are and what binds them together so as to form a proposition. For Frege, all propositional constituents are senses. For Russell, physical objects like a man or a mountain can be constituents of propositions, alongside universals. This view is due to Russell's theory of acquaintance. For instance, Frege claimed that the sense of the name 'Mont Blanc', and not Mont Blanc itself "with all its snowfields", occurs in the proposition/Thought that Mont Blanc is 4,000 metres high. Russell rejoined that in spite of all its snowfields Mont Blanc is itself a component part of what is actually asserted in the proposition expressed by, "Mont Blanc is more than 4,000 metres high" (for details see, e.g., King, 2001). This leads to consequences that do not tally with our intuitive use of the term 'proposition'; for instance, that propositions must be mind-friendly. Thus we would agree with Frege against Russell over whether Mont Blanc can be in any sensible way part of anything deserving the name 'proposition'.[3] Language-users understand many sentences without being acquainted with the concrete objects that the sentences talk about by means of abstract objects.

Russell distinguishes between two kinds of propositional constituents, things and *concepts*. Things are indicated by proper names, concepts by all other words. Russell's view on what binds the constituents of a proposition together is not clear. He says that the 'propositional contributions of verbs' hold the constituents together, but he does not say how it is that, for instance, the relation expressed by the verb 'to love' binds *a*, *b* together so as to form the proposition that *a loves b*. Russell pointed out that "every proposition has a unity which renders it distinct from the sum of its constituents" (Russell, 1903, p. 52). We agree, of course, that the mereological sum of *a, love, b* is not the proposition that *a loves b*. Russell suggested that the unity of a proposition is given by the fact that particular constituents stand in the relation expressed by the verb constituent. This seems, *prima* facie, to be a plausible idea. Yet it cannot be correct, because the constituents' standing in a particular relation is what makes a proposition *true*, not the proposition itself. If it were the proposition, then there would be *no false* propositions.[4] For instance, if *a* is in the relation of love with *b* then it is true that *a* loves *b*. But if *a* is *not* in the relation of love with *b*, we could not say that the proposition that *a loves b* is false, because there would be no such proposition.

Another attempt at a more fine-grained individuation of the objects of belief was Carnap's definition of *intensional isomorphism*. Since L-equivalent expressions sharing the same intension do not have to be synonymous, Carnap proposed intensional isomorphism as a criterion of the identity of belief. Roughly, two expressions are intensionally isomorphic if they are composed from expressions with the same intensions *in the same way*.

---

[3] Yet the term 'structured proposition' has been frequently used to refer to a semantic structure underlying a sentence. Thus in some theories of structured meanings 'structured propositions' stand in the place of *PWS* propositions as entities explicating the sense of a sentence.

[4] Cf. Russell's problem with negative facts.

However, Church in (1954) launched a counterexample involving two intensionally isomorphic sentences, one of which can be easily believed and the other not. The problem he tackled stems from Carnap's principle of tolerance (which itself is plausible). We are free to introduce into a language *syntactically simple* expressions which denote the same intension in *different ways* and thus are not synonymous. Yet they are intensionally isomorphic according to Carnap's definition. Church used as an example of such expressions two predicates P and Q, defined as follows: $P(n) = n < 3$, $Q(n) = \exists xyz(x^n + y^n = z^n)$, where x,y,z,n are positive integers. P and Q are necessarily equivalent, because for all n it holds that $P(n)$ if and only if $Q(n)$. For this reason P and Q are intensionally isomorphic, and so are the expressions "$\exists n(Q(n) \land \neg P(n))$" and "$\exists n(P(n) \land \neg P(n))$". Still one can easily believe that $\exists n(Q(n) \land \neg P(n))$ without believing $\exists n(P(n) \land \neg P(n))$. Criticism of Carnap's intensional isomorphism can be also found in Tichý (1988, pp. 8–9), where Tichý points out that the notion of intensional isomorphism is too dependent on the particular choice of notation.

Many other attempts at structured meanings can be encountered in the seventies and eighties. Lewis (1972), for instance, can be seen as an attempt to model structured meanings *via* trees (an idea shared with Bealer (1982)). The idea of structured meaning was propagated also by Cresswell in (1975) and (1985), in which meaning is defined as an ordered *n*-tuple. For instance, if M('+'), M('7'), M('5') are the meanings of '+', '7' and '5', respectively, Cresswell would construe the meaning of '7+5 = 12' as the triple

$$\langle \text{M('+')}, \text{M('7')}, \text{M('5')} \rangle.$$

That this is far from being a satisfactory solution is shown in Tichý (1994, 2004), Jespersen (2003). In brief, these tuples are set-theoretic entities. A tuple cannot be true or be known, hoped, etc., to be true, so it fails both as a truth-bearer and as the second relatum of an attitude relation. The above tuple is 'flat' from the procedural or algorithmic point of view. The *way* of combining particular parts together is missing here; i.e., the instruction to *apply* the function *plus* to a particular argument. And it is to no avail to add the operation of application to a tuple to somehow create propositional unity, since the operation would merely be an element alongside other elements. Sets, unlike procedures, are algorithmically simple, have no 'input/output gaps', and are flat mappings (characteristic functions). Similar attempts to model structured meanings as tuples can be found also in Kaplan (1989) and Soames (1987).[5]

---

[5] See also King (1995) and (2001) for critical comments on the tuple theory of sentential meaning.

2.3. The Pragmatic Turn

Quine, a contemporary of Carnap and his great opponent (as well as a friend), denies that there is anything such as the meaning of an expression: talk about 'meaning' must be understood metaphorically. Quine's famous slogan, "Don't ask for meaning, ask for using" concisely characterizes his philosophy of language. An expression does not have a meaning proper. Instead, it has a property of being *used* by some community in a certain way driven by some linguistic convention.[6] Thus a language does not serve as a system of 'names' representing entities beyond language. Quine's followers, the later Wittgenstein, Brandom, Peregrin and other inferentialists then study the rules of *language games* and the role particular expressions play in these games. The subject-matter of semantics is according to them the study of the *inferential structure* of language as a *social institution*.

Inferentialism raises a batch of questions deserving and demanding to be answered, such as, Which rules are correct? Can the correctness of the rules be recognized independently of meaning? Can the *a priori* character of logical rules (and of meaning) be explained by inter-subjective agreement? Is this intersubjective agreement not itself in need of explanation? Is so-called normative necessity really explained in terms of such pragmatic terms as 'commitment' or 'entitlement'? And shouldn't we ask whether there are objective and rigorous mathematical reasons for intersubjective agreement?[7]

We agree with the inferentialists that semantic structures and the allied principle of compositionality are central. And we also agree that the realm of *set-theoretical* entities is not rich enough to provide a plausible explication of meaning. The need for *structured, hyperintensional* entities serving as *meaning* explicatum has been spelled out by Carnap, Cresswell, Russell, and others. Our major objection to the pragmatic turn is, however, this. We disagree over whether the semantic role of expressions is exhausted by their role in a social language game. It is certainly a legitimate and interesting question to ask why and how it is that certain sequences of signs become meaningful in the language game played by a community. Just that this is not the subject-matter of an *a priori*, logical semantics. Thus Quine actually changed the subject under scrutiny: from logical *a priori* semantics to empirical investigation of language use.

---

[6] See, for instance, Quine (1953/1963).
[7] See, for instance, Materna (2007) for a critical survey of Quine's approach.

2.4. The Procedural Turn

Now we are going to show that there are entities beyond language that are suitable candidates as meanings. These structured entities are abstract *procedures*. Our procedural semantics has expressions refer to, or rather encode, procedures, which are higher-order structured objects. The parts of a procedure have to be other procedures and cannot be the objects themselves, though the procedure may lead up to an object as its final output. A procedure (including any procedure figuring as a constituent subprocedure) is a presentation of an object rather than a presented object. When knowing a procedure we need not know its output before actually executing it. We need to be acquainted with the procedure first before being able to execute it so as to arrive at the result. And some procedures may even fail to provide an output. A procedure is a different object than its product (if any), which is why exhaustive knowledge of the procedure does not include knowledge of its product. One thing is to know what to do (to know the instruction, i.e. procedure), quite another thing is to actually execute the procedure, and yet another thing is to know and understand what sort of object, if any, is the output.

Another proponent of procedural semantics is Yannis Moschovakis (1994; 2006). In Moschovakis (1994) the meaning of a term *A* is "an (abstract, idealized, not necessarily implementable) algorithm which computes the denotation of *A*". (Moschovakis, 2006, p. 27; see also 1994). The later version (2006) works with a formal language that extends the typed λ-calculus and so can accommodate, *per* Montague, reasonably large fragments of natural language. Moschovakis outlines his conception thus:

> ... when we view natural language with a programmer's eye, it seems almost obvious that we can represent the meaning of a term *A* by the algorithm which is expressed by *A* and which computes its denotation. (Moschovakis, 2006, p. 42)

Tichý began developing his theory as early as in the sixties. In 1968 (see 2004) he spelled out the idea that meaning can be viewed as a kind of a generalized *algorithm* or *algorithmic computation*.[8] He argued for the conception of language as a *code*. If expressions *encode* their meanings due to a linguistic convention, meaning itself must be an extra-linguistic entity defined independently of an expression. Tichý opted for procedures known

---

[8] In (1968) and (1969) Tichý explained his idea in terms of Turing machines. Later, especially in (1988), as TIL matured, his notion of *construction* became a general notion, defined independently of Turing machines.

nowadays as *constructions* and defined six kinds of them. We are going to reproduce these definitions in the next section. Below we summarise the semantic schema of TIL.

In sum, we explicate the sense of an expression as a TIL *construction* which is an abstract procedure. Our vision of semantics is informed by a top-down approach, descending from procedures to their products, which may be lower-order constructions or non-constructions such as possible-world intensions and extensions (individuals, truth-values, sets, etc). A Fregean sense differs from expressions and surely from denotations. It should explain the connection between an expression and the object denoted. A TIL construction is such an entity. Thus our neo-Fregean semantic schema is the adjusted version of Frege's diagram as visualized in Figure 2.

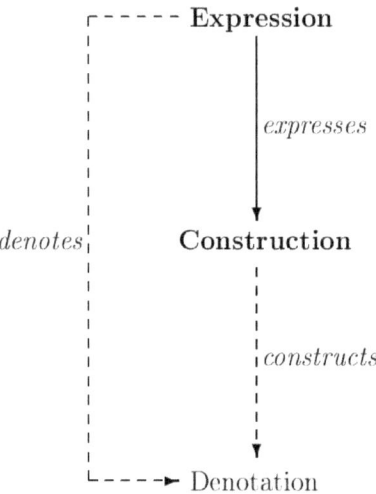

Figure 2: Semantic schema of TIL

The most important interconnection in this schema is between an expression and what is expressed by it: its meaning, i.e., a construction. Once we exactly define *construction*, we can logically examine it; we can investigate what (if anything) the construction constructs, what is entailed by it, etc. Thus constructions are semantically primary, denotations secondary. Once a construction is explicitly given, the entity (if any) it constructs is already implicitly given. When logically analysing an expression, we aim at discovering the construction encoded by the expression. As a limiting case, the logical analysis may reveal that the construction fails to construct anything; we will say that it is *improper*.

Thus constructions are invariably (and context-invariantly) assigned to empirical, analytical and mathematical expressions as their structured meanings. In the case of mathematical expressions there may be no denotation, yet there will always be a sense, *viz.* a construction. In such a case the construction is a 'blind alley' or a 'road to nowhere' failing to produce anything, like the sense of 'the greatest prime number'. Mathematicians had to first understand what to seek, only then could they prove that the construction encoded by 'the greatest prime number' does not construct anything, i.e. that there is no greatest prime.

Empirical expressions always have a denotation. It is the *PWS* intension constructed by their meaning, a construction. We shall often use Cresswell's term 'hyperintension' when talking about the construction of an intension. One advantage of the semantic schema of TIL immediately stands out: we can distinguish between the *reference* and the *denotation* of an empirical expression. A term's denotation is the *PWS* intension unambiguously singled out by the respective construction. Its reference is the value of this intension in the actual world at the present time. It might be tempting to say that while constructions are primary and denotations secondary semantic notions, the references of empirical terms and expressions were tertiary. But they are not. The relation of denotation is intra-semantic, but the relation of reference extra-semantic. Given a denotation, logical analysis cannot tease out its reference. So there is no room for reference in our semantic schema.

3. The Procedural Semantics of TIL

All the entities TIL works with, including constructions, receive a type. The ontology of TIL is organized in an infinite, bi-dimensional hierarchy of types. One dimension is made up of non-constructions, i.e., entities unstructured from the algorithmic point of view. TIL adopts the functional approach, because functions are 'procedure-friendly': a function can be applied to its argument or declared by the procedure of abstracting over its arguments. *Yet a function is not a procedure.* We view functions as set-theoretical mappings, and one and the same mapping can be produced by infinitely many procedures.

The other dimension of the type hierarchy is made up of structured, higher-order constructions which construct lower-order entities. Thus our definitions are inductive, and they proceed in three stages. First, we define the simple types of order 1 comprising non-constructions. Then we define constructions and, finally, the ramified hierarchy of types.

3.1. Simple Types

Definition 1 (Types of order 1)
Let B be a base, i.e., a collection of non-empty sets.
  i) Every member of B is a *type of order 1*.
  ii) Let $\alpha, \beta_1, \ldots, \beta_m$ be *types of order 1*. Then the set $(\alpha\beta_1\ldots\beta_m)$ of partial functions with values in α and arguments in $\beta_1, \ldots, \beta_m$, respectively, is a *type of order 1*.
  iii) *Nothing is a type of order 1* unless it follows from i) and ii).

The types ad ii) are *functional* types. They are sets of *partial functions*, i.e., functions that associate every m-tuple of arguments with *at most* one value. Thus total functions are a special kind of partial functions.

TIL is an open-ended system and the choice of base depends on the area and language we happen to be investigating. When investigating purely mathematical language, the base can consist of, e.g., two atomic types: o, the type of truth-values, and v, the type of natural numbers.

When analyzing an ordinary natural language, we usually use the *epistemic base* $\{o, \iota, \tau, \omega\}$. It is the collection of four atomic types:
$o = \{\mathbf{T}, \mathbf{F}\}$, the set of *truth-values*,
$\iota =$ the universe of discourse (members: *individuals*),
$\tau =$ the set of *real numbers* (or of *time moments*),
$\omega =$ the logical space, the set of *possible worlds*.

Since *function* is a primitive notion, we model *sets* and *relations* by their characteristic functions. Thus, for example, the set of prime numbers is a function of type $(o\tau)$ that associates any number with $\mathbf{T}$ or $\mathbf{F}$ according as the given number is a prime. The binary relation $>$ defined on numbers is a function of type $(o\tau\tau)$ that associates any couple of numbers with $\mathbf{T}$ or $\mathbf{F}$ according as the first number is greater than the second. The type of binary arithmetic functions is $(\tau\tau\tau)$. The set of such functions is an object of type $(o(\tau\tau\tau))$.

Intensions are entities of type $(\beta\omega)$: mappings from possible worlds to some type β. The type β is frequently the type of the *chronology* of α-objects, i.e., mapping of type $(\alpha\tau)$. Thus α-intensions are frequently functions of type $((\alpha\tau)\omega)$, abbreviated as '$\alpha_{\tau\omega}$'. Extensions are entities of a type α where $\alpha \neq (\beta\omega)$ for any type β.

Examples of frequently used intensions are:
Propositions (denoted by sentences) are of type $o_{\tau\omega}$; *properties of individuals* (usually denoted by nouns or intransitive verbs like 'is a student', 'walks') are of type $(o\iota)_{\tau\omega}$; binary *relations-in-intension* between individuals are of type $(o\iota\iota)_{\tau\omega}$; *individual offices/roles* (cf. Church's individual

concepts, usually denoted either by superlatives like 'the highest mountain' or terms with built-in uniqueness, like 'The President of the USA') are of type $\iota_{\tau\omega}$.

3.2. Constructions

Constructions are procedures, or instructions, specifying how to arrive at less-structured entities. For instance, the sense of an empirical sentence is an instruction on how to evaluate its truth-conditions in any possible world at any time. *Qua* procedures, constructions are algorithmically structured, unlike set-theoretical objects, which are devoid of structure. *Qua* abstract, extra-linguistic entities, constructions are accessible only *via* a verbal definition. The 'language of constructions' is a modified hyperintensional version of the typed λ-calculus, where Montague-like λ-terms denote, not the functions constructed, but the constructions themselves. Constructions *qua* procedures specify *partial functions* by assigning to input objects (of any type, even constructions of any order) output objects (or, in well-defined cases, fail to assign).

When claiming that constructions are algorithmically structured, we mean the following. A construction C consists of one or more particular steps, or *constituents*, that are to be individually executed in order to execute C. The objects a construction operates on are not constituents of the construction. Just like the constituents of a computer program are its sub-programs, so the constituents of a construction are again constructions. Thus on the lowest level of non-constructions, the objects that constructions work on have to be supplied by other (albeit trivial) constructions. The constructions themselves may occur not only as constituents to be executed, but also as objects that still other constructions operate on. Therefore, one should not conflate *using* constructions as constituents of compound constructions and *mentioning* constructions that enter as input/output objects into compound constructions. So we must strictly distinguish between *using* constructions as constituents and *mentioning* constructions as objects. Mentioning is, in principle, achieved by *using* atomic constructions. A construction C is atomic if it does not contain any other construction as a used subconstruction (a 'constituent of C') but C. There are two atomic constructions that supply entities (of any type) on which compound constructions operate: *Variables* and *Trivializations*.

Variables are constructions that construct an object dependently on *valuation*: they *v-construct*, where *v* is the parameter of valuation. With the important difference that we construe variables as extra-linguistic objects and not as expressions, our theory of variables is identical to Tarski's. Thus, in TIL variables construct objects of the respective types dependently on valuation in the following (Tarskian) way. For each type α there are countably infinitely many variables $x_1, x_2, \ldots$. The members of α (unless α is a singleton) can be organised in infinitely many infinite sequences. Let the sequences be given (as one is allowed to assume in realist semantics). The valuation v takes a sequence $\langle s_1, s_2, \ldots \rangle$ and assigns s_1 to the variable x_1, s_2 to the variable x_2; and so on.

When X is an object of any type (including a construction), the Trivialization of X, denoted '^{0}X', constructs X without the mediation of any other constructions. ^{0}X is the unique atomic construction of X that does not depend on valuation: it is a primitive, non-perspectival mode of presentation of X.

The other constructions are *compound*, as they consist of other constituents than themselves. They are *Composition*, *Closure*, *Execution* and *Double Execution*. Composition is the instruction to apply a function f to an argument A to obtain the value (if any) of f at A. Closure is the procedure of constructing a function by abstracting over variables in the usual manner of the λ-calculi. Finally, higher-order constructions can be used once or twice over as constituents of compound constructions. This is achieved by the constructions *Execution* and *Double Execution*.

Definition 2 (Constructions)

i) *Variables $x, y, z, p, q, \ldots$ are **constructions**.* They construct objects dependently on valuation, so they v-construct, where v is the parameter of valuations.

ii) Where X is an object whatsoever, the *Trivialization* ^0X is a **construction**. It constructs X without any change.

iii) *Execution* ^1X is a **construction**. It v-constructs what X v-constructs.

iv) *Double execution* ^2X is a **construction**. If X is a construction that v-constructs a construction Y, and Y v-constructs Z, then ^2X v-constructs Z; otherwise ^2X is v-improper, i.e., does not v-construct anything.

v) *Composition* $[XX_1\ldots X_m]$ is a **construction**. If X v-constructs a function f of type $(\alpha\beta_1\ldots\beta_m)$, and X_i, $1 \leq i \leq m$, v-construct objects b_i of types β_i, then the Composition $[XX_1\ldots X_m]$ v-constructs the value (if any) of f on $b_1, \ldots, b_m$. Otherwise the Composition is v-improper.

vi) *Closure* $[\lambda x_1\ldots x_m\, X]$ is a **construction**. If $x_1, \ldots, x_m$ are pairwise distinct variables v-constructing objects of types $\beta_1, \ldots, \beta_m$, respectively, and X v-constructs objects of type α, then the *Closure* $[\lambda x_1\ldots x_m\, X]$ v-constructs a function f of type $(\alpha\beta_1\ldots\beta_m)$ in the following way.

Let v' be a valuation identical with v at least up to assigning objects $b_1,\ldots,b_m$ of types $\beta_1,\ldots,\beta_m$, respectively, to variables $x_1,\ldots,x_m$. If Y is v'-improper (see iv), then f is undefined on $\langle b_1,\ldots,b_m \rangle$. Otherwise the value of f on $\langle b_1,\ldots,b_m \rangle$ is the α-entity v'-constructed by Y.

vii) *Nothing is a **construction** unless it so follows from* i)–vi).

Comments:
Ad i).
Since constructions are extra-linguistic, objective procedures (indeed, this is anathema for Quine *et alii*), variables (*qua* constructions) are not letters. Letters like 'x', 'y', ..., 'p', 'q', ... are chosen just as names of variables, while variables themselves are specific incomplete constructions. They do not construct anything unless being triggered by a valuation. Occurrences of variables can be *free* or *bound*. In TIL there are two kinds of binding: λ-bound variables are bound as in the λ-calculi; ^{0}bound (i.e. Trivialization-bound) variables are bound if they occur as a subconstruction of a construction ^0C. In this case a variable is just mentioned as an object, i.e., is not used as a constituent and not free for valuation.

Ad ii).
Trivialization is just as important as it seems at first blush to be redundant. As mentioned above, Trivialization makes it possible to supply input objects to operate on. A constituent of a hyperproposition is not the physical object Mont Blanc. Rather, it is a construction of Mont Blanc, for instance $^0Mont\_Blanc$, the simplest v-independent construction of the object. Moreover, lower-order objects that higher-order constructions operate on may in turn be constructions supplied by a Trivialization. In this way Trivialization makes it possible to distinguish between using constructions as constituents of compound constructions and mentioning constructions as input/output objects.

Ad iv).
Theoretically, constructions can be executed *n*-times over. Thus we could inductively define ^nX, if need be, to generate, e.g., *Triple* and *Quadruple Execution*.

Ad v).
In the λ-calculi a λ-term corresponding to Composition is usually called 'application'. Yet there is an important difference between the formal language of the λ-calculi and our language of construction. Whereas the *term* 'application' is interpreted as denoting the *result* of applying a function to one of its arguments, i.e., the respective value, the construction Composition is a compound procedure detailing particular steps. The respective term *encodes* this very *procedure*. Thus, for instance, the λ-term '$(+3,2)$'

is usually interpreted as denoting the number 5, but the term '$[^0+\ ^03\ ^02]$' is interpreted as denoting the construction $[^0+\ ^03\ ^02]$, i.e. a procedure consisting of the following steps:

(i) ^{03}: take the number 3;
(ii) ^{02}: take the number 2;
(iii) $^0+$: take the *function addition*;
(iv) $[^0+\ ^03\ ^02]$: apply the result of step (iii) to the objects obtained at (i) and (ii).

The result is the number 5. This bifurcation between a procedure and its product is a cornerstone of TIL.

Ad vi).

Similarly the λ-term in the λ-calculi corresponding to Closure is 'λ-abstraction'. Again, while the term is interpreted as denoting the resulting *function*, the construction Closure is the compound *procedure* whose product is the function. For example, the term '$\lambda x(x+1)$' is usually interpreted as denoting the *function* Successor, while the construction $\lambda x[^0+\ x\ ^01]$, where $x \to v$, the type of natural numbers, is the compound *procedure* whose *product* is the Successor function.[9]

Notation and abbreviations.
'X/α' means that the object X is (a member) of type α;
'$X \to_v \alpha$' means that the type of the object (v)-constructed by X is α.
We will standardly use the variables $w \to_v \omega$ and $t \to_v \tau$.
If $C \to_v \alpha_{\tau\omega}$, the frequently used Composition $[[Cw]t]$, the intensional descent of the α-intension v-constructed by C, will be written as 'C_{wt}'.

Examples.
Positive_number(s)/$(o\tau)$; $x,y \to_v \tau$; $[^0: xy] \to_v \tau$.
Higher_than/$(o\iota\iota)_{\tau\omega}$; Zugspitze/$\iota$; $x \to \iota$;
$\lambda w \lambda t\ \lambda x[^0Higher\_than_{wt}\ x\ ^0Zugspitze] \to (o\iota)_{\tau\omega}$.

However, as mentioned above, constructions themselves are objects and thus also receive a type. Only it cannot be a type of order 1, because a construction cannot be of the same type as the object it constructs.

Thus, for instance, the above Closure $\lambda w \lambda t\ \lambda x[^0Higher\_than_{wt}\ x\ ^0Zugspitze]$ constructs a property of individuals, which is an object of type $(o\iota)_{\tau\omega}$, of order 1. However, this Closure itself is an object belonging to a type of order 2. The Trivialization of this Closure, i.e., $^0[\lambda w \lambda t\ \lambda x\ [^0Higher\_than_{wt}\ x\ ^0Zugspitze]]$, constructs an object belonging to a type of

---

[9] Thus obviously constructions are *not* reducible to set-theoretic entities.

order 2 (*viz.* the Closure). Thus the Trivialization of this Closure is an object of a type of order 3. In general, constructions that construct entities of order 1 are constructions of order 1. They belong to a type of order 2, denoted '$*_1$'. The type $*_1$ serves as a base for the induction rule: any collection of partial functions, of type $(\alpha\beta_1\ldots\beta_n)$, involving $*_1$ in their domain or range is a *type of order 2*. Constructions belonging to a type $*_2$, constructing entities of order 1 or 2, and partial functions involving such constructions, belong to a *type of order 3*; and so on *ad infinitum*.

The definition of the ramified hierarchy of types decomposes into three parts. First, simple types of order 1 were already defined by Definition 1. Second, we define constructions of order n, and third, types of order $n+1$.

Definition 3 (Ramified hierarchy of types)

(T1) *Types of order* 1. See Definition 1.
(T2) *Constructions of order n.*
   i) Let α be a type of order n. If x is a variable then if $x \to_v \alpha$, then x is a *construction of order n*.
   ii) Let X be a member of a type of order n. Then ^0X, ^1X and ^2X are *constructions of order n*.
   iii) Let $X, X_1, \ldots, X_m$ be *constructions of order n*. Then $[XX_1\ldots X_m]$ is a *construction of order n*.
   iv) Let $x_1, \ldots, x_m, X$ be *constructions of order n*. Then $[\lambda x_1\ldots x_m\, X]$ is a *construction of order n*.
   v) Nothing is a *construction of order n* unless it so follows from i)–iv).
(T3) *Types of order $n+1$.* Let $*_n$ be the collection of all constructions of order n.
   i) $*_n$ and every type of order n are *types of order $n+1$*.
   ii) If $\alpha, \beta_1, \ldots, \beta_m$ are *types of order $n+1$*, then the collection $(\alpha\beta_1\ldots\beta_m)$ of partial functions from $\beta_1, \ldots, \beta_m$ into α is a *type of order $n+1$* as well.
   iii) Nothing is a *type of order $n+1$* unless it so follows from i) and ii).

Remark. As is common in type theories, entry (T3)-(i) is the application of type raising in order to make the type order of the functional type $(\alpha\beta_1\ldots\beta_m)$ homogeneous.

We furnish expressions with TIL constructions in all contexts, and constructions *qua* procedures are (algorithmically) structured. Their structure is near-identical to the syntactic structure of the respective expressions. However, as mentioned above, constructions are not syntactic structures. They are objectual procedures consisting of sub-procedures. Thus Russell's question, "What binds the constituents of propositions together?" is answered (see also King, 1995, 2001). It is the very procedure that generates a compound whole from its particular constituents. The meaning of an expression E is not a list of the meanings of the sub-expressions of E. Rather, it is the procedure detailing in what particular ways its atomic sub-procedures, or as the case may be, the objects they produce, are to be combined.

3.3. Method of Analysis

Uncovering the genuine meaning of a given expression, i.e., the construction encoded by the expression, is a non-trivial task. When performing an analysis, we adhere to the constraint on natural-language analysis dictated by Carnap's principle of subject matter and pursue an *admissible* and *literal analysis* of an expression E. Such an analysis terminates in a construction C such that C uses, as its constituents, constructions of just those objects that receive mention in E, i.e., the objects denoted by sub-expressions of E. The principle is central to our general three-step *method of logical analysis* of language:[10]

(i) *Type-theoretical analysis*. Assign types to the objects mentioned, i.e., *only* those that are denoted by sub-expressions of E, and do not omit any semantically self-contained sub-expression of E.

(ii) *Synthesis*. Combine constructions of these objects so as to construct the object D denoted by E.

(iii) *Type checking*. Use the assigned types for *control* so as to check whether the various types are compatible and, furthermore, produce the right type of object in the manner prescribed by the analysis.

As an example we are going to analyse Russell's famous, if not notorious, sentence "The King of France is bald". The sentence talks about the office of King of France, ascribing to the individual (if any) that occupies this office the property of being bald. Strawson's test reveals the presupposition that there is a king of France, i.e., that the office is occupied (see Strawson, 1952, pp. 173ff. and 1964). If not, then the proposition denoted by the sentence has no truth-value.[11] For, if it were false then it would be true that the King

---

[10] For details see Duží, Jespersen, and Materna (forthcoming) or Duží and Materna (2005).

[11] On our approach the sentence is not meaningless. The sentence has a sense, namely the instruction of how to evaluate, for any possible world w and any time t, its truth-condition. It

of France is not bald, which is not true either. This fact has to be brought out by our analysis. Here is how.

Ad (i) *King_of*/$(\iota\iota)_{\tau\omega}$—an attribute, i.e., a function that dependently on worlds and times associates an individual with at most one individual; *France*/ι; *King_of_France*/$\iota_{\tau\omega}$; *Bald*/$(o\iota)_{\tau\omega}$.

Ad (ii) Now we have to combine the *constructions* of the objects *ad* (i) in order to construct the proposition, of type $o_{\tau\omega}$, denoted by the whole sentence. The simplest constructions of the above objects would be their Trivialisations: $^0King\_of$, ^0France, ^0Bald. The attribute *King_of* has to be extensionalized first *via* the Composition $^0King\_of_{wt}$, and then applied to France. This yields

$$[^0King\_of_{wt}\,^0France] \to_v \iota$$

By abstracting over w, t we obtain the office of King of France:

$$\lambda w \lambda t\,[^0King\_of_{wt}\,^0France] \to \iota_{\tau\omega}$$

But the property of being bald cannot be ascribed to an individual office. Rather, it is ascribed to the individual occupying the office. Thus the office has to be subjected to intensional descent, i.e. extensionalized, to v-construct the individual (if any) occupying the office:

$$\lambda w \lambda t\,[^0King\_of_{wt}\,^0France]_{wt} \to_v \iota$$

The property itself has to be extensionalized as well: $^0Bald_{wt}$. Composing these two constructions yields a truth-value **T**, **F**, or nothing, depending on whether the office of King of France is occupied and, if so, whether its incumbent is bald:[12]

$$[^0Bald_{wt}\,\lambda w \lambda t\,[^0King\_of_{wt}\,^0France]_{wt}] \to_v o$$

Finally, abstraction over w, t constructs the proposition:

$$\lambda w \lambda t\,[^0Bald_{wt}\,\lambda w \lambda t\,[^0King\_of_{wt}\,^0France]_{wt}] \to o_{\tau\omega}$$

This Closure can be equivalently simplified to

$$\lambda w \lambda t\,[^0Bald_{wt}\,[^0King\_of_{wt}\,^0France]]$$

---

is just that if we evaluate this instruction in such a state-of-affairs where the King of France does not exist, i.e., the office of the King is vacant, the intensional descent of the office fails to pick up an individual of which baldness may be predicated. Thus the process of evaluation yields a truth-value gap.

[12] For further details on how to predicate properties, see Jespersen (2008).

Gloss: In *any* world (λw) at *any* time (λt) do this: First, find out who is the King of France by applying the extensionalized attribute *King_of* to *France* ($[{}^0King\_of_{wt}{}^0France]$). If there is none, then terminate with a truth-value gap, because the Composition $[{}^0King\_of_{wt}{}^0France]$ is *v*-improper. Otherwise, output **T** or **F** according whether the so obtained individual has the property of being bald ($[{}^0Bald_{wt}[{}^0King\_of_{wt}{}^0France]]$).

Ad (iii). By drawing a type-theoretical tree, we check whether the particular constituents of the above Closure are combined in the correct way (see Duží and Materna, 2005). The tree is depicted in Figure 3:

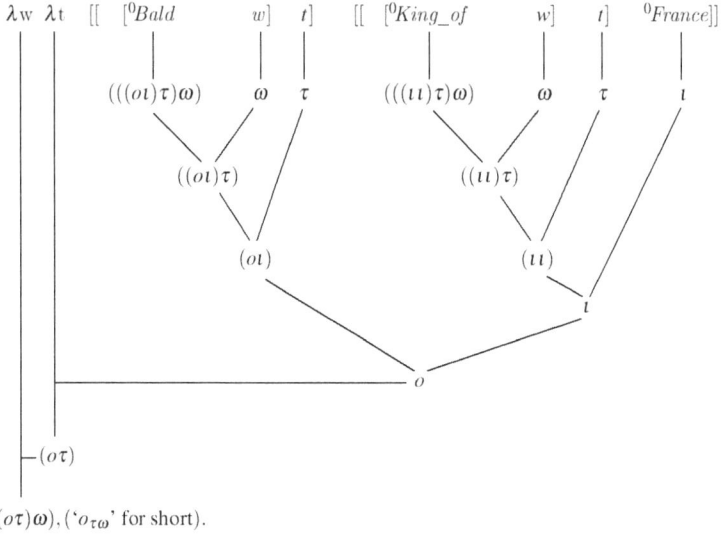

Figure 3: Type-theoretical semantic tree

The following sentence may serve as a simple example of a hyperintensional context in which a construction is merely mentioned as the object of predication:

"Charles calculates $3 + 5$".

Obviously, Charles does not calculate the number 8 denoted by '$3+5$', nor is he related to this particular notation. He is trying to execute the procedure encoded by '$3+5$', i.e., the construction $[{}^0+{}^03\,{}^05]/*_1$. Thus the types of objects that the sentence talks about are:

Charles/ι;[13] $3,5/\tau$; $+/(\tau\tau\tau)$; *Calculate*/$(o\iota*_1)_{\tau\omega}$.

---

[13] If a proper name functions as a non-descriptive label tagging an individual then the sense of the name is a Trivialization of the individual. For details on the semantics of proper names, see Duží et al. (forthcoming, Section 3.2).

The sentence expresses the construction

$$\lambda w \lambda t [^0 Calculate_{wt} {}^0 Charles {}^0[^0+{}^0 3 {}^0 5]].$$

Whoever evaluates the truth-conditions of this sentence need not execute the procedure $[^0+{}^0 3 {}^0 5]$. This is left to Charles to do. The evaluator only checks whether Charles is busy calculating the sum of 3 and 5.

TIL, thanks to its strong typing and procedural semantics, is capable of handling three levels of granularity: the extensional level of truth-functions, the intensional level of modalities and the hyperintensional level of attitudes and linguistic meanings. The sense of a sentence containing empirical expressions is an algorithmically structured *construction* of a proposition denoted by the sentence. The denoted proposition is a flat mapping with domain in possible worlds. Our motive for working top-down has got to do with anti-contextualism: any given unambiguous term or expression (even one involving indexicals or anaphoric pronouns) expresses the same construction as its sense in whatever sort of context the term or expression is embedded within. And the meaning of an expression determines the respective denoted entity (if any), but not *vice versa*.

We strictly distinguish not only between a procedure and its product, but also between a function and its value. A procedure C, i.e. a TIL construction, can be either used (to construct, e.g., a function f) or mentioned. If mentioned, C itself is an object of predication. If used, C can be used either intensionally or extensionally. If the former, the whole function f v-constructed by C is an object of predication. Using Scholastic terminology, we say that the construction C occurs with *de dicto supposition*. If C is used extensionally, the respective value (if any) of the function f v-constructed by C is an object of predication. In this case we say that C occurs with *de re supposition*.

If a sentence S contains indexicals, and S is used in isolation, then its meaning is an incomplete instruction, which is an open construction containing at least one free variable as a constituent. The construction can be completed, i.e. closed, in one of two ways. Either it is completed by a context of utterance providing a valuation of the free variables. Or the meaning of S is completed by a text supplying the meanings of the antecedents referred to by the anaphoric pronouns involved.

Thus, for instance, the sentence

(1)   "He is wise"

has an incomplete meaning. It expresses the open construction

(1') $\lambda w \lambda t [^0 Wise_{wt} he]$.

Types. $Wise/(o\iota)_{\tau\omega}$; $he/*_1 \to \iota$.

When embedded in a linguistic context, for instance in the sentence

(2)   "Charles believes of the Pope that *he* is wise"

the sentence (1) retains its meaning, (1'), yet the whole sentence (2) has a complete meaning, because 'he' now refers to the meaning of 'the Pope'. This meaning, i.e., the construction ^0Pope, occurs with *de re* supposition here to yield an individual. Of course, the office $Pope/\iota_{\tau\omega}$ cannot be wise, rather the individual occupying the office is believed by Charles to be wise. Thus the truth-value of the proposition denoted by the sentence at the given $\langle w,t \rangle$ depends only on the particular individual who occupies the office at this $\langle w,t \rangle$; it is irrelevant who occupies it at worlds/times other than $\langle w,t \rangle$.

On the other hand, in an 'oblique' context (*oratio obliqua*) we do not use the office in this manner, we just mention it, and the truth-value of the proposition is dependent on the occupancy of the role in all worlds at all times. For instance, in the sentence (3) ^0Pope occurs with *de dicto* supposition:

(3)   "Charles believes that the Pope is wise".

The reason is this. Charles is now related to the meaning of the embedded clause, i.e. the construction $\lambda w \lambda t [^0Wise_{wt} \, ^0Pope_{wt}]$. To be sure, wisdom is again predicated of the person occupying the papacy, thus ^0Pope occurs with *de re* supposition in the construction $\lambda w \lambda t [^0Wise_{wt} \, ^0Pope_{wt}]$. Yet the context of Charles' belief is either intensional or hyperintensional. Regardless of who occupies the papacy, if anybody at all, Charles believes the (hyper-)proposition that the Pope is wise. Charles may be ignorant enough to believe the (hyper-)proposition even when the office is vacant. The general rule being invoked here is that a higher context, whether intensional or hyperintensional, is dominant over an extensional one. However, the meaning of 'The Pope', namely the ^0Pope, remains the same. The shift concerns neither the meaning nor the denotation, but only the *supposition* with which the (same) meaning occurs.

Our top-down approach furnishing all expressions with a hyperintensional semantics—i.e., assigning *constructions* (of intensions) to (empirical) expressions as their meanings in all kinds of context—makes it possible to adhere to the compositionality principle. In a word, compositionality is saved without resorting to contextualism. In TIL, there is no such contextual thing as the *intension/extension of an expression*. Instead every expression either denotes an extension or an intension, independently of contextual embedding. What is dependent on context is the *supposition*, which comes in a *de dicto* and *de re* variant.

That the sentences (2) and (3) are logically independent should be obvious now. Whereas (3) can be true even if the Pope does not exist, (2) has *no* truth-value in such a situation, as is readily seen by applying Strawson's test. If the Pope does not exist, (2) is not true. Yet, if it were false, then the sentence "Charles does *not* believe of the Pope that he is wise" would be true. Yet this is not possible, either. Neither believing nor non-believing can be truly predicated of a non-existing Pope. Moreover, in order that (2) be true, Charles must have some 'intimate relation' to the person who happens to be the Pope (though Charles does not need to know of this person that he is the Pope). On the other hand, (3) can be true even when Charles has no relation to the individual, if any, occupying the office.

Yet standard *intensional* analyses are λ-convertible:

(*De re* belief)
$\lambda w \lambda t [\lambda x [^0 Believe_{wt} {}^0Charles\ \lambda w \lambda t [^0 Wise_{wt} x]] {}^0 Pope_{wt}]$
(*De dicto* belief)
$\lambda w \lambda t [^0 Believe_{wt} {}^0 Charles\ \lambda w \lambda t [^0 Wise_{wt} {}^0 Pope_{wt}]]$
Additional types:
$Charles/\iota;\ Believe/(o\iota o_{\tau\omega})_{\tau\omega}$.

Thus, either λ-convertibility does not amount to equivalence, or we need a more fine-grained hyperintensional analysis, or both. In fact, both are true, though for different reasons. First, β-reduction is not an equivalent transformation when partial functions are involved. Even if we rename the variables w, t to forestall collision of variables, the resulting construction is equivalent to neither (*De re* belief) nor (*De dicto* belief); the three construct three distinct propositions:

(*Reduced*)
$\lambda w \lambda t [^0 Believe_{wt} {}^0 Charles\ \lambda w' \lambda t' [^0 Wise_{w't'} {}^0 Pope_{wt}]]$.

The proposition constructed by (*Reduced*) can be true even at a $\langle w,t \rangle$ at which the Pope does not exist, causing $^0 Pope_{wt}$ to be v-improper. At such a $\langle w,t \rangle$ the proposition v-constructed by $\lambda w' \lambda t' [^0 Wise_{w't'} {}^0 Pope_{wt}]$ is a degenerate constant function that does not have a truth-value at any $\langle w',t' \rangle$. Still there is no logical reason why Charles could not be related to such a proposition. Thus (*Reduced*) is *not* equivalent to (*De re* belief). And it is not equivalent to (*de dicto* belief) either, because the propositions believed by Charles are different in such state-of-affairs $\langle w,t \rangle$ where the Pope does not exist. Whereas the proposition constructed by $\lambda w \lambda t [^0 Wise_{wt} {}^0 Pope_{wt}]$ is a non-constant function, the proposition v-constructed by $\lambda w' \lambda t' [^0 Wise_{w't'} {}^0 Pope_{wt}]$ is a degenerate constant function. For this reason (*Reduced*) is *not* an admissible analysis of (2), and *de re* belief is not reducible to *de dicto* belief, or *vice versa*.

So far so good—and the above constructions (*De re* belief) and (*De dicto* belief) could be considered adequate analyses of (2) and (3), respectively. However, these intensional analyses do not solve the problem of the identity of beliefs. For instance, the truth of (3) does not guarantee that (4) is true as well.[14]

(4) "Charles believes that the Pope is wise and no bachelor is married".

Yet the proposition denoted by "The Pope is wise" is the same as the proposition denoted by "The Pope is wise and no bachelor is married", because the proposition that no bachelor is married is the proposition *TRUE* taking the value **T** in all possible worlds at all times. Using Carnap's terminology, the sentence "The Pope is wise" is L-equivalent to "The Pope is wise and no bachelor is married"; but L-equivalence is too weak a criterion to govern the identity of beliefs. Since the two sentences are not synonymous, Charles can rationally assent to the former without assenting to the latter. And since intensional logics can individuate meanings only up to (L-)equivalence, we need a hyperintensional analysis.

Let $Believe^*/(o\iota^*_n)_{\tau\omega}$ be a hyperintensional belief, which is a relation-in-intension between an individual and a hyperproposition (a propositional construction). Then an adequate analysis of (3) is this.

(*De dicto* belief*)
$\lambda w \lambda t [^0Believe^*_{wt} {}^0Charles\, ^0[\lambda w \lambda t[^0Wise_{wt} {}^0Pope_{wt}]]]$.

The undesirable deduction of (4) is now blocked: the construction $\lambda w \lambda t[^0Wise_{wt} {}^0Pope_{wt}]$, i.e., the meaning of the embedded clause, is not identical to the construction $\lambda w \lambda t[[^0Wise_{wt} {}^0Pope_{wt}] \wedge [[^0No\, ^0Bachelor_{wt}] {}^0Married_{wt}]]$.

Additional types: $Bachelor, Married/(o\iota)_{\tau\omega}$; $No/((o(o\iota))(o\iota))$, a restricted quantifier modelled as the function that associates a given set M of individuals with the set of all sets that have an empty intersection with M. Since the population of bachelors v-constructed by $Bachelor_{wt}$ and the population of married people v-constructed by $Married_{wt}$ are disjoint in any world w at any time t, the second conjunct is necessarily true.

In an attempt to analyse the sentence (2) hyperintensionally, we encounter technical complications. Charles now believes *explicitly* of the Pope that *he* (*viz.* the incumbent of the papacy) is wise. If we analyzed the sentence similarly to how we analyze *implicit*, i.e. intensional, belief, we would obtain:

$\lambda w \lambda t [\lambda x [^0Believe_{wt} {}^0Charles\, ^0[\lambda w \lambda t[^0Wise_{wt} x]]]\, ^0Pope_{wt}]$.

---

[14] Some (intensional) epistemic logics nevertheless consider the inference from (3) to (4) valid. Such belief is then called *implicit* belief. The attributer does not reproduce Charles' internal perspective on what he (Charles) believes. Rather, what gets represented is the attributer's external perspective, reporting what Charles *could* consistently believe were he to deduce it.

However, this construction is *not* an adequate analysis of (2). The reason is that the *construction* $[\lambda w \lambda t [^0 Wise_{wt} x]]$ is now mentioned rather than used. Thus the variable x is not accessible to logical manipulation; it is ^{0}bound by Trivialization. Whoever the Pope may be, the result of the Composition $[\lambda x[^0 Believe_{wt} \, ^0 Charles^0 [\lambda w \lambda t [^0 Wise_{wt} x]]]\, ^0 Pope_{wt}]$ is that Charles believes the open construction $[\lambda w \lambda t [^0 Wise_{wt} x]]$. If, for instance, Joseph Ratzinger is the Pope, the Composition is v-congruent to $[^0 Believe_{wt} \, ^0 Charles^0 [\lambda w \lambda t$ $[^0 Wise_{wt} x]]]$rather than to$[^0 Believe_{wt} \, ^0 Charles^0 [\lambda w \lambda t [^0 Wise_{wt} \, ^0 Ratzinger]]]$.[15]

To adequately analyze the sentence, "Charles (explicitly) believes of the Pope that *he* is wise", we must complete the meaning of the embedded clause "*he* is wise", i.e., $[\lambda w \lambda t [^0 Wise_{wt} \, he]]$, *via* the meaning of 'The Pope'. Since this meaning occurs with *de re* supposition, the completion amounts to substituting a construction of the holder of the office of Pope for the variable *he*. If there is no such individual (the office being vacant) the substitution must fail, so that the resulting proposition lacks a truth-value. If the Pope is, for instance, Ratzinger, the resulting hyperproposition believed by Charles must be the construction $[\lambda w \lambda t [^0 Wise_{wt} \, ^0 Ratzinger]]$.

Using terminology known from programming languages, the above λ-transformation from (*De re* belief) to (*Reduced*) is a substitution *by name*. The *de re* occurrence of the construction of the office has been drawn into the intensional context of the embedded clause. In such a case the principle of dominancy of a higher context over a lower one applies: all the constituents occurring in a hyperintensional context occur hyperintensionally, and all the constituents occurring in an intensional context occur intensionally, i.e., *de dicto*. What is wanted here is a substitution *by value*. We have to test whether there is a holder of the Papacy, first of all; if there is, then the Trivialization of the *v*-constructed individual is to be substituted.

To this end we use the functions *Sub* and *Tr*. The function $Sub/(*_n *_n *_n *_n)$ applied to constructions C_1, C_2 and C_3 returns as its value the result of the substitution of C_1 for C_2 in C_3. The function $Tr/(*_1 \iota)$ associates an individual with its Trivialization. Thus, for instance, if $x \rightarrow_v \tau$, the Composition $[^0 Tr \, x]$ $v(2/x)$-constructs $^0 2$, $v(3/x)$-constructs $^0 3$, etc. Note that the variable x is now *free* in this Composition, unlike in $^0 x$ which constructs x independently of valuation.

Similarly, the Composition $[^0 Tr \, ^0 Pope_{wt}]$ $v(W/w, T/t)$-constructs $^0 Wojtyła$ at any world/time pair $\langle W, T \rangle$ at which Karol Wojtyła is the Pope, $v(W'/w, T'/t)$-constructs $^0 Ratzinger$ at a world/time pair $\langle W', T' \rangle$ at which Joseph Ratzinger is the Pope, and it is $v(W''/w, T''/t)$-improper with respect

---

[15] Constructions are *v*-congruent if they *v*-construct one and the same entity.

to $\langle W'', T'' \rangle$ lacking a pope or having more than one. Thus an adequate analysis of (2) is:

(*De re* belief*)
$\lambda w \lambda t [^0Believe^*_{wt}\ ^0Charles\ [^0Sub[^0Tr^0Pope_{wt}]\ ^0he\ ^0[\lambda w \lambda t [^0Wise_{wt}he]]]]$

If we want to consider also the implicit variant of the *believing* relation, the analysis using the substitution technique involving *Sub* and *Tr* is

$\lambda w \lambda t [^0Believe_{wt}\ ^0Charles\ ^2[^0Sub[^0Tr^0Pope_{wt}]\ ^0he\ ^0[\lambda w \lambda t [^0Wise_{wt}he]]]]$

Double Execution is called for here, because Charles is not related to the construction of the proposition, but to the so constructed proposition. The result of the substitution, i.e., the respective construction (if any), has to be executed again to yield the product of the procedure rather than the procedure itself.

In both cases undesirable inferences are blocked. For instance, if Charles believes (implicitly/explicitly) of the Pope that he is wise, we cannot deduce that Charles believes of the Pope that he is wise and that the first-order arithmetic is not recursively axiomatisable. Yet the question arises whether the hyperintensional approach just sketched is not too much of a good thing, by being too restrictive. Is perhaps the strict *identity* of constructions not too strong a criterion of synonymy and so of the identity of beliefs? Indeed it is: the identity of constructions is too restrictive to provide a workable criterion. In the next section we introduce a slightly coarser criterion of synonymy, which is the identity of *concepts*. We will show that there are constructions that differ so minimally as to be indistinguishable from the *procedural* point of view. In order to define *synonymy*, we introduce the relation of *procedural isomorphism* as defined over the set of constructions of a particular order, and define a concept as a representative of a given class of procedurally isomorphic closed constructions.

4. Concepts

4.1. Historical Background

If we want to explicate the notion of concept rigorously, then an explication (in Carnap's sense) should take into account the history of the ways the term 'concept' has been used till now. Thus we first summarise some important milestones in the development of theories of concept.

In the cognitive sciences, in particular in psychology, a concept is usually conceived of as a sort of mental object. In these empirical sciences topics like concept acquisition and concept evolution are being investigated. However, in what follows we are going to apply a *logical* theory of concepts to issues relevant from the semantic point of view. Thus the brief summary that follows concerns logically relevant features of semantic theories. Leaving aside the psychologistic tradition, concepts are often identified with expressions, or with sets. When comparing contemporary and traditional textbooks, one is likely to come away with the impression that contemporary logic is no more interested in studying concepts as a special category of interest.[16] Perhaps the only realist theories of concepts that are interesting from the logical point of view are those of Aristotle, Bolzano, Frege and Church.

For *Aristotle*, concepts were most likely structured entities, because he identified what we nowadays are prepared to call 'concept' with *definiens* ($^{c}o\rho\iota\sigma\mu o\varsigma$), which is not a simple entity (see Materna and Petrželka, 2008).

Bolzano, in (1837), worked out a remarkable realist theory of concepts as the meanings of the non-sentential components of a sentence. Bolzano noticed that *the semantics of the whole expression is not reducible to the semantics of particular subexpressions* (Bolzano, 1837, Vol. 1, p. 244). He distinguished a *concept* from the *content* (*Inhalt*) of a concept. The content of a concept is a collection (*Summe*) of the respective sub-concepts, whereas the concept itself consists in the *way these components combine* into a whole (*die Art, wie diese Theile untereinander verbunden sind*). Thus his ingenious insight led him to the conclusion that concepts are objective entities endowed with structure. Moreover, the classical Aristotelian conception of concept (= *definiens*) construed the structure of a concept merely as a conjunctive way of combining particular components, while Bolzano considered any logical way of forming a concept to be legitimate.[17]

Frege defined *concept* (*Begriff*) in (1891, 1892a). His definition is one of the rarely occurring attempts at defining *concept* in modern logic; but in a sense it is narrower than Bolzano's, and Frege's concepts are devoid of structure. Frege's key distinction between *function* and *object* locates concepts among functions. A concept is the *characteristic function* of a class assigning the value **T** to those elements of the universe that fall under the concept, and **F** to the rest.

---

[16] With the possible exception of the 'formal concept analysis' theory applied broadly in computer science; see Ganter and Wille (1999). A 'formal concept' is conceived here as a pair ⟨*content, extent*⟩, where the content is a cluster of properties (attributes) which are individually necessary and jointly sufficient conditions for the elements of the extent to fall under the formal concept.

[17] See Bolzano (1837), where a not uncontroversial criticism of 'the law of inverse proportion between the content and extension of a concept' is presented, using the example 'an adept in all European languages' vs. 'an adept in all living European languages'.

There are many problems connected with this conception. The way Frege defines *function* is *not unambiguous*. This has been convincingly shown in Tichý (1988), where Frege's oscillation between taking functions as mappings and something like rules for forming mappings is documented. Roughly, this oscillation is between functions-in-extension and functions-in-intension. Frege places concepts in the position of *denotation* (*Bedeutung*) in his semantic schema, which suggests the interpretation of functions as mappings. Yet, one would rather suppose concepts to be identified with *senses*, because just as there are more expressions with different senses connected with the same entity denoted by these expressions, so there are more concepts of the same entity. One would expect concepts to be *ways to an object* rather than the object itself. Put differently, concepts are reasonably expected to be *conceptualizations* of objects rather than the conceptualized objects themselves.

Frege himself demonstrates in (1892a) the dichotomy between construing a function as a mapping (*Wertverlauf*) and an 'unsaturated entity'. If concepts are Fregean 'unsaturated entities', then the expression (*Begriffswort*) that denotes a concept should never stand in the position of a grammatical subject. A grammatical subject should stand for an *object* (*Gegenstand*), whereas concepts are functions that are applied to objects. Hence, a concept is not an object from the viewpoint of Frege's distinction between object and function (*Begriff*). As Frege's opponents pointed out,[18] there are counterexamples to this viewpoint. Consider the sentence

"The concept of horse is a zoological concept".

Any instance of concepts being *mentioned* (in order to figure as object of predication) seems to refute Frege's claim. Frege's answer is interesting, though. If the respective *Begriffswort* stands in subject position it no more denotes a concept; it denotes an object. Yet the intuition that a concept remains a concept whether *used* or *mentioned* has much to be said for it, since it seems that Frege's concept of horse vacillates between being a concept and being an object only relative to a flawed theory of concepts.

Another objection concerns concepts identified with classes. Empirical concepts cannot be (characteristic functions of) *classes*. For instance, the predicate 'is a dog' does not denote a class, because the denotation would be dependent on the variability of the population of dogs over worlds and times. If the concept of dog were identified with the actual/current population of dogs, the concept itself would change with every new dog seeing the

---

[18] Not least Benno Kerry, whose criticism is responded to in Frege (1892a).

light of day and every dog drawing its last breath. This objection does not apply to mathematical concepts, because in mathematics worlds and times are irrelevant. For instance, the set of primes is the same independently of possible worlds and times; it cannot change its population. Still there is a serious problem with Frege's definition even in the case of mathematical concepts. Intuitively, we would certainly say that the predicates 'is a number with just two factors' and 'is a number greater than 1 and divisible just by 1 and itself' *express* two *different* (albeit equivalent) concepts. However, according to Frege these expressions *denote* one and the same concept.

Frege considers only general concepts as denoted by *predicates*. In Bolzano, by contrast, also other kinds of expression express concepts, with the exception of sentences. Thus, for instance, while 'is a mountain' does denote a concept according to Frege, 'the highest mountain' either does not, or denotes a so-called singleton concept. But then saying, for instance, that the highest mountain is in Asia would amount to saying that the respective singleton is in Asia, which makes no sense.

Church was one of the most significant of Frege's followers. Church realized that Frege's explication of the notion of concept was untenable. Church's proposal slots concepts into the position of Fregean *senses*. Consequently, concepts should be associated not only with general ('predicate-like') expressions but with any kind of expression possessing a sense. Even sentences express concepts; in the case of empirical sentences the concepts are concepts of propositions. Church's concept is now a *way* to the denotation rather than a special kind of denotation. More concepts can identify one and the same object. Actually, as Church maintains, *the sense* of an expression E should be *a concept* of what E denotes. This is a very subtle formulation: any expression possesses just one sense but there are other expressions that express another concept of the same denotation.

Moreover, as mentioned above, in (1954) Church critically examined Carnap's intensional isomorphism, and proposed replacing it by *synonymous isomorphism* as the criterion of the identity of beliefs. Later, in (1993), he proposed three alternatives of the definition of synonymy. The degree to which concepts should be fine-grained was of the utmost importance to Church. When summarising Church's Alternatives, Anderson (1998, p. 162) defines three options considered by Church. Senses are identical if the respective expressions are (A0) 'synonymously isomorphic', (A1) mutually λ-convertible, (A2) logically equivalent. (A2), the weakest criterion, was refuted already by Carnap in his (1947), and would not be acceptable to Church, anyway. (A1) is surely more fine-grained; however, partiality throws a spanner in the works. As we showed above, β-reduction is not guaranteed to be an equivalent transformation as soon as partial functions are involved. On the other hand, (A1) could serve as a kind of explication of (A0) (see Church, 1993, p. 143).

What would we, as semantic realists, say about this connection between *sense* and *concept*? Accepting Church's version as an intuitively appealing one, we claim that *senses are concepts*. In TIL we define synonymy on the basis of *procedural isomorphism*, which is close to Church's Alternative (A1) as soon as β-reduction is excluded.[19]

4.2. A Procedural Theory of Concepts

A closed construction is a structured, abstract procedure, an objectual instruction that, when executed, yields an object, or in well-defined cases fails to. Moreover, expressions express constructions as their senses. Expressions with complete meanings express closed constructions. Expressions that contain indexicals do not have a complete meaning; they express an open construction and thus do not express a concept. To illustrate our conception, consider the following expressions:

(a) "A prime that is greater than any other prime"
(b) "The prime that is greater than any other prime"
(c) "A brother of"
(d) "The brother of"
(e) "The brother of Albert Einstein"
(f) "His only brother"

Obviously, (a)–(e) express concepts unlike (f), as the following analyses reveal.

Types *ad* (a) and (b): Let v be the type of natural numbers. Then $Prime/(ov); \neq, >/(ovv); \forall/(o(ov)), the/(v(ov))$: a *singularizer*, namely a function that associates a singleton S with the only element of S, and it is undefined if S is not a singleton; $x, y \rightarrow v$.

(a') $\quad \lambda x [^0 \wedge [^0 Prime\, x] [^0 \forall \lambda y [^0 \supset [^0 Prime\, y] [^0 \supset [\neq xy] [^0 > xy]]]]] \rightarrow (ov)$

This construction constructs an object, *viz.* an empty class. Thus (a) expresses a concept of an empty class of natural numbers. We also say that (a') is an *empty concept*.

(b') $\quad [^0 the\, \lambda x [^0 \wedge [^0 Prime\, x] [^0 \forall \lambda y [^0 \supset [^0 Prime\, y] [^0 \supset [\neq xy] [^0 > xy]]]]]] \rightarrow v$

---

[19] As for a procedural interpretation of Church's hierarchy, see Tichý (1988, Chap. 10, esp. p. 159).

This construction is improper and thus fails to identify an object. The class constructed by $\lambda x[\ldots]$ is empty, so it is not a singleton, and thus the function *the* is undefined on this empty class. Yet (b) expresses a concept, *viz.* the construction (b'). We say that (b') is a *strictly empty concept*. The parallel between (strictly) empty concepts and the metaphor of 'roads to nowhere' is that, though a road to nowhere lacks a destination, it is no less a road for that.

Types ad (c)–(f): $Brother\_of/(o\iota\iota)_{\tau\omega}$; $The/(\iota(o\iota))$; $Einstein/\iota$; $x, y \to \iota$.

(c') $\quad {}^0Brother\_of \to (o\iota\iota)_{\tau\omega}$

This Trivialization identifies an empirical relation between individuals. Thus (c) expresses a concept of brotherhood. We say that (c') is a *simple concept* (of brotherhood), because it conceptualizes brotherhood without the mediation of other concepts.

(d') $\quad \lambda w \lambda t [\lambda y [{}^0 The [\lambda x [{}^0 Brother\_of_{wt}\, xy]]]] \to (\iota\iota)_{\tau\omega}$

This Closure identifies the following function f. Dependently on a world and time, f associates every individual y with at most one individual x such that x is the only brother of y. Thus (d) expresses a concept of the only brother of somebody.

(e') $\quad \lambda w \lambda t [{}^0 The [\lambda x [{}^0 Brother\_of_{wt}\, x\, {}^0 Einstein]]] \to \iota_{\tau\omega}$

This Closure identifies an individual office, of type $\iota_{\tau\omega}$. It is an empirical function g that in those worlds and times at which Einstein happens to have just one brother has a value, *viz.* Einstein's only brother. In other worlds and times, i.e., those where Einstein has no brother or more than one, g has no value and, so, is undefined. Thus (e) expresses a concept of the only brother of Albert Einstein.

(f') $\quad \lambda w \lambda t [{}^0 The [\lambda x [{}^0 Brother\_of_{wt}\, x\, his]]] \to_v \iota_{\tau\omega}$

This construction contains the free variable $his \to_v \iota$. Since the evaluation of *his* is extrinsic (i.e., not determined by the meaning of the expression 'His only brother' itself) and dependent on pragmatic factors like the context of utterance, no object is conceptualized by (f'). So it does not express a concept.

Our first preliminary characterisation of *concept*, then, is:

Concepts are closed constructions.

Yet, constructions are a bit too fine-grained from the procedural point of view. Some closed constructions differ so slightly that they are *virtually* identical. In natural language we cannot even render their distinctness, which is due to the role of λ-bound variables that lack a counterpart in natural language.

Compare these two constructions of the property of being taller than 170 cm:

$$\lambda w \lambda t\, \lambda x[{}^0Taller_{wt}\, x\, {}^0170];\ \lambda w' \lambda t'\, \lambda y[{}^0Taller_{w't'}\, y\, {}^0170].$$

Types: $x, y \to \iota$; $Taller/(o\iota\tau)_{\tau\omega}$.

The procedural difference between them cannot be distinguished in an ordinary natural language. We will say that they are α-equivalent.

A similar case is the objectual version of η-reduction in the λ-calculus. Compare the η-equivalent constructions

$${}^0Believe;\ \lambda w[{}^0Believe\, w];\ \lambda w \lambda t\, {}^0Believe_{wt};\ \lambda w \lambda t\, \lambda xy[{}^0Believe_{wt}xy].$$

Though the number of steps to be executed is increasing, the additional instructions to compose a construction with variables and abstract over these same variables make for insignificant differences in terms of what procedure is prescribed. All four constructions share the common property of constructing the relation-in-intension *Believe* without the mediation of any other closed constructions. Again, the four constructions above are procedurally isomorphic.

These considerations motivate the following definition.

Definition 4 (Procedural isomorphism)
Let C, D be constructions. Then C, D are α-equivalent, denoted "${}^0C \approx_\alpha {}^0D$", $\approx_\alpha/(o *_n *_n)$, iff they v-construct the same entity and differ at most by using different λ-bound variables. C, D are η-equivalent, denoted "${}^0C \approx_\eta {}^0D$", $\approx_\eta/(o *_n *_n)$, iff one arises from the other by η-reduction or η-expansion. C, D are *procedurally isomorphic* iff there are constructions $C_1, \ldots, C_n$ ($n > 1$) such that ${}^0C = {}^0C_1$, ${}^0D = {}^0C_n$, and each C_i, C_{i+1} are either α- or η-equivalent.

It still remains a problem, however, how *concept* is to be defined. Materna, in (1998) and (2004), defines a concept as a *class* of procedurally isomorphic constructions (using the term 'quasi-identical constructions'). The obvious disadvantage of this definition is that a concept is reduced to a class, i.e., a set-theoretical entity. Materna proposes a remedy in distinguishing between using and mentioning a concept. When a concept is used, any representative of the class of quasi-identical constructions is used. Only when mentioning a concept is the whole concept, i.e. the class of quasi-identical constructions mentioned. However, this solution is too close to Frege's contextual dichotomy between concept and object. Horák proposed, in (2002), a remedy that we recapitulate here. Horák defined the unique *normal* form of a *construction* C: *NF(C)*. The normal form *NF(C)* is the *simplest* member of the class of constructions procedurally isomorphic to C. The simplest member is defined as the alphabetically first, non-η-reducible construction.

For instance, the following constructions are procedurally isomorphic:

$$\lambda x[^0 + x^0 1];\ \lambda y[^0 + y^0 1];\ \lambda z[^0 + z^0 1];\ \lambda x[\lambda x[^0 + x^0 1]x];\ \lambda y[\lambda x[^0 + x^0 1]y], \ldots$$

The normal form of these constructions is $\lambda x[^0 + x^0 1]$.

Thus we now define *concept*.

Definition 5 (Concept)
A *concept* is a closed construction in its normal form.

By way of summary, we say that procedurally isomorphic constructions, i.e., α- or η-equivalent constructions, *represent one and the same concept*. The meaning of an expression is a construction. If an expression contains indexicals its meaning is an open construction; the meaning of a non-indexical expression is a concept.

Thus our definition is close to Church's alternative (A1), but we do not include β-equivalence, because our logic is a logic of *partial functions*. As explained above, β-reduction is not an equivalent transformation when partial functions are involved.

Having defined *concept*, we can now easily distinguish between the *synonymy* and mere *equivalence* of expressions. Expressions are *equivalent* if they denote one and the same object.[20] Expressions are *synonymous* if their meanings are procedurally isomorphic and thus represent one and the same concept. Whereas equivalent expressions can be mutually substituted *salva veritate* in intensional contexts, in hyperintensional contexts of explicit attitudes only synonymous expressions are substitutable.[21]

5. Conclusion

In this paper we introduced a realist procedural semantics, which is at variance with denotational semantics (such as model theory) and pragmatist

---

[20] Notice that all true mathematical sentences are equivalent. The 'Great Fact' problem arises when only denotation is taken to be semantically salient, neglecting the category of concept.

[21] Notice that in (1968) Tichý took standard model theories to task for indulging in the illusion that Fregean sense could be defined in terms of the synonymy of expressions. He argued that *prior* to the definition of the notion of synonymy the notion of *sense* has to be defined.

semantics (such as inferentialism). Having a notion of objective, extra-linguistic procedure at our disposal, we illustrated the fact that this robust notion of semantic structure makes it possible to analyze critical fragments of natural language in compliance with key semantic principles, such as compositionality and referential transparency in all kinds of context.

Whereas denotational semantics assigns only procedural products to expressions as their meanings and pragmatist semantics eschews meanings in favour of socially constituted rules, the procedural semantics of TIL has expressions express constructions, which are higher-order structured procedures. Our semantics belongs to the neo-Frege-Churchian programme characterized by the sense/denotation dichotomy. We proposed a substantial adjustment of Frege's semantic schema explicating the sense of an expression E as the TIL construction encoded by E and the denotation of E as the product of the sense. Anti-contextualism is obtained by assigning constructions to expressions as their structured, context-invariant meanings. Compositionality is obtained through a strict demarcation between procedures and their products. In extensional or intensional contexts we *use* a construction to construct a product, whereas in hyperintensional contexts we *mention* a construction which features itself as an object of predication.

The question then cropped up of just how fine-grained the senses of expressions should be. Since constructions are a bit too finely individuated from the procedural point of view, the question remains unsettled. Our proposal at this point, though, is this. Any two senses are identical just when they are *procedurally indistinguishable*. We defined the relation of *procedural isomorphism* on the set of constructions, and introduced a slightly coarser category of entities to individuate senses, *viz. concepts*. Our conclusion is that any two expressions are *synonymous*, hence substitutable even in hyperintensional contexts of explicit attitudes, if they express one and the same concept.

Acknowledgements. Parts of this paper were taken from the manuscript Duží et al. (ms.). The authors, together with Bjørn Jespersen, also presented their programme of procedural semantics at the Joint Paris-Arché conference *Abstract Objects in Semantics and the Philosophy of Mathematics*, February 28–March 1, 2008. We are grateful to Bjørn Jespersen for his valuable comments that improved the quality of the paper. This research has been supported by the Grant Agency of the Czech Republic (Project No. 401/07/0451, *Semantization of Pragmatics*) and the Czech Academy of Sciences (Project No. 1ET101940420, *Logic and Artificial Intelligence for Multi-agent Systems* within the programme *Information Society*).

References

Anderson, C. A. (1998). Alonzo Church's contributions to philosophy and intensional logic. *The Bulletin of Symbolic Logic*, 4(2), 129–171.

Bealer, G. (1982). *Quality and Concept*. Clarendon Press, Oxford.

Bolzano, B. (1837). *Wissenschaftslehre*, vols. I, II. Sulzbach.

Burge, T. (2005). *Truth, Thought, Reason: Essays on Frege*. Oxford University Press, Oxford.

Carnap, R. (1947). *Meaning and Necessity*. Chicago University Press, Chicago.

Church, A. (1954). Intensional isomorphism and identity of belief. *Philosophical Studies*, 5, 65–73.

Church, A. (1956). *Introduction to Mathematical Logic*. Princeton University Press, Princeton.

Church, A. (1993). A revised formulation of the logic of sense and denotation. Alternative (1). *Noûs*, 27, 141–157.

Cresswell, M. J. (1975). Hyperintensional logic. *Studia Logica*, 34, 25–38.

Cresswell, M. J. (1985). *Structured Meanings*. MIT Press, Cambridge.

Duží, M., Jespersen, B., and Materna, P. *Procedural semantics for hyperintensional logic. Foundations and applications of TIL*. Springer. (Forthcoming).

Duží, M. and Materna, P. (2005). Logical form. In G. Sica (Ed.), *Essays on the Foundations of Mathematics and Logic*, volume 1 (115–153). Polimetrica International Scientific Publisher, Monza.

Frege, G. (1891). *Funktion und Begriff*. H. Pohle, Jena. (Vortrag, gehalten in der Sitzung vom 9. Januar 1891 der Jenaischen Gesellschaft für Medizin und Naturwissenschaft, Jena, 1891).

Frege, G. (1892a). Über Begriff und Gegenstand. *Vierteljahrschrift für wissenschaftliche Philosophie*, 16, 192–205.

Frege, G. (1892b). Über Sinn und Bedeutung. *Zeitschrift für Philosophie und philosophische Kritik*, 100, 25–50.

Ganter, B. and Wille, R. (1999). *Formal Concept Analysis. Mathematical Foundations*. Springer, Berlin, Heidelberg and New York.

Horák, A. (2002). *The Normal Translation Algorithm in Transparent Intensional Logic for Czech*. PhD thesis, Masaryk University, Brno. URL = http://www.fi.muni.cz/~hales/disert/.

Jespersen, B. (2003). Why the tuple theory of structured propositions isn't a theory of structured propositions. *Philosophia*, 31, 171–183.

Jespersen, B. (2008). Predication and extensionalization. *Journal of Philosophical Logic*, 37, 479–499.

Kaplan, D. (1989). Demonstratives. In J. Almog, J. Perry, and H. Wettstein (Eds.), *Themes From Kaplan* (481–563). Oxford university Press, Oxford.

King, J. C. (1995). Structured propositions and complex predicates. *Noûs*, 29(4), 516–535.

King, J. C. (2001). Structured propositions. In E. N. Zalta (Ed.), *The Stanford Encyclopedia of Philosophy*. URL = http://plato.stanford.edu/entries/propositionsstructured/.

Kirkham, R. L. (1997). *Theories of Truth*. The MIT Press, London and Cambridge. (Original work published 1992).

Lewis, D. (1972). General semantics. In D. Davidson and G. Harman (Eds.), *Semantics of Natural Language* (169–218). Reidel, Dordrecht.

Materna, P. (1998). *Concepts and objects*. Acta Philosophica Fennica, 63. Tummavuoren Kirjapaino Oy, Helsinki, Vantaa.

Materna, P. (2004). *Conceptual Systems*. Logos, Berlin.

Materna, P. (2007). Once more on analytic vs. synthetic. *Logic and Logical Philosophy*, 16, 3–43.
Materna, P. and Duží, M. (2005). The Parmenides principle. *Philosophia*, 32, 155–180.
Materna, P. and Petrželka, J. (2008). Definition and concept. Aristotelian definition vindicated. *Studia Neoaristotelica*, 5(1), 3–37.
Montague, R. (1974). *Formal Philosophy: Selected Papers of R. Montague* (R. Thomason, Ed.). Yale University Press, New Haven.
Moschovakis, Y. N. (1994). Sense and denotation as algorithm and value. In J. Väänänen and J. Oikkonen (Eds.), *Lecture Notes in Logic*, volume 2 (210–249). Springer, Berlin.
Moschovakis, Y. N. (2006). A logical calculus of meaning and synonymy. *Linguistics and Philosophy*, 29, 27–89.
Quine, W. V. O. (1963). Two dogmas of empiricism. In *From a Logical Point of View* (20–46). Harper et Row, Publisher, New York and Evanston. (Original work published 1953).
Russell, B. (1903). *Principles of Mathematics*. Norton, New York, second edition.
Soames, S. (1987). Direct reference, propositional attitudes and semantic content. *Philosophical Topics*, 15, 47–87.
Strawson, P. F. (1952). *Introduction to Logical Theory*. Methuen, London.
Strawson, P. F. (1964). Identifying reference and truth values. *Theoria*, 96–118.
Tichý, P. (1968). Smysl a procedura. *Filosofický Časopis*, 16, 222–232. Translated as 'Sense and procedure' in: Tichý (2004), pp. 77–92.
Tichý, P. (1969). Intensions in terms of Turing machines. *Studia Logica*, 26, 7–25. Reprinted in: Tichý (2004), pp. 93–109.
Tichý, P. (1978). Questions, answers and logic. *American Philosophical Quarterly*, 15, 275–284. Reprinted in Tichý (2004), pp. 293–304.
Tichý, P. (1986). Indiscernibility of identicals. *Studia Logica*, 45, 251–273. Reprinted in Tichý (2004), pp. 649–671.
Tichý, P. (1988). *The Foundations of Frege's Logic*. De Gruyter, Berlin and New York.
Tichý, P. (1994). The analysis of natural language. Reprinted in: Tichý (2004), pp. 801–841.
Tichý, P. (2004). *Collected Papers in Logic and Philosophy*. Filosofia, Czech Academy of Sciences, Prague and University of Otago Press, Dunedin.

Hyperintensions and Procedural Isomorphism: Alternative (½)*

Bjørn Jespersen

> *Ich finde aber, daß eine Frage, für deren Beantwortung es kein objectives Kriterium gibt, überhaupt in der Wissenschaft keine Stelle hat.*
> Frege to Husserl, 1906

Abstract

It is a thrice-told tale in contemporary philosophical logic, especially epistemic logic and formal semantics, that at least the logical objects figuring as complements of explicit attitudes, not least sentential senses, need to be hyperintensionally individuated. As early as 1947, Carnap realized that a sentence like "John believes that D" constitutes neither an extensional nor intensional context. This prompted him to develop the notion of *intensional isomorphism*, which Church found wanting and urged be replaced by *synonymous isomorphism*. This paper, in turn, recommends the notion of *procedural isomorphism* as the principle governing the individuation of hyperintensions. The basic idea is that any two hyperintensions are identical as soon as they are two near-identical procedures, a procedure being an instruction that details what operations are to be applied to what entities in what order to produce a particular kind of product. Near-identity, or isomorphism, is couched in terms of α- and η-conversion as known from the

---

* Versions of this paper were read at Formal Philosophy Seminar, Leuven, 5 March 2009; *Propositions: Ontology, Semantics, and Pragmatics*, Venice, 17–19 November 2008; *ECAP 6*, Kraków, 21–26 August 2008. I am indebted to Marie Duží, Sebastian Sequoiah-Grayson and the participants for valuable comments, and to the Editor, Katarzyna Kijania-Placek, for granting me the extra time needed to write up this paper after the Leuven talk.

λ-calculi. The resulting granularity of hyperintensions-as-procedures slots in between Church's Alternatives (0) and (1) and is, for this reason, dubbed 'Alternative (½).'

1. Why Go Hyperintensional?

A *hyperintension* is an intension whose principle of individuation is finer than logical equivalence: any two merely logically equivalent hyperintensions are two and not one. In general, anything finer than logical equivalence qualifies as hyperintensional. But how much finer? What is their upper bound? How hyper are hyperintensions? Cresswell, who coined the phrase 'hyperintension,' offers only a negative definition of hyperintensionality (in terms of context of substitution):

> Hyperintensional contexts are simply contexts which do not respect logical equivalence. (Cresswell, 1975, p. 25)

That is, it is not logically necessary that two distinct, logically equivalent hyperintensions are going to be mutually substitutable when occurring in an hyperintensional context.

In general, hyperintensionality is required in order to distinguish between, e.g., inverse relations such as Jumbo being larger than Mickey and Mickey being smaller than Jumbo, a pint glass being half-full and being half-empty, and knowing (believing, hoping, predicting, etc.) that if it rains then the streets are wet and knowing (etc.) that if the streets are not wet then it is not raining. Thus, while in a particular case there may be only one relation, one set or one truth-condition, there will be at least two hyperintensions converging in the same relation, the same set or the same truth-condition, respectively. For instance, Prawitz (1968) points out that, even though it is often assumed that logically equivalent propositions are identical,

> [A] person may not realize that a logical equivalence holds, and when analyzing certain senses of knowing, one may prefer to state that all [algebraically generated] propositions ... are to be different. (Prawitz, 1968, p. 138)

The introduction of hyperintensions—hyperpropositions, hyperproperties, hyperrelations (relations-in-hyperintension), etc.—serves to draw distinctions at a higher level than that at which logical equivalence holds sway. Otherwise it is not possible to maintain that one thing is to know that if it is raining then the streets are wet and another thing is to know that if the streets are not wet then it is not raining, for there would be but one thing to know. Just as it is part of logical lore that there are contexts, or operators, that are non-extensional (e.g., "If A had been true, then B would have been true," or "Probably, A"), so the insight is gaining ground that some contexts or operators are non-intensional, because they are hyperintensional. Thus, where A, B are propositions and = the relation of identity between propositions, the inference

$$\frac{\Box A \quad A = B}{\Box B}$$

is valid for intensional contexts like modal ones.[1] But it is not going to be valid for hyperintensional contexts. For illustration, if K is a relation-in-intension between a knower and a hyperproposition, and K' the operator 'it is known hyperintensionally that,' then the following two arguments are, first of all, not well-formed:

$$\frac{K_a A \quad A = B}{K_a B} \qquad \frac{K'A \quad A = B}{K'B}$$

In the first premises A, B are supposed to be hyperpropositions; in the second premise, possible-world propositions. This makes the notation 'A,' 'B' ambiguous, which is anathema to logic. Secondly, if the notation is tidied up, e.g. along the lines of '$Ext(A)$,' '$Ext(B)$' to denote the propositions that the hyperpropositions A, B yield when extensionalized, it is obvious that $Ext(B)$ cannot validly be inserted into a context governed by K_a or K'. For then a lower-order entity (a proposition) is substituted for a higher-order entity (a hyperproposition). K_a and K' require the admissible substituends for A to be identical or near-identical to A; A, B share the same extension and are, therefore, *equivalent*, but they are not *identical*. Let me amplify this point. For unambiguous notation, let C, D be two distinct hyperpropositions. Let ≈ be the equivalence relation defined between hyperpropositions, such that it obtains between arbitrary C, D iff C, D yield the same proposition. If we rewrite the above twin inference along these lines, we get:

$$\frac{K_a C \quad C \approx D}{K_a D} \qquad \frac{K'C \quad C \approx D}{K'D}$$

---

[1] If A', B' are *formulae* denoting propositions, then in case A' ⇔ B' (i.e., if A', B' are *co-entailing* formulae) it follows that they co-denote the same proposition in the logic of total functions.

There is no logical guarantee that if C is known so is D, since $C \neq D$. If $C = D$ then the two arguments come out valid, of course.

The distinction between identical and equivalent intensions is lost in possible-world semantics. If the variables f, g range over intensions (functions from possible worlds) and w over possible worlds then if f, g return the same values (sets, truth-values, individuals, or whatever) for the same arguments (worlds) then $f = g$. That is, co-intensionality, the principle of individuation of intensions, is identical to necessary co-extensionality:

$$\forall fg(\forall w(f(w) = g(w)) \rightarrow f = g)$$

By contrast, the *raison d'être* of hyperintensions is that they are capable of being co-intensional and yet distinct.

In general, what is wanted is a means to distinguish between identical and merely equivalent pieces of knowledge, objects of belief, objects of mathematical proof and calculation, sentential senses, conceptualizations or descriptions of sets, etc. For instance, when calculating $7 + 5$ you are certainly not calculating 12, so the complement of *calculating* should not be a number, but must be a hyperintension issuing in a number. Nor are you calculating just any old hyperintension issuing in 12. You may master addition while having no clue of how to calculate the square root of a number. So these two arguments need to come out invalid:

(I)
$$\frac{a \text{ calculates } 7+5; \quad 7+5 = 12}{a \text{ calculates } 12}$$

(II)
$$\frac{a \text{ calculates } 7+5; \quad 7+5 = \sqrt{144}}{a \text{ calculates } \sqrt{144}}$$

As an aside, the respective conclusions are different in principle, in that the former is arguably nonsensical for want of an operation: calculating means applying operations to numbers, but "*a* calculates 12" mentions no operator; unless, of course, 12 is itself interpreted as an operation, e.g. the application of the successor function to 0 twelve times over. (Further discussion of this point would take us into philosophy of mathematics proper.) "*a* calculates $\sqrt{144}$," for its part, does mention both an operation and a number.

A logical analysis blocks (I), (II) when '$7 + 5$' in the first premises is shown to occur *de dicto*, thus picking out a hyperintension, and *de re* in the second premises, thus picking out an extension. The substitution of '12' and '$\sqrt{144}$' for '7 + 5' in the context "*a* calculates..." fails because a term for a lower-order entity, such as a number, cannot replace a term for a higher-order entity, such as a number-producing hyperintension. Similarly in (IV) below. Such a logical analysis presupposes, however, an ontology of logical objects rich enough to encompass hyperintensions.

As for the difference between possible-world propositions and hyper-propositions as attitudinal complements, consider:

(III)
$$\frac{a \text{ knows that } b \text{ is a half-}empty \text{ glass;}}{a \text{ knows that } b \text{ is a half-}full \text{ glass}} \quad b \text{ is half-empty if, and only if, } b \text{ is half-full}$$

Valid? *Yes*, provided *knowing* is an implicit attitude, which is defined to be logically closed. Agents entertaining implicit attitudes are idealized *à la* Hintikka (1962): they know all the logical equivalences and implications of their original pieces of knowledge and are insofar logically omniscient.[2] *No*, in case *knowing* is an explicit attitude, which is defined not to be logically closed and which requires the agent to be able to actively apply rules of inference and other logical operations to the originally known proposition(s).[3]

For a non-empirical example, consider:

(IV)
$$\frac{a \text{ knows that } 7 + 5 = 12;}{a \text{ knows that } \sqrt{144} = 12}$$

Valid? As above, logic adjudicates neither way: all logic does is charting the logical consequences of a philosophical decision to settle either for validity or invalidity. What recommends settling for validity is that it catalogues the propositions, i.e. pieces of knowledge, that are, so to speak, waiting for the knower to pick them up: if you know *A*, then if *A* entails *B*, *B* is true and follows from what you already know. So you cannot go wrong by including *B* into your stock of knowledge. What speaks against going with validity is that it cannot be a matter of logical necessity that every, or any, agent

---

[2] Interestingly, whereas epistemology is haunted by skepticism, i.e. the spectre of knowing too little or nothing at all, epistemic logic is haunted by omniscience, i.e. the spectre of knowing too much.

[3] For the implicit/explicit distinction, see (Levesque, 1984) and (Fagin, Halpern, Moses, and Vardi, 2003, §9.5). Agents are said to be *aware* of their explicit attitudes. For a logic of explicit knowledge, see Duží, Jespersen and Müller (2005).

knowing *A* also knows *B*, as soon as *A* entails *B*, in (III) by being equivalent to *B*. Logic should not grant a knower a piece of knowledge for free.

Let me mention in passing that I am *not* recommending *non-normality* if this would mean emulating the effect of blocking the arguments (I) through (IV) by relating agents to *impossible*-world propositions interpreted in a parallel universe of impossible, or non-normal, worlds rather than relating agents to hyperpropositions.[4] One complaint is that going non-normal presupposes that laws of logic be indexed to (im-)possible worlds. But laws of logic, being formal, ought to be independent of worlds altogether: a law of logic is not 'at a world.' Possible worlds are exclusively in the business of modelling empirical variability. (This is what makes possible-world semantics a valuable addition to model-theoretic semantics, which Tarski created to model mathematical and logical language.) The invocation of impossible worlds to avail oneself of an alternative logical space in which to evaluate propositions and other intensions is a 'sideways' move rather than an 'upwards' move. Whereas we ought to be ascending into a sphere of hyperintensions, the possible-world semanticist who finds himself running out of worlds to model, e.g., inconsistent beliefs and flawed reasoning attempts to apply the same old model-theoretic techniques by conjuring up a(n) (il-)logical space next door. The overall goal is to furnish various operators with an interpretation that makes them at most 'semi-penetrating,' in the parlance of Dretske (1977). Thus, if argument (III) is interpreted in that second space, there is going to be a model in which the conclusion is false. Such an analysis comes across as being philosophically shallow. For one thing, it fails to make good sense of the nature of the intentional object that an epistemic or doxastic agent is intellectually related to (see Duží, Jespersen and Materna, ms., §5.1.3).

As for hyperpropositions, Cresswell himself recommends

> [A]n analysis of propositions which assumes that they are *structured* entities, and that the clue to their structure is found in the sentences which express them. And we want to do this while preserving the highly desirable connection with the possible worlds approach to semantics. (*Ibid.*, p. 27)

This suggests a general three-storey conceptual edifice:

(1) *hyperintensional storey* (functions-in-intension)

(2) *intensional storey* (functions-in-extension/mappings)

(3) *extensional storey* (functional values: $f(a) = f(b)$)

---

[4] See Priest (1992) on non-normal worlds.

The top storey will cover, e.g., explicit attitudes, linguistic (e.g., sentential) senses, conceptualizations of equivalences in mathematics and logic, and perhaps also probabilities. The middle storey will cover, e.g., implicit attitudes and alethic modalities (such as logical possibility and counterfactuals, perhaps also laws of nature). The bottom storey will cover, e.g., numbers, sets, individuals, truth-functions, and truth-values.[5] While we know what co-intensionality and co-extensionality amount to (unless $f(a)$, $f(b)$ are hyperintensions), we do not know what principle of individuation ought to apply at the top level since, as we saw, Cresswell's definition of hyperintensionality is the negative one that logical equivalence does not entail identity.

Notice, however, that the three-tiered conceptual edifice just sketched may well turn out to be importantly insufficient by not speaking explicitly about hyperintensional store*ys*. For the question, "How hyper are hyperintensions?" may conceivably turn out to rest on the false assumption that it must be susceptible to exactly one answer. Maybe there ought to be multiple degrees of hyperintensional individuation. Already in 1941, when distinguishing between functions-in-extension and functions-in-intension, had Church pleaded for functions-in-intension of different degrees:

> It is possible, however, to allow two functions to be different on the ground that the *rule* of correspondence is *different in meaning* in the two cases although always yielding the same result when applied to any particular argument. When this is done we shall say that we are dealing with *functions in intension*. The notion of difference in meaning between two rules of correspondence is a vague one, but . . . can be made exact in various ways. We shall not attempt to decide what is the true notion of difference in meaning but shall speak of functions in intension in any case where a more severe criterion of identity is adopted than for functions in extension. There is thus not one notion of function in intension, but many notions; involving various degrees of intensionality. (Church, 1941, pp. 2–3. Emphasis mine)

In a similar vein, Myhill recommends that

> *[W]e should never identify [hyper-]intensions unless we are forced to*; for the fewer identifications we make the more flexible will be our analysis of belief-sentences. (Myhill, 1963, p. 306. Quoted from Anderson (2001), p. 404. My insertion)

And after a remark on attitudes Kripke vents this afterthought on propositions:

---

[5] See (Sullivan, 1998) for an attempt to accommodate fine-grained Fregean and coarse-grained Russellian propositions within the same theory.

> How this relates to the question what 'propositions' are expressed by these [attitude-reporting] sentences [and] whether these 'propositions' are objects of knowledge and belief... are vexing questions. I have no 'official doctrine' concerning them, and in fact I am unsure that the apparatus of 'propositions' does not break down in this area. ... Of course there may be more than one notion of 'proposition,' depending on the demands we make of the notion. (Kripke, 1980, p. 21; *ibid*.: 21, n. 21. My insertions)

Before proceeding, though, it is important to bear in mind that the question of the individuation of hyperintensions is, strictly speaking, a *functional* and not also a *structural* one. The functional question is *what* function(s) hyperpropositions are supposed to fulfil. Hyperpropositions must be fine enough in order to block various arguments and coarse enough so as not to draw distinctions without a difference. The structural question would be what interior properties hyperintensions need to possess in order to fulfil their function(s), these properties bearing not least on the enabling structure of (various kinds of) hyperintensions, or more loosely put, on what hyperintensions 'look like.' If one answers the structural, but not the functional, question one displays one or more logical objects and explains how they 'work,' without explaining what they are good for. If one answers the functional, but not the structural, question one explains what kind(s) of logical objects would be needed to solve one or more particular questions, without explaining how they would do that.

Thomason (1980), for instance, introduces the primitive type p of hyperpropositions. The introduction of p acknowledges the very need for hyperpropositions; but Thomason neither defines a measure of hyperintensionality, nor does he describe the elements of p, i.e. how they are structured so as to fulfil their function(s). Thomason's p is a placeholder for the category of hyperproposition within a semantic edifice, leaving it to others to fill that place with properly defined hyperpropositions. Filling the slot comes down to deciding on and motivating one or more calibrations of hyperintensionality. One may proceed to specify the make-up of different kinds of hyperintension (*in casu* hyperpropositions) to provide a structural account of hyperintensions as well. Below I suggest *procedural isomorphism* as the general criterion of hyperintensional individuation, leaving it open whether procedural isomorphism ought to form a spectrum of different calibrations. Procedural isomorphism is achieved by going hyperintensional in the logic of functions, the (typed) λ-calculus, the λ-terms of which are re-interpreted as denoting procedures.

2. How Hyper?

The very fact that intensionality has had to be qualified as 'hyper' goes to show that something has gone awry. Indeed, there may well be several degrees of intensionality; but the thing is that possible-world intensions are, strictly speaking, not intensional entities at all, if for no other reason than that they are extensionally individuated. Rather, we ought to refer to them as 'hyperextensions' to set these fine-grained extensional entities apart from coarse-grained ones (sets, etc.) 'Intension' would then be reserved for today's hyperintensions. In fact, this proposed linguistic revision would not be a revolutionary but retrograde move. Prior to the advent of possible-world semantics, intensions—under whatever name and whether construed objectually, mentalistically or syntactically—were understood to be fine-grained. Gaskin (2008, pp. 11–13) enumerates several historical examples of fine-grained propositions, running the gamut from Stoic *axiômata* through Leibniz's *propositiones* to Bolzano's *Sätze an sich* and Frege's *Gedanken*.[6] We should, uncontroversially, see modern-day theories such as the following as being part and parcel of this millennium-long research project:

- Carnap: intensional isomorphism (1947)
- Church: synonymous isomorphism (1954)
- Lewis: annotated trees and ordered *n*-tuples (1972)
- Cresswell: ordered *n*-tuples (1975, 1985)
- Zalta: propositions *encoding* lower-level propositions (1988)
- Bealer: metaphysically simple propositions, analyzed as trees (1982)
- Moschovakis: idealized algorithms (1994, 2006)
- Tichý: constructions (1988, 2004)

But the respective exact calibration of these various hyperintensions is often a big, open and contentious question. One notorious example is Frege. In the current stand-off between 'neo-Russellian' and 'neo-Fregean' approaches in analytic philosophy of language, the former moniker has come to stand for extensionalist theories, and the latter for intensionalist theories. The monikers are inspired not least by Russell and Frege's famous 1904 clash over whether Mont Blanc is a constituent of the proposition/*Gedanke* that Mont Blanc is over 4,000 metres high. For a Russellian proposition to be about Mont Blanc, it must contain that very mountain as a constituent.

---

[6] I shan't consider quotational, inscriptional and other sententialist/meta-linguistic theories; first, because pieces of notation are too fine-grained as objects of attitude; second, because the Church-Langford argument from translation counts decisively against them.

For a Fregean *Gedanke* to be about Mont Blanc, it must contain a mode of presentation of Mont Blanc. (Whether Russell and Frege spoke at cross-purposes in that exchange is another matter.) However, Frege himself was never a clear-cut intensional logician. In fact, he thought of himself as an *Umfangslogiker*, which goes toward explaining why he was puzzled that "$a = b$" and "$a = a$" differ in cognitive value (*Erkenntniswert*) even when both are true. His solution has the two sentences express two different *Gedanken* that are modes of presentation of the same truth-value, a *Gedanke* being a sentential *Sinn*. (Similarly, Church would say that the meaning of a sentence is a *concept of a truth-value*.) But Frege never made the step from characterizing *Sinn* and listing the various functions it was to fulfil to providing a rigorous definition. Hence, *Sinn* remained a marginal notion in Frege's theory proper. One key question thus left unanswered was the granularity of *Sinn*. Frege was torn, throughout his career, between two rival criteria, one coarse-grained, the other fine-grained; namely, logical equivalence and cognitive value. For want of a definition of *Sinn*, he was unable to decide either way; or conversely, for want of a criterion winning out, he was unable to provide a definition. He would occasionally even try to dodge the issue by dismissing it as being unsusceptible to scientific treatment (*vide* the motto of this paper culled from Frege's 1906 correspondence with Husserl). Hodes (1982), Sundholm (1994) and Penco (2003) detail the story of Frege's oscillation between the two rival criteria. For instance, as Hodes rightly wonders,

> Why do "$2+5 = 7$" and "not (not $(2+5 = 7)$)" express the same thought, whereas "$2+5 = 7$" and "$3+5 = 8$" do not? (*Ibid.*, p. 176)

For surely $2+5 = 7$ and *not* (*not* $(2+5 = 7)$) ought to count as two different *Gedanken*, considering the fact that the latter contains (two occurrences of) an operation (negation) absent in the former. Van Heijenoort (1977a, p. 105) adduces an example that is equally awkward for Frege.[7] The following biconditional is provable in Frege's logicistic system:

$$(2^2 = 4) \equiv (2+2 = 4)$$

---

[7] Van Heijenoort suggests, (1977a, pp. 99–100), that the Fregean *Sinn* of a formula T is to be identified with a tree T', whose semantic structure will be isomorphic to the syntactic structure of T. The suggestion is *prima facie* appealing, not least because the diagrammatic structure of trees is in the vicinity of the syntactic structure of Frege's *Begriffsschrift* notation. However, as Van Heijenoort himself points out, "a tree is a mapping. ... Thus, in Fregean terms, a tree would be the object that is the *Werthverlauf* of a certain function. This conclusion may seem quite odd." Indeed. But, worse, if a Fregean *Sinn* is to be sliced in terms of cognitive significance rather than merely logical equivalence, then a mapping won't do as *analysans* due to the coarse individuation of mappings.

So by the same token, "$2^2 = 4$" and "$(2+2 = 4)$" are predicted to express the same *Gedanke*. But surely this prediction must be wrong: again, different operations are involved, so it is going to take more than mere linguistic competence to establish this equivalence. The extra that is needed is mathematical computation.

In general, what is appealing about logical equivalence as a criterion of individuation is that it is exact and well-entrenched. Furthermore, Frege wanted his *Begriffsschrift* to take note of all and only *logically* relevant factors of an inference, and logical equivalence seems insofar the obvious choice; anything beyond logical equivalence would be of mere poetic or psychological significance. Yet Frege's *Begriffsschrift* comes with the category of *judgement* (*Urteil*), which relates reasoning agents to *Gedanken*. Where '⊢' is Frege's judgement stroke (*Urteilsstrich*) and not the symbol of validity, Frege's question in (1892) is whether the *judgement* ⊢ Fa is identical to the *judgement* ⊢ Fb, provided $a = b$. A predominantly extensionalist logician, Frege would have expected an answer in the affirmative. Still the two judgements are not identical, and he recognizes the manifest difference in cognitive value between making the one rather than the other. (The difference between *judgements* rather than *Gedanken*, or propositions or sentences, is the specifically *Fregean* source of what has become known as *Frege's puzzle*.)

The difference in cognitive value between ⊢ Fa and ⊢ Fb suggests that Fregean hyperpropositions ought rather to be sliced according to the amount of knowledge they convey. What speaks against, though, is that it is hard to bring on a logically rigorous footing. Apart from having to stave off psychologistic intrusions, much epistemological footwork needs to be completed before arriving at a worked-out concept of cognitive value. What is appealing about such a criterion, however, is that logical differences in the complements of ⊢, such as the occurrence of negation, ought to be reflected at the level of *Gedanken*. An even number of occurrences of negation will cancel each other out at the level of truth-conditions, for sure, but the logical procedure of calculating the truth-condition of $\neg\neg A$ is different from the logical procedure of calculating the truth-condition of A, and establishing their concurrence in the same truth-condition requires a non-trivial calculation. Frege, in 1906, proposed the following. If S, S' are two different sentences, then $Gedanke(S) = Gedanke(S')$ if anyone accepting one must, for logical reasons, immediately accept the other.[8] Frege points out that what goes nowadays by the name 'transparency' is required: grasp-

---

[8] Cited from Sundholm (1994, pp. 305–306).

ing *Gedanke*(*S*), *Gedanke*(*S'*) forms no obstacle, since the relevant agents are linguistically competent, i.e. understand each item belonging to the language(s) that *S*, *S'* are part of.⁹ Therefore, it is not an option for the neo-Fregean semanticist to follow Richard (2001, pp. 546–547) in abandoning this principle (as formulated by Richard):

> Transparency: It is impossible for a (normal, rational) person to understand expressions which have identical senses but not be aware that they have identical senses (or be such that he would immediately come to know this if he were to reflect on it).

If Transparency is preserved, the outline of the solution to Church's *paradox of analysis*, as commonly exemplified by pairs of predicates like {'is a brother,' 'is a male sibling'} or {'lasts a fortnight,' 'lasts fourteen days'}, becomes clear. If the predicates are synonymous then they are mere notational variants, hence one of them is semantically redundant. To understand one predicate is *ipso facto* to understand the other, since there is but one thing to understand: their identical sense. The solution consists in preventing the paradox of analysis from arising in the first place: since there are two predicates and one sense, there is no traffic between a simple (non-composite) and a composite sense to analyze. But the relation between the two predicates and their shared sense must be transparent to the language-user. Alternatively, the predicates are not synonymous, but only co-intensional, hence are not mere notational variants, and so there are two senses for the agent to understand. For instance, 'is a brother' may have a simple sense and 'is a male sibling' a composite sense. This opens up the possibility that an agent competent with respect to one predicate and its sense may fail to be competent with respect to the other predicate/sense pair, without being guilty of failing to recognize a pair of synonyms. Since the paradox of analysis requires a pair of synonyms to get off the ground, this second scenario also prevents the paradox from arising. As is seen, the para-

---

⁹ Cf. Tichý (1966, p. 364; 2004, p. 55, n. 1.) In current analytic philosophy of language, however, the transparency constraint is often lifted or relaxed. This is how, for instance, the spurious Kripkean category of necessary propositions *a posteriori* (as distinct from nomological necessity) has been generated. If 'Hesperus,' 'Phosphorus' are rigid designators then, if true, "Hesperus = Phosphorus" expresses a necessary truth (and if false, a necessary falsehood). Yet if the reference relation of 'Hesperus,' 'Phosphorus' is not transparent, it follows that the truth that Hesperus is Phosphorus must be established *a posteriori* rather than *a priori* and exclusively by means of semantic analysis. Since linguistic competence with respect to 'Hesperus,' 'Phosphorus' need not amount to perfect or ideal competence, it is not incoherent that a language-user may qualify as competent and still be unable to recognize that they rigidly co-designate.

dox of analysis points to the need for a distinction between co-intensionality and hyperintensionality.[10]

The general hypothesis I extract, tentatively, from Frege's idea of cognitive significance and Church's concerns over the paradox of analysis is that the hyperintensional tuning of cognitive significance should be the same as for synonymy. Recall, however, that both Frege's and Church's respective semantic theories come with *indirect senses*. The Churchian logician Anderson (2001, p. 146) admits that, "One will then need to maintain that two synonymous expressions may nevertheless differ in indirect sense." Hence, synonymy is not absolute, but relative to orders ($\alpha_1, \alpha_2, \ldots, \alpha_n$, in Church's notation), yielding an upward-branching hierarchy of senses. To put it melodramatically, this hierarchy forms a towering hell of indirect senses, which is first and foremost an unfortunate artefact of Frege-Churchian semantics.[11] Says Tichý,

> [I]f someone is to be able to understand the term 'the author of *Waverley*' in all possible contexts [i.e., whatever the degree of nesting], he must internalize an infinite sequence of distinct presentations [senses]. This seems an unrealistically tall order. (Tichý, 1988, p. 126, my insertions)

To avoid erecting a tower of indirect senses in a non-*ad hoc* manner, the most rewarding strategy is no doubt to construe all contexts as *transparent* and none as opaque or oblique.[12]

What I intend by 'synonymy' in the following is, thus, absolute in the sense that a pair of expressions are synonymous *tout court* or not at all. To resume the thread, operating with the same calibration for cognitive significance and synonymy comes nowhere near to defining either, of course, but it does give us a critical platform from which to intuitively assess various problematic cases. Here are two.

The first case is due to Anderson (2001, p. 417). He says,

> Notice that if '+' is assigned a binary function concept as its meaning, it will not be up to us to stipulate that the true equation:
>
> $$x + (y+z) = (x+y) + z$$

---

[10] So does the *paradox of inference*. See Duží et al. (ms., §5.4.3.1).

[11] See Salmon (1993) for a dilemma for Churchian semantics, as discussed in Anderson (1998, pp. 145–146). The dilemma is that Church's solutions to the paradox of analysis and Mates' puzzle rule each other out. As a consequence, either the sense hierarchy goes by the board, or in Mates' puzzle the two indirect senses in question come out distinct.

[12] For statements of the programme of a theory of transparent hyperintensional semantics, see Tichý (1986b, §0) and Duží et al. (ms., §1.1). For technical details, see Tichý (1988, Chap. 12) and Duží et al. (ms., Chapters. 1 and 2, esp. §2.6).

expresses a synonymy. This is slightly more evident if we use a notation whereby the function is written before the arguments, say:

$$\text{Plus}\,(x, \text{Plus}\,(y,z)) = \text{Plus}\,(\text{Plus}\,(x,y), z).$$

Indeed, it is to be observed that something is explicitly denoted on the left side of the equation which need not be explicitly denoted on the right, namely, y plus z. It could still be maintained that the entity is 'implicitly denoted' by way of being determined by a concept which must be grasped if the right hand side is understood, and hence that we do not really have distinctness of denotations. But the claim is strained, at best.

Strained it is, and no doubt too crude, as soon as cognitive significance has got anything going for it as a criterion of hyperintensional individuation. As Anderson points out, 'Plus (y,z),' which occurs in the left-hand formula, denotes something that goes unmentioned in the right-hand formula. Only the question is what that something is. The question comes down to whether 'Plus (y,z)' denotes the algebraic *calculation* of adding the value of y to the value of x, or the *result* of this addition. From an epistemological point of view, there are manifest differences between comprehending, and executing, the calculation encoded by the left-hand formula and comprehending, and executing, the calculation encoded by the right-hand formula. One thing is to add the value of x to the sum of adding the values of y and z; another thing is to add the sum of the values of x and y to the value of z. This is so even when the value of all three variables should be 0, because the steps differ. So non-trivial distinctions can be drawn at the level of associativity, the scopes brought about by the brackets making for procedural differences. The most straightforward way of accommodating those differences in one's formal semantics is to not to

> [Push] the connection between [hyper-]intensions and extensions to the metalevel [but instead] to explicitly have this connection at the object level in order to be able to give it a computational interpretation. (Muskens, 2005, p. 486, my insertions)

The gist of the idea of a procedural, or computational, semantics according to which procedures receive mention already at the object-level (rather than merely at the meta-level) can be exemplified thus. '7 + 5' does *not* denote (refer to, name, express, pick out, ...) a *sum*; rather it denotes (etc.) a sum-producing *procedure*. So '7 + 5' and '5 + 7' are *not* two semantically identical (though notationally slightly distinct) terms for one and the same sum; rather they denote two distinct procedures issuing in the same sum. This claim is superficially at odds with Tichý's claim about '9 − 2' that this term names 7, rather than the procedure of subtracting 2 from 9 (Tichý,

1986a, p. 514). However, Tichý's general stance is that an expression represents "a definite intellectual journey to an entity" (Tichý, 1988, p. 284). Thus in (1986a) he hastens to point out that '3 + 4' names 7 in a different way than does '9 − 2.' In (1988, p. 224) it is stated unequivocally what, say, '7 + 5' and '$\sqrt{144}$' denote or name, namely two different procedures converging in the same product, the number 12. So neither term, or name, is semantically redundant. Nor is either of Anderson's 'Plus $(x, \text{Plus}(y, z))$' and 'Plus $(\text{Plus}(x, y), z)$.'

Anderson's strong reservations about the notion of an entity being 'implicitly denoted' and abolishing distinctness of denotations is, in my view, also borne out by another consideration. When an agent is related to a hyperintension, by entertaining an explicit attitude or comprehending a piece of language, it is essential that the agent not be also related to what the hyperintension yields. It may yield either a lower-order hyperproposition, a possible-world intension, or an extensional entity (or nothing at all, if we include 'vacuous' hyperpropositions). If a relation to, say, a hyperproposition would also relate the agent to what it yields then the agent would, *ipso facto*, also be related to a truth-condition or a truth-value. To be sure, it is internal to a hyperintension what it yields, so in some intuitive sense it is true to say that when a hyperintension is given, so is its product. But, it must be possible for us mortals to grasp a hyperintension and yet fail to realize what it yields. Computation of some sort ought to be required to intellectually proceed from a hyperintension to what it yields. Otherwise much of the rationale for distinguishing between hyperintensions and coarser intensions will be undermined, reinstating logical omniscience and blurring the line between equivalent and synonymous expressions.

The second case, credited to Richmond Thomason, is also due to Anderson, at (2001, p. 406, n. 28), and is no less complicated than the first. Are "The yellow car ran down the street" and "The car, which was yellow, ran down the street" synonymous? Neither Anderson nor Thomason mentions this, but it may well make a semantically salient difference that in the former sentence the adjective 'yellow' occurs in attributive position and in the latter in predicative position. For when occurring in attributive position it becomes an option to construe 'yellow' as denoting a modifier. And in this case it becomes definitely non-trivial whether a pair of sentences moulded in the vein of, "The yellow car ran down the street" and, "The car, which was yellow, ran down the street" are synonymous. If the first adjective ('yellow,' in the example at hand) denotes a modifier then it is sensible to consider whether the two sentences are synonymous at all only if the modifier that the adjective denotes is intersective or subsective rather than privative. The distinction between these three kinds of modification is the following, in rudimentary notation, where A represents a modifier, B a property, and A^* the property formed from A by means of pseudo-detachment.[13]

> *Intersective.* From $(AB)a$ infer $A^*a \wedge Ba$
> *Subsective.* From $(AB)a$ infer Ba
> *Privative.* From $(AB)a$ infer $\neg Ba$
> *Pseudo-detached.* From $(AB)a$ infer A^*a

The semantic status of the modifier involved affects whether one sentence is, potentially, synonymous with another. If the modifier in question is privative then the grammatical subject cannot be, 'The B, which was A^*,' because 'The AB' cannot denote an individual which is a B. For instance, "The forged banknote was spent on a toy car" does not mean that a particular with the property of being a banknote had the two further properties of being forged and being spent on a toy car, for a forged banknote is not a banknote (as little as a toy car is a car). Instead it means that the particular with the property of being a forged banknote (hence, with the property of being a forged something, as per the rule of pseudo-detachment) had the further property of being spent buying a toy car. Therefore, "The forged banknote was spent on a toy car" does not have the same truth-condition, let alone the same meaning, as "The banknote, which was forged, was spent on a toy car." The latter has a truth-condition which cannot possibly be satisfied, as it requires a particular to be both a banknote and a forged banknote. An additional point is that "The car, which was yellow, ran down the street" is not synonymous with, "The car which was yellow ran down the street." The former means that some unique car had two properties: being yellow and running down the street. The sentence may be equivalently rephrased as, "The car, which ran down the street, was yellow." The latter means that among several cars the only yellow one ran down the street; "The car which ran down the street was yellow" would not be an equivalent rephrasing.

---

[13] For discussion and justification of the rule of pseudo-detachment of a property from a property modifier, see Duží et al. (ms., §4.4). The bare bones of the rule is this. Let $[AB]$ be the property resulting from applying A to B, and let $[AB]_{wt}$ be the result of applying the property $[AB]$ to the world and time variables w, t to obtain a set, in the form of a characteristic function, applicable to a. Further, let $=$ be the identity relation between properties, and let p range over properties, x over individuals. Then the proof of the rule is this:

1. $[[AB]_{wt}\, a]$ assumption
2. $\exists p[[Ap]_{wt}\, a]$ 1, EG
3. $[\lambda x \exists p[[Ap]_{wt}\, x]\, a]$ 2, β-expansion
4. $[\lambda w' \lambda t'[\lambda x \exists p[[Ap]_{w't'}\, x]]_{wt}\, a]$ 3, β-expansion
5. $A^* = \lambda w' \lambda t'[\lambda x \exists p[[Ap]_{w't'}\, x]]$ definition
6. $[A^*_{wt}\, a]$ 4, 5, Leibniz' Law.

3. Going Hyperintensional in the λ-Calculus

The notion of procedural semantics contrasts with denotational semantics and goes back to the early 1970s, when procedural semantics was a research programme in AI.[14] One of its advocates, W. A. Woods, sums up

> two extreme interpretations of procedural semantics—a black-box approach in which the internal structure of a meaning function is inaccessible (only the input-output relations are available), and a low-level detail approach in which every detail of the operation of the meaning function procedure is considered a 'part of the meaning.' The former gives rise to a sense of equivalence between meaning functions that is too weak . . ., in that it counts as equivalent meaning functions whose input-output relations are the same (in all possible situations) regardless of the means by which those extensions are determined [thus identifying, e.g., tautologies]. The low-level detail interpretation is at the opposite extreme of this spectrum. Its sense of equivalence is so strong that it counts two meaning functions as different if they differ in any detail of their operation regardless of the extent to which they effectively do the same thing. The notion of abstract procedure that is required for the characterization of meaning functions appears to lie somewhere between these extremes—providing a degree of internal structure that is considered significant, while leaving certain low-level details unspecified (or specified with suitable don't-care conditions). (Woods, 1981, p. 329, my insertion)

So which low-level details are semantically insignificant? For instance, to use a standard example, if there is any point in distinguishing between

$$a = b$$

and

$$b = a$$

then what would that be? Well, one reason might be to have hyperintensional structure match syntactic structure perfectly, on a one-to-one basis. But what would be the *semantic significance* of the sequence in which the *relata* of a reflexive relation occur? Another reason might be that someone were so thick ('intellectually challenged') that they would agree that a was identical to b and deny that b was identical to a. Ostensibly, a fine-tuned doxastic logic would want to be able to draw a distinction between a being identical to b and b being identical to a. But is this bound to be a distinction without a difference? The jury is out on that one, so the more careful strategy is to incorporate very fine, perhaps excessively fine, distinctions into

---

[14] For a clash between the two approaches, see Johnson-Laird (1977) and Fodor (1978).

one's semantics; if need be, distinctions without a difference can be soaked up by imposing a slightly coarser criterion of individuation.

My suggestion is that the upper bound of hyperintensions must be cast in terms of procedural isomorphism. A hyperintension should, accordingly, be seen as an instruction of what operations to apply to what objects in order to arrive at an outcome. When an agent entertains an explicit attitude, the agent is *ipso facto* intellectually related to an instruction, or prescription, detailing how to form a lower-order entity by applying particular operations to particular (kinds of) entities. For instance, to grasp a hyperproposition is to understand a procedure, which is a structured instruction. Therefore, to grasp a sentential meaning is not to grasp a truth-condition, but a procedure whose product is a truth-condition.[15] Since we can distinguish between (fine-grained) procedures and (coarse-grained) truth-conditions, it becomes possible to model explicit attitudes in terms of the former and implicit attitudes in terms of the latter.

A procedural semantics demands a strict demarcation between the very procedure of calculating the square root of 144 and the result of the calculation (a functional value). (I) as well as (II) and (IV) above showed, first, that mathematical attitudes require hyperintensional complements and, second, that a hyperintension, or a term for it, must be capable of occurring in two different manners. $\sqrt{144}$, or '$\sqrt{144}$,' when taken out of context, exemplifies the bifurcation between procedure and product. $\sqrt{144}$ is always a procedure—to wit, the procedure of applying the square root function to 144—but in some contexts this very procedure (the very application of a function) is what matters, and in other contexts what matters is the product of this application (the number 12). What is needed, but won't be offered here, is a worked-out theory of the different manners in which hyperintensions, or terms denoting them, may occur.[16]

Duží and Materna (this volume) motivates the use of Tichý's Transparent Intensional Logic as the right hyperintensional logic to hook up with a procedural semantics. The key point is that TIL is based on the (typed) λ-calculus, whose two operations of functional abstraction and application are procedure-friendly. Thus, abstraction may be recast as the very proce-

---

[15] This holds for *empirical* sentences. It is interesting to note that a *realist* (as opposed to *anti-realist* or *idealist*) procedural semantics is not truth-conditional and, therefore, flouts Dummett's tenet that semantic realism is inherently truth-conditional. This suggests that Dummett's conception of semantic realism is too narrow, as it is not obvious how it would accommodate hyperpropositions.

[16] See Duží et al. (ms., §§2.6, 2.7) for exact logical rules governing how constructions occur *mentioned* or *used*, the latter dividing into *intensional* and *extensional* occurrences.

dure of forming a function, while application may be recast as the very procedure of applying function to argument. I refer to Duží and Materna (this volume) for the relevant definitions. To illustrate the basic idea, though, here is how Anderson's Plus example in 'Polish notation' translates into TIL. (Bear in mind that variables are TIL constructions and as such qualify as constructional constituents; I am leaving out Trivialization, since we do not need to *mention*, but are only *using*, hyperintensions.)

$$[= [Plus\,[x[Plus\,[yz]]]]\,[Plus\,[Plus\,[xy]]z]]$$

If x, y, z range over, say, the natural numbers, then $=$ is the identity relation between pairs of natural numbers. $[Plus\,[x[Plus\,[yz]]]]$ constructs (determines, presents, yields, etc., in the manner defined by the definition of the TIL construction Composition) a natural number, and so does $[Plus\,[Plus\,[xy]]z]$. The application of $=$ to these two numbers yields a truth-value: True, if the two sums are identical numbers; False, if not. So the overall procedure is a truth-value-constructing one. This procedure is the meaning of the formula, "$[= [Plus\,[x[Plus\,[yz]]]]\,[Plus\,[Plus\,[xy]]z]]$." If we λ-bind x, y, z to obtain the Closure

$$\lambda xyz[= [Plus\,[x[Plus\,[yz]]]]\,[Plus\,[Plus\,[xy]]z]]$$

what we get is a procedure producing a mapping from triples of natural numbers to truth-values. The procedure $\lambda xyz[= [Plus\,[x[Plus\,[yz]]]]\,[Plus\,[Plus\,[xy]]z]]$ is the meaning of the closed formula "$\lambda xyz[= [Plus\,[x[Plus\,[yz]]]]\,[Plus\,[Plus\,[xy]]z]]$." The question arises, however, whether, e.g., $\lambda xyv[= [Plus\,[x[Plus\,[yv]]]]\,[Plus\,[Plus\,[xy]]v]]$, swapping v for z, is also the meaning of "$\lambda xyz[= [Plus\,[x[Plus\,[yz]]]]\,[Plus\,[Plus\,[xy]]z]]$." The question, cast in the vernacular of the λ-calculi, is whether α-conversion preserves meaning. To place the question in a wider context, recall that the lambda-calculus has three rules of conversion: α-, β-, and η-conversion. These three rules may be characterized informally as follows.

α-*conversion* expresses the idea of 'renaming,' i.e. replacing one λ-bound variable x by another λ-variable y, typically in order to avoid collision of variables (i.e. a free occurrence of a variable becoming bound).

β-*conversion* expresses the ideas of functional application (β-reduction where a formal parameter x is substituted by a particular value a) and functional abstraction (β-expansion where a is replaced by a λ-bound x). In particular, the former provides the logic of predication: $(\lambda w(\lambda x(F_w x)a))$ is the procedure of predicating the property F (i.e. its extension at w) of a at w, and its product is a possible-world proposition. (In TIL the notation would be, "$[\lambda w[\lambda x[^0F_w x]^0a]]$," '$^0F_w$' abbreviating '$[^0F_w]$.')

η-*conversion* expresses the idea that two functions are identical if, and only if, they take the same arguments to the same values: $\lambda xFx = \lambda xGx$ if $(\forall x(F(x) = G(x)))$. The rule often serves to trim λxFx to simply F. (In TIL the notation is, "$[^0\forall \lambda x[= [^0Fx][^0Gx]]]$.")

Having launched the philosophical notion of synonymous isomorphism, Church went in search of a rigorous logical definition, dubbing the various candidate definitions 'Alternatives.' Here is a list of Church's Alternatives, with the TIL-based Alternative (½) thrown in:

- Alternative (0). α-conversion (synonymous isomorphism)
- Alternative (½). α- and η-conversion (procedural isomorphism)
- Alternative (1). α- and β-conversion
- Alternative (1'). α-, β-, and η-conversion
- Alternative (2). Logical equivalence

If Alternative (½) is embraced, then the answer to the question whether swapping v for z in the Plus example above preserves meaning is Yes. For a natural-language example, the following conversion also preserves meaning:

$$\lambda w \lambda t [^0Elephant_{wt} \, ^0a] \approx_\eta \lambda w \lambda t [[\lambda x [^0Elephant_{wt} x]]^0a]$$

According to Alternative (½), there is no semantic difference between obtaining the extension of the property of being an elephant at $\langle w,t \rangle$ by means of $[^0Elephant_{wt}]$ or $[\lambda x[^0Elephant_{wt}x]]$. However, β-conversion, at least in its unqualified version, goes too far (see Duží and Materna (this volume) and Duží et al., ms., §2.7).

The tentative thesis, then, is that hyperintensions are isomorphic modulo α- or η-convertibility, the idea being that α- or η-convertible procedures are semantically indistinguishable.[17] Identify hyperintensions with TIL constructions. Then, where $\approx$ is the relation of equivalence, since $\lambda x[...x...] \approx_\alpha \lambda y[...y...]$ it holds, in the mathematical case, that $\lambda x[^0+ x \, ^01] \approx_\alpha \lambda y[^0+ y \, ^01]$ and, in the empirical case, that $\lambda w \lambda t [\lambda x[^0F_{wt} x]] \approx_\alpha \lambda w' \lambda t' [\lambda y[^0F_{w't'} y]]$. Further, since $^0F \approx_\eta \lambda x[^0F x]$ it holds, for instance, that $\lambda x[^0+ x \, ^01] \approx_\eta \lambda y[\lambda x[^0+ x \, ^01]y]$ and that $\lambda w \lambda t[\lambda x[^0F_{wt} x]] \approx_\eta \lambda w \lambda t[^0F_{wt}]$.

Hence, the transitive closure of procedurally isomorphic hyperintensions is this. Any two hyperintensions C, D are *procedurally isomorphic* iff there are hyperintensions $C_1,...,C_n (n \geq 1)$ such that $C = C_1, D = C_n$ and any two C_i, C_{i+1} are either α- or η-equivalent.

---

[17] The origin of the current notion of procedural isomorphism is *Quid* (for 'quasi--identity'), as defined in Materna (1998, §5.3).

Conclusion

The present advocacy of Alternative (½) is as inconclusive and tentative as Anderson's of his own Alternative (1*) in (2001 p. 403). However, the shared conviction is that Frege and Church have succeeded in pointing in the right general direction of where to find an answer to the question of the calibration(s) of hyperintensions. In spite of the current prevalence of denotational semantics, some of us just never Russelled our Frege-Church and want instead to reconstrue it as an explicitly procedural semantics.

References

Anderson, C. A. (1998). Alonzo Church's contributions to philosophy and intensional logic. *The Bulletin of Symbolic Logic*, 4, 129–171.

Anderson, C. A. (2001). Alternative (1*): a criterion of identity for intensional entities. In C. A. Anderson and M. Zelëny (Eds.), *Logic, Meaning and Computation* (393–427). Kluwer Academic Publisher, Dordrecht.

Bealer, G. (1982). *Quality and Concept*. Clarendon Press, Oxford.

Bolzano, B. (1837). *Wissenschaftslehre*, volume 1–6. J. Seidel, Sulzbach.

Carnap, R. (1947). *Meaning and Necessity*. Chicago University Press, Chicago.

Church, A. (1941). The calculi of lambda conversion. In *Annals of Mathematical Studies*. Princeton University Press, Princeton.

Church, A. (1954). Intensional isomorphism and identity of belief. *Philosophical Studies*, 5, 65–73.

Church, A. (1993). A revised formulation of the logic of sense and denotation. Alternative (1). *Noûs*, 27, 141–157.

Cresswell, M. J. (1975). Hyperintensional logic. *Studia Logica*, 34, 25–38.

Cresswell, M. J. (1985). *Structured Meanings*. MIT Press, Cambridge.

Dretske, F. (1977). Epistemic operators. *The Journal of Philosophy*, 67, 1007–1023.

Duží, M., Jespersen, B., and Materna, P. (Manuscript). *Procedural Semantics for Hyperintensional Logic. Foundations and Applications of Transparent Intensional Logic*. Forthcoming as vol. 17 of *Logic, Epistemology and the Unity of Science*. Springer-Verlag.

Duží, M., Jespersen, B., and Müller, J. (2005). Epistemic closure and inferable knowledge. In L. Běhounek and M. Bílková (Eds.), *The Logica Yearbook 2004* (125–140). Filosofia, Czech Academy of Sciences, Prague.

Duží, M. and Materna, P. (2010). Concepts as structured meanings. This volume.

Fagin, R., Halpern, J. Y., Moses, Y., and Vardi, M. Y. (2003). *Reasoning About Knowledge*. MIT Press, Cambridge and London.

Fodor, J. A. (1978). Tom Swift and his procedural grandmother. *The Journal of Philosophy*, 6, 229–247.

Frege, G. (1892). Über Sinn und Bedeutung. *Zeitschrift für Philosophie und philosophische Kritik*, 100, 25–50.

Gaskin, R. (2008). *The Unity of the Proposition*. Oxford University Press, Oxford.

van Heijenoort, J. (1977a). Frege on sense identity. *Journal of Philosophical Logic*, 6, 103–108.

van Heijenoort, J. (1977b). Sense in Frege. *Journal of Philosophical Logic*, 6, 93–102.

Hintikka, J. (1962). *Knowledge and Belief*. Cornell University Press.

Hodes, H. T. (1982). The composition of Fregean thoughts. *Philosophical Studies*, 41, 161–178.
Johnson-Laird, P. N. (1977). Procedural semantics. *Cognition*, 5, 189–214.
Kripke, S. A. (1980). *Naming and Necessity*. Basil Blackwell, Oxford.
Levesque, H. J. (1984). A logic of implicit and explicit belief. In *AAAI-84: Proceedings of the Fourth National Conference on Artificial Intelligence* (198–202).
Lewis, D. (1972). General semantics. In D. Davidson and G. Harman (Eds.), *Semantics of Natural Language* (169–218). Reidel, Dordrecht.
Moschovakis, Y. N. (1994). Sense and denotation as algorithm and value. In J. Väänänen and J. Oikkonen (Eds.), *Lecture Notes in Logic*, volume 2 (210–249). Springer, Berlin.
Moschovakis, Y. N. (2006). A logical calculus of meaning and synonymy. *Linguistics and Philosophy*, 29, 27–89.
Muskens, R. (2005). Sense and the computation of reference. *Linguistics and Philosophy*, 28, 473–504.
Myhill, J. (1963). An alternative to the method of extension and intension. In *The Philosophy of Rudolf Carnap*, volume 11 of *Library of Living Philosophers* (299–310). Open Court and Cambridge University Press, La Salle and London.
Penco, C. (2003). Frege: two theses, two senses. *History and Philosophy of Logic*, 24, 87–109.
Prawitz, D. (1968). Propositions. *Theoria*, 34, 134–146.
Priest, G. (1992). What is a non-normal world? *Logique et Analyse*, 139–140, 291–302.
Richard, M. (2001). Analysis, synonymy and sense. In C. A. Anderson and M. Zelëny (Eds.), *Logic, Meaning and Computation: Essays in Memory of Alonzo Church*, volume 305 of *Synthese Library* (545–571). Kluwer Academic Publisher, Dordrecht.
Salmon, N. (1993). A problem in the Frege-Church theory of sense and denotation. *Noûs*, 27, 158–166.
Sullivan, A. (1998). Singular propositions and singular thoughts. *Notre Dame Journal of Formal Logic*, 39, 114–127.
Sundholm, G. (1994). Proof-theoretical semantics and Fregean identity criteria for propositions. *The Monist*, 77, 294–314.
Thomason, R. (1980). A model theory for propositional attitudes. *Linguistics and Philosophy*, 4, 47–70.
Tichý, P. (1966). K explikai pojmu 'obsah věty'. *Filosofický časopis*, 14, 364–372. Published as "On explicating the notion 'the content of a sentence'" in Tichý (2004).
Tichý, P. (1971). An approach to intensional analysis. *Noûs*, 5, 273–297. (Reprinted in Tichý, 2004).
Tichý, P. (1986a). Constructions. *Philosophy of Science*, 53, 514–534. (Reprinted in Tichý, 2004).
Tichý, P. (1986b). The indiscernibility of identicals. *Studia Logica*, 45, 251–273. (Reprinted in Tichý, 2004, pp. 293–304).
Tichý, P. (1988). *The Foundations of Frege's Logic*. De Gruyter, Berlin and New York.
Tichý, P. (2004). *Collected Papers in Logic and Philosophy* (V. Svoboda, B. Jespersen, and C. Cheyne, Eds.). Filosofia, Czech Academy of Sciences, Prague and University of Otago Press, Dunedin.
Woods, W. A. (1981). Procedural semantics as a theory of meaning. In A. K. Joshi, I. A. Sag, and B. L. Webber (Eds.), *Elements of Discourse Understanding* (300–334). Cambridge University Press, Cambridge.
Zalta, E. (1988). *Intensional Logic and the Metaphysics of Intentionality*. MIT Press, Cambridge and London.

Against Propositional Radicals*

Marián Zouhar

1. Introduction

There is a semantic tradition according to which a syntactically complete sentence, S, expresses a proposition, P, such that

(i) all simple constituents of P correspond to some simple constituents of S;
(ii) the structure of P mirrors the syntactic structure of S;
(iii) if S is indexical-free, every simple constituent of P is supplied exclusively on the basis of the semantic conventions associated with simple expressions;
(iv) if S is not indexical-free, some constituent of P corresponds to the indexical expressions involved in S and is determined, *inter alia*, by the context of use.[1]

---

* The paper has been supported by the grant project VEGA No. 2/0162/09.

[1] Since the aim of the paper is mainly critical and thus negative, I will use the notion of proposition usually adopted by the proponents of views to be discussed in this paper. According to this notion, propositions are structured entities and can be represented by—though not necessarily identified with—ordered tuples involving the semantic contents expressed or denoted by the constituent phrases of the sentences used to express these propositions. Though this view faces serious problems of its own, I ignore them in the present paper (for details, see Jespersen 2003). It would be better to take propositions as unstructured entities, as 'flat' mappings from possible worlds (or possible world/time pairs) to truth-values, and insert the structure corresponding to the syntactic structure of sentences at another level of semantic content (for an insightful suggestion along these lines, see Tichý's Transparent Intensional Logic; cf. Tichý 1988, 2004).

If points (i) through (iv) are satisfied, *P* is what may be called the proposition expressed by *S* or what *S* said (in a given context).

This is just a rough sketch of one basic principle of semantic theories dating from Frege, Russell, the early Wittgenstein and Carnap onwards. It is by no means complete because there are some very important queries that are to be resolved before any full-blooded semantic theory can be obtained. For example, this is just a sample of the questions to be answered:

What are propositions?
What can be counted as a simple propositional constituent?
Are propositions structured entities?
If they are structured, what is their structure?
If propositions are not structured, which entities (if any) are structured in their stead?

For some reasons, which will emerge later, this tradition is sometimes called *semantic minimalism*.

Nowadays many philosophers find minimalism outdated. They have gathered a lot of syntactically complete sentences which either fail to express complete propositions or express inappropriate propositions that cannot be identified with what is said, even if the above points (i)–(iv) are followed. The rival research program has it that these drawbacks can be removed if we take into account also the context in which sentences are used and admit that the context of use contributes somehow to (determining) the propositions expressed. This is called *contextualism*. Here are some examples used to support contextualism. The sentences

(1) "Everyone was sick"
(2) "The table is covered with books"
(3) "You're not going to die"
(4) "I've had breakfast"
(5) "This apple is red"
(6) "This melon is red"
(7) "Peter is tall"
(8) "Steel is strong enough"

may express, relative to some context of use, the following respective propositions (possible completions supplied by the context of use are suggested in square brackets):

(1a) *that everyone [who attended A's birthday party] was sick*
(2a) *that the table [in Room No. 221] is covered with books*

(3a) *that Bob is not going to die [from the cut]*
(4a) *that Bob has had breakfast [on August 24, 2008]*
(5a) *that Abe is red [on the outside]*
(6a) *that Mel is red [on the inside]*
(7a) *that Peter is tall [for a jockey]*
(8a) *that steel is strong enough [to support the roof]*.[2]

Bob in (3a) is the contextually determined referent of 'you' in an utterance of (3), Bob in (4a) is the contextually determined referent of 'I' in an utterance of (4) and Abe and Mel are the contextually determined referents of 'this apple' and 'this melon,' respectively, in utterances of (5) and (6), respectively.

Here is an illustration of an argument that is typical for contextualism:

> For a bird to be red (in the normal case), most of the surface of its body would have to be red, though not its beak, legs, eyes, and of course inner organs. Furthermore, the red color should be the bird's natural color, since we normally regard a bird as being 'really' red even if it is painted white all over. A kitchen table, on the other hand, is red even if it is only painted red, and even if its 'natural' color underneath the paint is, say, white. . . . Similarly, a red apple needs to be red only on the outside, but a red hat needs to be red only on its external upper surface; a red crystal is red both inside and outside, and a red watermelon is red only inside. A red star only needs to appear red from the earth, a red glaze needs to be red only after it is fired. A red pen need not even have any red part (the ink may turn red only when in contact with the paper). In short, what counts for one type of thing to be red is not what counts for another. (Lahav, 1989, p. 264)

The quotation suggests that 'is red' makes different contributions to propositions expressed by sentences involving this predicate. This is an important fact and Lahav derives from it various conclusions, in particular that the semantic behavior of 'is red' cannot be described in a *systematic* way and that the principle of compositionality breaks down as a result of the first conclusion. Perhaps, we might add that the above quotation suggests that there is no simple property of *being red*; instead, there are various other properties like *being red on the inside* or *being red when fired*, etc.

There are various versions of contextualism. Some contextualists claim that the sentences listed above cannot express anything truth-evaluable, if considered outside their context of use; in particular, it is only utterances of sentences that can be supposed to express truth-evaluable content. There are also contextualists who claim that some of the above sentences express

---

[2] Italicized 'that'-phrases will be used to denote propositions throughout the paper.

propositions that are inappropriate in a given situation; the completions are necessary to get appropriate propositions that can be taken as what is said in a given context. And according to some contextualists (at least some of) the above sentences express incomplete propositions or *propositional radicals* and the completions are necessary to obtain complete propositions.[3] To illustrate this idea, let's quote R. Buchanan and G. Ostertag:

> [T]he linguistic meaning of ['The F is G'] is a template of the form ... '[the x: Fx & $\_x$](Gx).' (Buchanan and Ostertag, 2005, p. 903)

The sign '_' indicates the empty place that is to be filled with something contributed by the context of use. It is this kind of contextualism I'll be concerned with here.

To illustrate the difference between minimalism and the last kind of contextualism, consider sentences (5) and (6) as uttered in a particular context. Since these sentences are syntactically complete, minimalism assumes that they express the following complete propositions (relative to the context of use fixing the referents for 'this apple' and 'this melon,' respectively):

(5b) *that Abe is red*
(6b) *that Mel is red*

where Abe and Mel are the contextually determined referents of 'this apple' and 'this melon,' respectively.[4] Importantly, 'is red' makes the same propositional contribution—it is the very same property *being red* that enters both propositions. However, according to the contextualists, 'is red' makes different contributions in different contexts, but some admit that there is a common core in its meaning appearing in all contexts. They deny that (5b) and (6b) are propositions; rather, they recognize that propositional radicals like

(5c) *that Abe is red _*
(6c) *that Mel is red _*

are what (5) and (6) express on basis of semantic conventions and compositionality alone. The empty places indicated by '_' are to be filled with con-

---

[3] Of course, there are also more fine-grained versions of contextualism according to which both (i) the completion of a propositional radical as well as (ii) the supplementation of the resulting proposition are required to get the proposition expressed by a given utterance. For example, Kent Bach suggests something along these lines in Bach (1994).

[4] For the sake of simplicity, I assume that complex demonstratives like 'this apple' or 'this melon' are referring expressions rather than quantifiers. In spite of this, I am sympathetic to the view defended by J. King (2001), for example, that they are, in fact, quantifiers of some kind.

textually determined constituents prior to any propositions being expressed relative to the context at hand.[5]

It should be plain that my usage of the label 'contextualism' is rather broad. For it covers also philosophers like Kent Bach who directed heavy criticism at various versions of contextualism.[6] Bach claims that every sentence expresses some semantic content independently of the context of use. Anyway, there are sentences that cannot express complete propositions, if considered outside their context of use (in spite of the fact that these sentences are indexical-free); they express, at most, what Bach calls 'propositional radicals' (cf. Bach, 1994, p. 127) and the context is necessary to fill empty places in the propositional radicals. In this way, Bach disagrees with the semantic tradition developed by Frege *et al.*; he does not admit that any well-formed sentence of a given language expresses a proposition. In many cases, Bach claims,

> a constituent is needed to specify what completes the proposition that is only incompletely expressed: some relevant class, respect, or contrast with respect to which the utterance is intended to be understood. (Bach, 1994, p. 128)

This is the reason why I take Bach to be a contextualist. Anyway, Bach is at most a mild contextualist because he tries to preserve the sharp boundaries between semantics and pragmatics (cf., for example, Bach, 1987, 2005). In

---

[5] It should be stressed that sentences like "Bob is tall" are elliptic in a certain sense (and, thus, is not a syntactically complete sentence). For expressions like "tall" are modifiers and there has to be something for them to modify. This sentence may be taken as shorthand for "Bob is a tall F"; in particular contexts it can be elliptic for "Bob is a tall man" or "Bob is a tall villager" or "Bob is a tall individual," etc. Something similar holds also for sentences like "This apple is red"; if "red" is viewed as a modifier, we may offer a deeper analysis. In particular contexts "This apple is red" may be elliptic for "This apple is a red apple" or "This apple is a red fruit" or "This apple is a red individual," etc. Anyway, these completions are not the ones invoked by the contextualists. The reason is that even if "This apple is red" is taken as shorthand for "This apple is a red fruit," it still remains open in which manner the object referred to by "this apple" is red. So, "This apple is a red fruit" should also express a propositional radical, if "This apple is red" expresses one. Analogously, if "Bob is tall" shortens "Bob is a tall man" in a particular context, the class of men need not be the comparison class invoked by contextualism. The contextualist might require "Bob is a tall man to be a jockey" as a proper completion; in this sense, the proper comparison class is that of jockeys. So, "Bob is a tall man" should express a propositional radical, if "Bob is tall" expresses one. In what follows, I shall ignore this kind of ellipsis. Thanks to Bjørn Jespersen for pointing this out to me.

[6] I have in mind here Bach's discussion with F. Recanati that is of particular importance; see Bach (2001), for example. Recanati wrote numerous papers and books in defense of a version of contextualism; for a recent formulation of his position see Recanati (2004).

fact, the idea of propositional radicals may be interpreted as resulting from his attempts to preserve the distinction.[7]

2. Contextualism and the Metaphysics of Properties

Observe that it is a necessary condition for the existence of the propositions *that Abe is red* and *that Mel is red* that the property *being red* exists. The dispute between minimalists and contextualists can be, in fact, reduced to the dispute concerning the very existence of properties like *being red, being tall, having had breakfast, being strong enough*, etc. The assumption that such properties do not exist can be ascribed both to mild contextualists as well as more radical ones. If this is so, the mild contextualists might say that every sentence of the form "X is red" expresses a propositional radical rather than a proposition. On the other hand, a more radical contextualist might maintain that such sentences fail to express any semantic content outside their context of use.

Observe that if we admit that the property *being red* does exist we would need an extra argument to demonstrate that this property plays no role in the semantics of 'is red' and that the predicate does not contribute this property to propositions expressed. Consider a sentence of the form "X is red" again and suppose that X is the semantic content of 'X' and *being red* is the semantic content of 'is red.' Given these suppositions, it is pretty natural to admit that, relative to a context of use, the sentence expresses the proposition *that X is red*. It would be quite difficult to block this conclusion. We would need very persuasive arguments which would demonstrate that "X is red" fails to denote *that X is red* in spite of the assumptions that 'X' denotes X and 'is red' denotes *being red*. I am afraid that such arguments would be unconvincing because any attempt to formulate them would be overridden by the truism that the semantic content of "X is red" is composed somehow of the semantic contents of 'X' and 'is red.' So, there is one possibility remaining for the contextualists—to claim that 'is red' fails to express any

---

[7] Bach distinguishes between two different processes: If a sentence expresses, on basis of semantic conventions and compositionality, a propositional radical, a *completion* of a conceptual constituent supplied by the context of use is required to get a proposition. On the other hand, if a sentence expresses a so-called *minimal* proposition the context of use is used to *expand* this proposition in order to get what was communicated by the speaker. The two processes lead to what is said by a given utterance. This is the reason why Bach distinguishes what he calls conversational implicitures (resulting either from completions or from expansions) from Grice's conversational implicatures (cf. Bach, 1994, p. 140).

property or, rather, that 'is red,' when used in a given context, expresses various properties, e.g., *having red skin, having red pulp, being red in such and such light*, etc.[8] Thus, before discussing some problems connected with propositional radicals, I'll touch briefly the metaphysical issue concerning the existence of such properties.

The contextualists are accustomed to reasoning as follows: The speaker utters a sentence, S, in a particular context, C. Now imagine a verification procedure that is to be undertaken in order to evaluate S (with respect to C). To put it in a simplified way, the constituent phrases of S pick out entities (individuals, properties, relations, etc.) that are to be involved in the verification procedure. However, in many cases the verification procedure has to pay attention also to some specific features of C that are not depicted by any constituent parts of S. So, given the assumption that the semantic content of a sentence should suggest what the verification procedure for the sentence looks like, it is plain that the semantic content of the sentence—i.e., the proposition it expresses—in the context cannot be determined solely by the semantic contents of its constituent phrases and their mode of composition. Therefore, the semantic content of S with respect to C is determined both by the semantic conventions associated with the constituent phrases of S as well as some features of C.

Now, as far as I can see, this sort of argument runs two different things together. We should distinguish (i) the property instantiated by an individual from (ii) the standard criteria for evaluating whether the individual in question does instantiate the property or not. Properties are usually represented as functions from possible worlds (or possible world/time couples) to sets of individuals (or of some other kinds of objects).[9] In particular, the property *being red* is represented as a function mapping possible worlds to sets of red individuals. However, this function is silent on what counts as red for a particular individual—this is a metaphysical matter having nothing to do with the logician's tools used to represent properties. It should be taken for granted that there are some standard criteria laying down what counts as being red for particular kinds of individuals. Anyway, these criteria are, so to speak, external to the property *being red* as such. Of course, these criteria may differ when different kinds of things are taken into account. But this should not lead us to embrace the conclusion that these different kinds of things instantiate different properties. For example, this apple and this

---

[8] The proponents of propositional radicals might say that 'is red' expresses the two-place relation *(. . .) is (a) red (. . .)* rather than the one-place property *being red*.

[9] For the sake of simplicity, I take into account only individual properties.

melon are both red despite the fact that they are red in different manners. In other words, being red in this or that way does not imply not being red.[10]

Lahav's quotation from the previous section strongly suggests that different kinds of things are associated with different standards of what is to be red. However, he implies that these kinds of things instantiate different properties. To put it differently, Lahav's evidence having something to do with standards of what is to be red for this kind of individual or other is interpreted as evidence for instantiating different properties. In a sense it is true that those different kinds of things possess different properties—this melon has the property *having red pulp*, this apple has the property *having red skin*, this glaze has the property *being red after being fired*, etc. Anyway, this does not go to show that all these things do not instantiate the property *being red*. What this suggests is merely that

melons are taken to instantiate the property *being red* provided they instantiate the property *having red pulp* as well;
apples are taken to instantiate the property *being red* provided they instantiate the property *having red skin* as well;
glazes are taken to instantiate the property *being red* provided they instantiate the property *being red after being fired* as well;
etc.[11]

It means that these kinds of objects are taken to have the property *being red* conditionally, i.e., they have to instantiate also some other property over and above *being red*. When I was speaking about standard conditions of property ascription or property instantiation, I had in mind conditions of the above sort. Certain properties can be instantiated by individuals only provided some more specific properties are instantiated by them. And the standard conditions of property ascription record these dependencies.[12]

---

[10] It is a truism that both melons and apples are said to be red. It means that they co-instantiate the very same property *being red*. This is an empirical piece of evidence undermining the contextualist's hypothesis that the two kinds of fruits instantiate different properties as far as their color is in question. This point has been highlighted by H. Cappelen and E. Lepore (2005).

[11] Observe that the above co-instantiation rules are not biconditionals. It means that a melon, say, may have the property *being red* even when it has some other property instead of *having red pulp*; when unripe, it may be taken to be red because it is painted red, for example.

[12] Pavel Tichý introduced a handy notion of *requisites* that can be used in this context. A property P is a requisite of a property Q *iff* for any individual o, possible world w and time t it holds that if o instantiates Q in w and t, then o has to instantiate P in w and t as well (cf., Tichý 2004, p. 360, for example; my definition is more general than Tichý's). So it

The contextualists seem to make an incorrect move. Instead of recognizing the above co-instantiation dependence, they conclude that objects cannot have the property *being red* (*simpliciter*) and, therefore, sentences of the form "*X* is red" cannot express propositions of the form *that X is red*. Anyway, this move is not sufficiently substantiated.

I shall further explain my point by an uncontroversial example; or so I hope. Some individuals have the property *being a composer*. I hope that contextualists are willing to admit that there is such a property and that sentences of the form "*X* is a composer" do express propositions of the form *that X is a composer*. This property is instantiated by Richard Wagner, for example. Now the reason why Wagner instantiates this property is because he composed several musical works. Had he devoted his life solely to literature or sheep farming, he could not instantiate this property. This fact points to a criterion of what counts as being an instance of the property *being a composer*—the property can be instantiated only by those individuals who produced some kind of work. Wagner composed several operas and is, thus, an operatic composer. Anyway, when people utter

(9) "Wagner is a composer"

they do not express propositions such as

(9a) *that Wagner is an operatic composer*
(9b) *that Wagner is the composer of* Tristan und Isolde
(9c) *that Wagner is the composer of* Tristan und Isolde *and* Der Ring des Nibelungen

Rather, they express the proposition

(9d) *that Wagner is a composer.*

End of story. Of course, to evaluate (9) is to test whether Wagner is a composer. And it means that we should assess whether he composed at least one piece of music. We find that there are about 13 operas to his credit and this would be the reason to evaluate (9) as true. In spite of this it would hardly be controversial to say that (9) expresses (9d) rather than any of (9a) through (9c). Of course, it holds that

---

can be said that the property *being red* is a requisite of the properties like *having red skin* or *being red on the outside*. In this sense, there is a necessary connection between properties; but this is not to say that the properties like *being red* do not exist. I am indebted to Marie Duží for pointing this out to me.

Wagner is taken to instantiate the property *being a composer* provided he instantiates the property *being an operatic composer* as well;

Wagner is taken to instantiate the property *being a composer* provided he instantiates the property *being a composer of Tristan und Isolde* as well;

Wagner is taken to instantiate the property *being a composer* provided he instantiates the property *being a composer of Tristan und Isolde and Der Ring des Nibelungen* as well;

etc.

Anyway, this is not to say that the property *being a composer* does not exist and that Wagner cannot instantiate this property.

There is a clear analogy between *being a composer* and *being red*. An individual is a composer if s/he satisfies certain conditions and the same holds for *being red*. An individual is a composer only if s/he authored at least one musical work; an individual is red only if it is red somewhere. In both cases an individual has to co-instantiate an additional property. To put it differently, in both cases there are some standard conditions concerning what counts as instantiating the property in question that should be distinguished from the property instantiated.

Given these standards of property instantiation, it is easy to imagine how standard verification procedures for propositions should look like. When we want to evaluate whether the proposition *that Mel is red* is true or false, we should investigate the pulp of this melon rather than any other part of it (this is not to say that in some non-standard cases it is the skin of the melon that is under scrutiny). It is also commonly known that when we say about this apple that it is red, the standard verification procedure of the proposition expressed is that we investigate its skin rather than its pulp or any other part of it (again, this is not to say that in non-standard cases we may not be interested in the color of its pulp). To stress the point, this does not mean that apples and melons instantiate, instead of the property *being red*, only properties like *having red skin* and *having red pulp*, respectively. Thus, an utterance of (5) ("This apple is red") need not express, relative to a particular context of use, the proposition *that Abe is red on the outside*; it expresses, relative to the context in question, the proposition *that Abe is red*. Of course, when we evaluate (5), we should test whether this apple is really red and this means that we should find out whether it is red *somewhere*. This apple is red because it is red somewhere. And the standard conditions of property instantiation associated with the property *being red* tell us that when an apple is under discussion the property is possessed by its skin rather than by its pulp. Anyway, it would be completely unjustified to demand that this fact should be reflected in the semantics of 'red.'

Given the above considerations, there are properties such as *being red*, pure and simple. And if this is the case, we should better admit that these properties are what constitute the propositional contributions of predicates like 'is red.' Otherwise, very strong arguments are required in order to demonstrate that the semantics of these predicates is much more complicated and much more dependent on various contextual features. Arguments purporting that speakers may use sentences of the form "*X* is red" to say something more specific, e.g. that *X* is red on the inside, are by no means persuasive.[13]

3. Propositional Radicals Questioned

We now return to propositional radicals. Of course, when we admit that there are properties like *being red* one should work very hard to demonstrate that the predicate 'is red' fails to contribute this property to propositions expressed by sentences involving this predicate. What the proponent of propositional radicals seems to suggest is that 'is red' contributes a two-place relation rather than a one-place property to the propositions expressed by utterances of "*X* is red." However, the very idea of propositional radicals seems to be rather suspicious. Let us look more closely at propositional radicals and discuss some problems connected with them.

Those authors who recognize propositional radicals as self-contained semantic entities do not say much about them. We only come to know that they are proposition-like entities, but cannot be evaluated as true or false. Propositional radicals involve empty places and become propositions when the empty places are filled with appropriate propositional constituents. We are told that propositional constituents filling empty places in propositional radicals are supplied by the context of use rather than by semantic conventions. We are also told that propositional radicals are expressed by syntactically complete sentences and that they are what some sentences express on the basis of semantic conventions and syntactic composition alone. Now, several questions remain open and I will address some of them, in particular:

Where does the empty place occurring in a propositional radical come from? How is it possible to distinguish sentences expressing propositions from those expressing propositional radicals?

---

[13] I am quite sympathetic to the position recently developed by E. Corazza and J. Dokic in their joint paper (2007). They argue that the semantics of "*X* is red" remains constant across various utterances in various contexts; what changes are the situations with respect to which these utterances are evaluated.

How many empty places are there, or could there possibly be, in a propositional radical?

Propositional radicals are usually represented as self-contained semantic entities involving (at least one) empty place. It is also agreed that propositional radicals are expressed by certain sentences on the basis of the semantic conventions associated with the constituent expressions and their composition alone. These sentences are supposed to be syntactically complete while propositional radicals are deemed to be semantically incomplete. This should be explained (or explained away) somehow. How is it possible that expressions occurring in a syntactically complete sentence provide for this essential incompleteness at the level of the semantic content? There seems to be something in the propositional radical that cannot correspond to any sentential constituent, namely the empty place itself. The reason is that the sentence is syntactically complete and, therefore, cannot contain any empty place. So, it is to be explained how it is possible that propositional radicals involve empty places corresponding to nothing in the sentences; in fact, no semantic convention relevant for expressions involved in a given sentence may introduce the empty place into its semantic content.

Of course, there are no empty places that might be visible at the level of surface structure. So if there are some empty places, they may occur, at most, at the level of logical form. In fact, this idea is suggested by Buchanan and Ostertag in the short quotation in Section 1. However, this option is no less dubious. One reason is that it is hard to imagine any syntactic process that would give rise to an empty place at the level of the sentence's logical form. Another reason is that if a sentence is syntactically complete, it has to be such both at the level of its surface structure and its logical form, provided we adhere to some well-established theories of sentences. For there are independent reasons for treating sentences as ordered couples ⟨SS, LF⟩, where SS represents the surface structure of the sentence and LF represents its logical form.[14] It is easy to see that if there is an empty place at the level of the logical form, the sentence identified with the couple ⟨SS, LF⟩ is by no means syntactically complete. And, conversely, if the sentence identified with the couple ⟨SS, LF⟩ is complete, there can be empty places neither at the level of its logical form nor at the level of its surface structure. So, we do not know how to answer the question concerning the origin of empty places in propositional radicals.

The proponents of propositional radicals may insist that the above kind of reasoning is mistaken through and through. They may argue that syntac-

---

[14] For some evidence supporting this theory of sentences see, for example, Neale (1993).

tically complete sentences do express propositional radicals, provided only the sentences involve n-place predicates expressing $n+1$-place relations. They may claim, for example, that 'is red' expresses, on the basis of semantic conventions alone, the two-place relation *(. . .) is (a) red (. . .)* and, when uttered in a particular context, expresses the property resulting from this relation by filling the second argument place. But it is open to me to object that this cannot be taken for granted. The idea that syntactically complete sentences may express incomplete semantic entities is something that is in need of vindication; or so I tried to show in the previous section. For, given that there is a property *being red*, we need extra arguments demonstrating that the predicate 'is red' fails to express this property. To put it differently, it should be explained why an n-place predicate is capable of expressing an $n+1$-place relation; for we naturally assume that an expression has as many empty places as has the relation it is used to express.

Anyway, let's suppose that some sentences do express propositional radicals, even when it is rather unclear how they do that. Then another problem comes to the fore: we have no clear syntactic criterion to distinguish between sentences expressing propositions from those expressing propositional radicals only. For note that in both cases only syntactically complete sentences are involved. So why should we admit that some of them express propositions while others express propositional radicals?

In fact, there is one criterion alluded to by the proponents of propositional radicals, but it is based on semantic-*cum*-pragmatic considerations. It deals with the verifiability of sentences. A sentence expresses a proposition (relative to the given context of use), if its literal meaning represents a procedure that would lead us to a particular truth-value (with respect to a particular possible world and/or time), if properly executed. On the other hand, if a sentence does not represent any procedure leading to a truth-value (with respect to a particular possible world and/or time), it has to express a propositional radical only. To put it differently, if the procedure encoded by the sentence determines a truth-value (with respect to a particular possible world and/or time), the procedure can be identified with a proposition; otherwise it is just a propositional radical. Of course, this is just a rough and inaccurate description of what is to be done in order to decide whether a sentence expresses a proposition, but this formulation is still sharp enough to make it clear that it cannot work properly.

Note that some sentences are hard to classify; they might fall in either of the two categories. Consider (4) ("I've had breakfast"). Some might claim that, when uttered in a particular context, it expresses the proposition *that Bob has had breakfast* (where Bob is the contextually determined referent of 'I'). But it might be argued that if anybody has had breakfast, they have had it sometimes in the past; so there are propositions such as *that Bob has had breakfast today*, *that Bob has had breakfast this morning* or *that Bob has had breakfast sometimes in the past*, and utterances of (4) can be used to express propositions of this kind. Consequently, *that Bob has had breakfast* is not a proposition but, at most, a propositional radical. So, if (4) expresses anything at all on the basis of semantic conventions and compositionality alone, it is to be a propositional radical rather than a proposition. On the other hand, Bach would say that (4) expresses a proposition because additions such as *today* or *this morning* or *sometimes in the past* result from so-called expansion (rather than completion); and it holds that what is expanded is a proposition and this proposition is supplemented by additional material such that a more comprehensive proposition results.

Analogous situations can be devised for other examples as well. And since we do not know how to handle them, the semantic-*cum*-pragmatic criterion is not impeccable. It would be better to have a syntactic criterion distinguishing sentences expressing propositions from those expressing propositional radicals. But we have none.

Be that as it may, let us assume, for the sake of further illustration, that (4) does express a propositional radical rather than a complete proposition.[15] Now, another problem with propositional radicals is that it is not quite clear how many empty places are to be found in the propositional radical expressed by a particular sentence. This problem is an extension of the above worry. The former problem dealt with the question whether the meaning of a given sentence involves at least one empty place or else none; now we have the problem of whether the meaning of a given sentence involves one empty place or more. What I have in mind here is that we have no clear syntactic criterion to help decide how many empty places are to be found in the propositional radical expressed by a particular sentence. Consider an example. Imagine that someone utters (4) in a particular context. Of course, this context is rich enough to make various completions possible. Thus, when uttered in this context, (4) may express any one of the following propositions:

(4a) *that Bob has had breakfast on August 24, 2008*

(4b) *that Bob has had breakfast in Bob's hotel on August 24, 2008*

(4c) *that Bob has had breakfast in Bob's hotel on August 24, 2008 while reading newspapers*

---

[15] Nothing hinges on this assumption; if you think (4) actually expresses a proposition, just choose another example.

In fact, there is no upper bound that would prevent us from incorporating further features of the context into the proposition expressed by (4) with respect to the context at hand. We might say that (4a) results from filling one empty place in the propositional radical

that Bob has had breakfast _,

while (4b) results from filling two empty places in the propositional radical

that Bob has had breakfast _ _

and (4c) results from filling three empty places in the propositional radical

that Bob has had breakfast _ _ _.

Alternatively, we might say that (4a)–(4c) all result from filling one empty place in the propositional radical

that Bob has had breakfast _.

Now, which one of the two hypotheses is to be preferred? I'm afraid that since we have no syntactic evidence justifying the number of empty places in a given propositional radical, we can substantiate neither the former nor the latter. In this connection, an air of arbitrariness is lurking behind the whole idea of propositional radicals. What we can do is just to decide by *fiat* that propositional radicals involve, for example, only one empty place and no more or that they involve exactly two empty places, etc.

* * *

There are at least two sorts of problems connected with propositional radicals. One of them deals with internal problems having to do with the fact that they are not well defined; as a result, there are doubts concerning their structure and correspondence with the sentences purportedly expressing them. The other sort of problems detracts from their explanatory power; it is not clear what they are good for. Since predicates like 'is red' contribute properties like *being red* to the propositions expressed, there seems to be no job propositional radicals might eventually perform.[16]

---

[16] I am indebted to Marie Duží and Bjørn Jespersen for their numerous comments that helped me to clarify and improve argumentation.

References

Bach, K. (1987). *Thought and Reference*. Clarendon Press, Oxford.
Bach, K. (1994). Conversational implicitures. *Mind and Language*, 9, 124–162.
Bach, K. (2001). You don't say? *Synthese*, 128, 15–44.
Bach, K. (2005). Context ex machina. In Z. G. Szabó (Ed.), *Semantics Versus Pragmatics* (15–44). Oxford University Press, Oxford.
Buchanan, R. and Ostertag, G. (2005). Has the problem of incompleteness rested on a mistake? *Mind*, 114, 889–913.
Cappelen, H. and Lepore, E. (2005). *Insensitive Semantics. A Defense of Semantic Minimalism and Speech Act Pluralism*. Blackwell, Oxford.
Corazza, E. and Dokic, J. (2007). Sense and insensibility or where minimalism meets contextualism. In G. Preyer and G. Peter (Eds.), *Context-Sensitivity and Semantic Minimalism* (169–193). Oxford University Press, Oxford.
Jespersen, B. (2003). Why the tuple theory of structured propositions isn't a theory of structured propositions. *Philosophia*, 31, 171–183.
King, J. C. (2001). *Complex Demonstratives. A Quantificational Account*. MIT Press, Cambridge, MA.
Lahav, R. (1989). Against compositionality: The case of adjectives. *Philosophical Studies*, 57, 261–279.
Neale, S. (1993). Grammatical form, logical form, and incomplete symbols. In A. D. Irvine and G. A. Wedeking (Eds.), *Russell and Analytic Philosophy* (97–139). University of Toronto Press, Toronto.
Recanati, F. (2004). *Literal Meaning*. Oxford University Press, Oxford.
Tichý, P. (1988). *The Foundations of Frege's Logic*. De Gruyter, Berlin and New York.
Tichý, P. (2004). *Pavel Tichý's Collected Papers in Logic and Philosophy* (V. Svoboda, B. Jespersen, and C. Cheyne, Eds.). Filosofia, Czech Academy of Sciences, Prague and University of Otago Press, Dunedin.

Beyond Minimalism

Truth's Role in Understanding

Filip Buekens

1. The Explanatory Role of Truth-Involving Habits

The *concept* of truth, in tandem with other concepts, plays an essential role in habits we employ when aiming at understanding others and being understood. This accounts for the indispensability of the explanatory role of the concept of truth. Like other central concepts such as *identity* or *cause*, the concept of truth is primitive hence unanalyzable, but application of a primitive, unanalyzable concept may be indispensable in explanations, and its centrality is accounted for when it is shown that the proposed explanations are ineliminable in our cognitive lives. Overlooking the centrality of these truth-involving habits might be excusable due to a confusion between the role of the truth-*predicate* in stating a semantic theory for a language and the role of applying the *concept* of truth in cognitive strategies that aim at understanding others—what they say, assert or implicate. From its central role other and better-known features of the concept follow: its uniformity, its language-independency, but also its notorious irrelevance in other explanatory projects, such as explanations of the success of science (Horwich, 1998a). I sketch the main lines of this novel approach, suggest some applications and compare the project with minimalism and some recent criticisms of the original Davidsonian program.

The proposed account is not well marked on contemporary maps. Traces can be found in Davidson, but the argument doesn't rest on the distinctively Davidsonian claim that a theory of meaning for a language L (with variable L) must take the form of an empirically adequate theory of truth for L. The position is consistent with the minimalist insight that truth cannot be analyzed, and so an account of the explanatory role of applying the concept will not follow deductively from its analysis.

To get the idea, consider how H. P. Grice reminds us of the fact that communicative exchanges are purposeful and cooperative enterprises governed by the *Cooperative Principle* and further maxims:

(Cooperative Principle)

Make your conversational contribution such as is required, at the stage at which it occurs, by the accepted purpose or direction of the talk exchange in which you are engaged. (Grice, 1989a, p. 29)

The maxim that will interest us is the *Maxim of Quality*:

(Quality)

Try to make your contribution one that is true. Specifically: Do not say what you believe to be false; do not say that for which you lack adequate evidence. (Grice, 1989a, p. 29)

The role of the concept of truth in the Maxim of Quality—an explanatory principle—cannot be accounted for by minimalist approaches to truth for its role is not confined to *expressing* an explanatory principle. More specifically, *the maxim does not summarize applications of the form "if X contributes to the conversation that p, then assume that p etc."* but states a principle of practical reason—a central habit—applied in understanding others. The reason is straightforward: the Maxim of Quality helps us understanding what a speaker contributes to the conversation, under the assumption that what he says is true, but understanding what he, under that assumption, *says* or *implicates* cannot be *assumed* in stating the maxim. We do not first understand what a speaker says or implicates and *then* generalize over what speakers tend to say; the role of the maxim is to get us into a position in which we come to understand various aspects of utterances by *assuming* them to be true. They need not be true, but the assumption that they have that property (i.e. fall under the concept of being true) is part of the process that yields understanding of what speakers say and implicate. Speaking from the point of view of the applier of the maxim: he is assumed to have a tacit grasp of the concept for grasping the maxim is assumed in Gricean rational reconstructions of the ability of speaker-hearers to identify what is said, asserted and conveyed.[1]

I suggest the concept of truth plays its explanatory by figuring in explanatory principles which, following P. F. Ramsey, I'll call *habits*:

---

[1] That the role of the Gricean principles is not confined to identifying what is conversationally implied, is forcefully argued by Carston (2002), Recanati (2004) and Soames (2008).

The human mind works essentially according to general rules or habits, a process of thought not proceeding according to some rule would simply be a random sequence of ideas; whenever we infer A from B we do so in virtue of some relation between them. We can therefore state the problem of the ideal as *"what habits in a general sense would it be best for the human mind to have?"* This is a large and vague question which could hardly be answered unless the possibilities were first limited by a fairly definite conception of human nature. (Ramsey, 1990, p. 90)

A complete argument for the indispensability of *truth-involving habits* would consist in a defense of at least two different, interlocking claims:

(i) **Indispensability**: Understanding others—their language and, more broadly, their mental economy—is an essential characteristic of the kinds of mind that characterize us. We wouldn't have the same mental economy if we were not constantly involved in the process of understanding others, and in trying to be understood. Other projects, like knowledge transmission and extended cooperative actions, rely on it.[2]

(ii) **Essentiality**: rational reconstructions of understanding, an indispensable practice, rely on the exercise of cognitive habits in which the concept of truth occurs essentially. It cannot be substituted *salve explicatione* for other concepts (like warranted assertibility, or being justified) in those habits without explanatory loss.

That the concept of truth as it figures in explanatory principles is expressed by the predicate ('is true') that figures in an explicit statement of a semantic theory for a language L illustrates that the concept that figures in explanatory principles *can* also figure in semantic descriptions, but the concept's explanatory role is not *derived from* the use of the predicate in expressing semantic properties of sentences. Moreover, the concept the familiar predicate expresses *also* guides our coming to understand implicatures, new parts or extensions of the language, stipulations, etc. Understanding is a state of mind that has a specific causal history *constitutive* to it, in the sense that no state of mind a person would amount to understanding another speaker were the truth-involving habits not involved in the explanatory process that brings about that state of understanding. (Compare: X performing

---

[2] The approach does not exclude that there can be a continuum from principles and habits involved in understanding others to habits and explanatory principles governing other practices that depend on understanding, as those governing knowledge acquisition via testimony, or the role of experts. Indispensability excludes an account of the explanatory role of (the concept of) truth through practices we are merely *contingently* connected to.

φ skillfully constitutively requires a causal history of X in which φ-ing was learned.)[3] We argue that in the process that leads to understanding, truth-involving explanatory principles are essential. If both claims are true, it follows that possession of the concept if truth is a presupposition of understanding others.

The approach assumes that the concept of truth figures in what are sometimes called *transcendent* modes of thinking.

Gila Sher (2004) identifies the *immanent mode of thinking* as a mode of direct engagement with an external subject matter, normally a structure of objects possessing properties and standing in relations, where the engagement is typically attributive, *i.e.* consists in the attribution of properties and relations to the objects in the structure. It is a mode of thinking in which we look *through* truth, not *at* truth. The minimalist claim that our understanding of truth is *fully* captured by accepting instances of the scheme

(M) $<p>$ is true iff p

(read as: the proposition that p is true iff p) assumes an immanent mode of thinking: believing that p is true is tantamount to believing that p. You are thereby thinking about a subject matter, and that subject matter allows for thinking about it *by* thinking that the proposition that p is true.

An application of the transcendent mode of thinking involving the concept of truth is exemplified by reflecting (explicitly or not) on an utterance not yet, or not yet fully understood with a view to coming to understand it. When at work in that transcendent mode of thinking the application of the concept of truth doesn't assume that one understands the specific type of content of the utterance it is applied to, and it is applied with a view to understand a specific type of content, with the further intention of acquiring knowledge of the world and/or the mind of the speaker. Were our cognitive lives confined to immanent uses of the concept of truth, it would be meaningless to even ask what the further role of the concept of truth would be, other than its 'disappearance' function nicely evoked by W. H. Auden in his poem *Words*:

> A sentence true makes a world appear
> where all things happen as it says they do.

That ascribing the property of being true figures in ineliminable cognitive habits involving our coming to understand others, suggests that what are

---

[3] The comparison is due to T. Williamson. This metaphysical point about understanding as requiring the 'right' causal history seems to figure behind Davidson's famous Swampman-intuition.

sometimes called *blind* truth ascriptions are possible, and that such ascriptions cannot be paraphrased away. (We assume that a sequence of sounds *is* true, without knowing what it says or implies about the world). But the point must be stated carefully. Frege (1918/1999) noticed that the truth-predicate is redundant when asserting *that it is true that I smell the scent of violets* (Frege, 1918/1999, p. 88), but he didn't defend a redundancy theory since he allowed blind truth ascriptions. (I can say of a scientists's hypothesis that it is true, without understanding the hypothesis.) Azzouni (2005) takes ineliminable blind truth ascriptions to be "the *raison d'être* for the presence of the truth idiom in the vernacular"—that is, "that it's because of them, and them alone, that ordinary language craves the truth idiom" (Azzouni, 2005, p. 275). This, he contends, is a crucial argument against minimalism, for ineliminably blind truth ascriptions cannot be paraphrased by the speaker into statements he understands yet do not contain the truth predicate. But this is not exactly right, for a minimalist can reply that he need not require that it is the *speaker himself—the user of the truth-predicate*—who must be in a position to give a truth-less paraphrase of a singular statement or generalization involving the predicate 'is true.'[4] The concept of a blind truth transcription remains interesting however, because the project in which ascribing the property of being true plays an explanatory role involves ascriptions in which it is assumed that what it is applied to is true, and this with a view to come to understand (part of) the utterance's communicative content.[5] The point of view from which one recognizes (M) as being true assumes that one is already immersed in the language, hence knows which propositions are semantically or pragmatically expressed in speech acts, while the point of view assumed in the process constitutive for understanding a speaker doesn't. Our interpreting practices always assume the possibility of rising to a transcendent level, in order to come to understand what speakers say and mean.

Theories about what grasping the meaning of the truth-predicate in one's home language consists in—accepting sentences like (M) as true, for example—can remain silent about the role of the *concept* of truth the predicate 'is true' expresses. It is conceptually coherent to assume that understanding the truth *predicate* is learned long after truth-involving habits of which the speaker has implicit, practical grasp, were already at work in the process of

---

[4] Azzouni (2005, p. 274): a blind truth ascription is ineliminable if the ascriber cannot replace it with an explicit one by using the sentence(s) the term in the blind truth ascription picks out. Explicit truth ascriptions are eliminable.

[5] It is not an option that the speaker explains, in his language, what the utterance means to the audience.

coming to learn her first language, and, by extension, coming to understand what other persons say, mean, think and do.[6] On the other hand, that very same concept is adequately expressed by the truth-predicate in (M). The predicate 'is true' means the same in (M) *and* Grice's maxim of quality.[7]

The possibility that truth ascriptions are not (and in an important sentence cannot be) confined to sentences of the home language is an essential mark of the transcendent mode of thinking involving truth and was acknowledged by Tarski (1944/1999), for he took assignments of truth conditions to be resulting in empirical statements (under the supposition that the axioms they were derived from were interpreted as empirical generalizations over uses of terms and predicates in a language).[8] If truth conditions can be assigned on the basis of empirically correct reference—and satisfaction axioms, we can ask what the role of the concept of truth—the concept that is expressed by the predicate 'is true' that figures in assignments of truth conditions—might be in the process of collecting empirical evidence for the correctness of the axioms. The fact that truth conditions for sentences of a language L can be empirically testable statements need not show *by itself* that the concept of truth plays a role in collecting evidence for their truth, but the possibility observed by Tarski shows that there is conceptual space for that question—a space not left open by the minimalist's *tautological* assignments of the property of being true to propositions, as in (M). And the fact that truth figures in Grice's maxims strongly suggests that the truth-involving explanatory principles are not confined to yielding knowledge of the literal meaning of what speaker/hearers communicate.

What, then, are the truth-involving indispensable explanatory habits? We cannot draw up a *list* of truth-involving platitudes for such a list would be *unprincipled*. No list tells us why an alleged principle or habit should belong to it (apart from the trivial fact that they have in common that the predicate 'is true' is required to state them). The alternative proposal is to see a central organizing principle in an exploration of the role of the T-concept in

---

[6] *Pace* Field (1972) and Horwich (1998a). What cannot be denied is that, as Tarski suggested, the T-schema is a good test for what we are inclined to accept when we understand *our* familiar truth predicate.

[7] Quine's (1970) subtle formulation makes this distinction: "for the sake of a generalization, we have restored to semantic ascent" (Quine, 1970, p. 12). Semantic ascent is taken up by the predicate 'is true'; the generalisation is take care of by quantifiers.

[8] Pointing to a French sentence, I can say that it is true if and only if snow is white. The resulting statement is an empirical truth. When the property of being true iff snow is white—in lambda notation: $\lambda x(x$ is true iff snow is white)—is being assigned to the *proposition* that snow is white, the resulting sentence is analytically and definitionally true. This grounds truth's transparency, the former reflects transcendent modes of using the predicate 'is true.'

explanatory inferences involved in understanding, and to think of habits that *should* figure on the list as constitutive of that explanatory practice. Crispin Wright's proposal for such a list (Wright, 1992/1999, p. 227) contains the following items:

(i) the transparency of truth,

(ii) the opacity of truth,

(iii) the conservation of truth-aptitude under embedding,

(iv) the correspondence platitude,

(v) the timelessness of truth,

(vi) the contrast of truth with justification.

Wright's heterogeneous list is a mix of logical properties of the truth predicate, remarks about the relation between truth and propositions and hints at substantial theories of truth (the correspondence principle). But what we want are not *philosophical* platitudes but insight in habits that govern and unify an explanatory practice. Take again Grice's maxim of quality: this seems to be a crucial and—if Grice is right—central principle involved in explanations (what does the speaker *mean* by saying what he does?) and its outcome is an understanding of what the speaker implicates.

If our model of how to approach the explanatory role of truth makes sense, a distinctive feature of classical minimalism—that "(the concept of) truth be 'independent of other concepts'" (Davidson, characterizing Horwich's 1998a position)—must be rejected, for the central habits involve important and motivated connections of truth with other concepts. On the other hand, if 'independence' simply means: "not analyzable into other, more primitive concepts," no minimalist should feel offended. But our position will raise an objection to, for example, Paul Horwich's proposal for a use-account of meaning (Horwich, 1998b)—an account of meaning intended as one in which the concept of truth does not figure in the analysis of meaning—for it remains to be seen whether such an account will not in the end have to appeal to truth-involving habits, i.e. habits involved in coming to *understand* uses of bits of language by speakers. But Horwich's point might simply be that there is no direct *route* from an *analysis* of the concept of truth to insight in meaning, the use of words, or understanding, and that is fully in line with the proposal presented here. The route proposed makes the relevant connections via truth's role in habits constitutive of understanding, not truth's supposed role in the *conceptual analysis* of the concept of *meaning*.

The approach is congenial to minimalism in other ways. Take the suggestion that recommendations like "it is best to believe the truth" and "the truth will out" are principles that cannot be explained or justified by minimalists and that, therefore, minimalism is false. To this a minimalist—in this case: John Burgess—replies as follows: "(I)t should be observed that these maxims (or 'general tendencies,' as Burgess puts it) are the sort of thing that, to the extent that they hold, are quite naturally thought of as bits of worldly wisdom *learned by sad experience*, or in other words, inferred inductively from an accumulation of unfortunate examples. There seems to be no reason why one should expect to be able to derive *deductively* from the *concept* of truth the conclusion that in general the truth becomes known, and it is best to believe it." (Burgess, 2002, p. 53). This reply seems right: going beyond minimalism doesn't mean that we have to reject Burgess' 'deductive requirement,' *for the relevant habits in which they figure are not discovered by inspecting the concept of truth.* It may well be that it is best to believe the truth (which truths, one might ask. All truths?), but why should that recommendation be motivated by an analysis of the concept? Moreover, note that such principles are *believed* by many, and proposed by many when asked what properties to ascribe to truth, and that they are often thought of as constitutive of truth. But why should a case for the indispensability or centrality of a truth-involving habit rest on empirical facts about general tendencies in what people believe about the role of true beliefs and the concept of truth in our lives? A similar suggestion is made by Michael Lynch, when he claims that "they (minimalists, FB) are no longer trying to simply to capture all things people normally want to say about truth" (Lynch, 2004a, p. 513). A philosophical account of what it means to grasp the concept of truth—thought of as getting a grip on central habits in which the concept figures essentially—must surely go beyond what people "normally want to say about truth," and why should the role of such habits be fully transparent to us? (After all, it took luminaries like Grice to make explicit one very specific habit involving the understanding of implicatures.)

Moreover, many claims presented as platitudes about truth are controversial if not downright wrong. Take the claim that we have a disinterested interest in truth. The trouble with this is, as Heal (1988/1989) pointed out, that a subject's interest in truth is better described as a subject's interest—perhaps a profound, unconditional interest—in certain *subject matters* (the behaviour of goldfish in a fishbowl, say), and that such interests are best *realized* by acquiring knowledge about goldfish-behaviour in a bowl. It does not follow from the fact that we have a disinterested interest in subject matter *S*, that we have a disinterested interest in epistemic states—knowledge—about *S*. People interested in goldfish-behaviour do not usually have an interest in *another* subject matter, i.e. states of knowledge involving goldfish. Those who are interested in the latter are sociologists of goldfish-amateurs.

The proposed project has a connection with a strategic reversal Davidson proposed early on in his work. "(A)ssuming translation, Tarski was able to define truth: the present idea is to take truth as basic and to extract an account of translation or interpretation" (Davidson, 1984, p. 134). Later, in his *Dewey Lectures* Davidson proposed "an approach (to the study of truth, FB) that makes the concept of truth an essential part of the scheme we all necessarily employ for understanding, criticizing, explaining and predicting thought and action" (Davidson, 1990, p. 282). Here, Davidson clearly alludes to the concept of truth as being at work in a transcendent mode of thinking: attributing a structure of truth to thoughts and utterances with a view to understanding speakers. Davidson developed the project via a thought experiment (radical interpretation), but I contend that the role of the concept might also, and more plausibly, be explored by investigating *micro-interpretation*. This suggests that the project should not be explored by pursuing the classic Davidsonian semantic program: it is not because the concept of truth figures in explicit statements of knowledge of what is sufficient for understanding speakers (knowledge of an empirically adequate theory of truth, according to him) that its explanatory role is revealed. The *same* concept the predicate 'is true' expresses is at work in language- and speaker-independent principles that govern the process of coming to understand speakers, and arguments for the indispensability of those habits should focus on the process, rather than the ways of stating parts—according to Davidson: the semantic, compositionally-determined parts—of their outcome. This is surely in the spirit of Davidson—we explore the role of truth-ascriptions in explanatory projects—but widely unfaithful to the letter of his original project.

Coming to understanding others is an essentially *explanatory* practice. Such practices must be governed by habits that structure the data, tell us what counts as evidence, and what they are supposed to explain. Since understanding a person's utterances is the result of a process in which we are necessarily involved—we wouldn't have the kind of mind we actually have were we not constantly interacting with other minds, which presupposes that we understand what these minds produce—we will have at least a *principled* basis for deciding which habits should and shouldn't be on a list of truth-involving habits. Nothing thus excludes that the envisaged project will capture certain *intuitions* behind the principles and platitudes on Wright's list discussed earlier. For example, the uniformity of truth reflects the fact that imposing a structure on the data cannot assume that assertions about different subject matters require the employment of *different* concepts of truth, for, *ex hypothesi*, when trying to understand what agents mean the subject matter he is thinking and talking about may still be obscure to us.⁹

In an attempt to recruit Davidson in the deflationist camp, Michael Williams makes a recommendation to the Davidsonian: "Why not give truth to the deflationists and let interpretation stand on its own feet? The only objection would have to be that the canons of interpretation themselves make use of a rich notion of truth." (Williams, 1999, p. 559). The interesting point now is that Williams rejects that the canons of interpretation—he has in mind the well-known Davidsonian principle of charity—do involve a *rich* notion of truth. "The relevant interpretive maxim involves a generalizing rather than an explanatory use (of truth, FB)" (Williams, 1999, p. 562). But now an objection mounted earlier applies: if the explanatory habits were mere *generalizations (that "most of the speakers beliefs are true"* is the generalization Williams has in mind), they would have to be *logically equivalent* with (perhaps infinite) conjunctions (or disjunction of conjunctions) the applier of the truth-predicate *understands*. But that is problematic. First, it neglects the transcendent mode of thought the application of the concept of truth requires in the habits: the habits are used to come to understand utterances, but one can hardly sum up (disjunctions or conjunctions of) propositions communicated by utterances one doesn't (yet) understand. Secondly, those generalizations have *counterfactual* force, which is not captured by infinite conjunctions (Gupta, 1993/1999). Thirdly, even if (*per impossibile*) there *were* an infinite conjunction (or disjunction) logically equivalent with the generic principle, it would not have the same explanation: a generalization might be explained on different grounds than its instances (see again Gupta, 1993/1999).¹⁰ That we should take speakers not to say what they

---

⁹ Understanding is a specimen of what Lipton (2004, p. 3) called *self-evidencing explanations*. These are explanations where what is explained provides an essential part of our reason for believing that the explanation itself is correct. We often infer that a hypothesis H is correct precisely because it would, if correct, provide a good explanation of the evidence. In our case, the evidence is that we *do* understand and learn from each other; we infer the presence of truth-involving habits, which, in turn, explain *why* we understand and ultimately learn from each other.

¹⁰ As Gupta (1993/1999, p. 288) points out: explaining why everyone in the boat died, need not entail an explanation why X, Y and Z (the victims) died. If they all died because the boat capsized, X might have died of a heart attack, while Y might have drowned. Not only can logically equivalent statements have different explanations, they can also have different explanatory roles. The explanatory asymmetry between truth-involving generalisations and an infinite list of instances is not particularly tied to T-involving generalisations. The argument is therefore not *ad hoc*. Minimalists may have neglected this point by focussing too quickly on trivial generalisations ("everything the pope says is true") or generalisations that are, for independent reasons, controversial or implausible ("truth is good," "truth is the goal of inquiry").

think is false is a constitutive explanatory principle (it yields, when applied in tune with other principles, understanding), it is justified by its outcomes (occurrences of understanding a speaker), but the principle need not and cannot explain or justify why a particular belief (say, the belief that snow is white) is true.

A related objection against Davidson was raised by when he stated that "(a)ll we need to know about truth is encapsulated in the way the truth predicate solves a simple syntactic problem. . . . The usefulness of Tarski's methods in a compositional theory of meaning does not, therefore, indicate any deep connection between the notion of truth and that of meaning"(Kölbel, 2001, p. 634). This objection doesn't apply to the project proposed here, for the concept of truth's explanatory role is not encapsulated by its role in stating the *outcome* of a theory of meaning but rather by its role in habits that yield understanding—and not just understanding the semantic content of sentences, but also what is communicated via non-semantic means.

The dialectical situation seems to be this: (i) A standard objection from those defending the Davidsonian program was that the minimalist is challenged *not* to explain meaning in terms of truth (a challenge Horwich picked up in *Meaning*). (ii) The minimalist challenge to the Davidsonian program was to give a non-substantial account of truth, given that (as the Davidsonian holds) that instances of the T-schema don't capture everything there is to say about truth. The way out was to accept what the minimalist concedes anyway—that can occur can occur in quantified sentences or in statements where a singular term is flanked by the truth predicate (the truth-predicate's expressive power is conceded and being relied on)—and then to make the further claim that some of these generalizations or generic statements reflect central, ineliminable cognitive habits involved in the process of coming to understanding others. The role of the *concept* of truth is thus not derived from the role of the truth-predicate in *stating* those principles; assuming that utterances are true plays a role in rational reconstructions of whatever understanding involves. A rational reconstruction of understanding others is therefore a theory that assumes (tacit) grasp the concept of truth in speakers and their audiences, and not a theory that merely appeals to the expressive powers of a language that contains a truth predicate to state the theory.

2. Examples

Grice's Maxim of Quality—I have already suggested that Grice's maxim of Quality is an emblematic example of a habit that helps us understand what speakers mean over and above what they say:

(Maxim of Quality)
Do not say what you believe to be false.
Do not say that for which you lack adequate evidence.

It is hard to see how the Maxim of Quality could be explanatory equivalent with, or be a mere generalisation of an infinite list of instances. And even if you would derive the Maxim of Quality from an infinite list of mini-norms ("do not say that snow is white if you believe that snow is not white,"...), wouldn't we want to *justify* acceptance of the mini-norms on the basis of our antecedent inclination to accept the *general* principle (Lynch, 2004b)?[11]

Implicitures—Linguists are familiar with the phenomenon of impliciture, cases where what is *said* by a speaker is an extension or expansion of the proposition (or propositional skeleton) of the sentence semantically expressed (Bach, 1994). Typical examples are "He lives two blocks away" (⇒ *he lives two blocks away from here*) and "You are not going to die" (⇒ *you are not going to die from that injury*). In theories that develop pragmatic accounts of the processes underlying these forms of pragmatic enrichment (Recanati, 2004; Carston, 2002), it is often argued that such expansions are part of the truth conditions of the utterance produced by the speaker because only the expanded versions are either truth-evaluable (as in the first example) or are not patently false (as in the second example). These 'contextualist' or 'pragmatic' strategies are plausible insofar as they rely on explanatory principles such as this: interpret a speaker's utterance so that what he *says* (over and beyond what he or his words semantically express) is (according to the audience's standards) true. (There are, of course, other constraints that steer the explanatory practice, such as principles of informativity and accuracy.) Here too it is plausible to describe the pragmatic enrichment process as involving a truth-involving habit. Note that the explanatory principle still allows the communicated content to be false, for it does not amount to the stronger principle that the speaker's beliefs must be true.

Communicated content and testimony—A family of theories connects the *presumption of truthfulness* in understanding what the speaker intends to communicate with explanations and justifications of why we believe what

---

[11] The derivation deduces the maxim from (i) the equivalence schema (<*p*> is true iff *p*), and (ii), 'for each *p*, do not say that *p* if you don't believe that *p*.' But, as Lynch (2004b, p. 111) remarks, to say that the instances are basic, is to say we don't need to justify our acceptance of them. But doesn't the maxim *explain* why we accept the instances? And, secondly, wouldn't we regard speakers we act according to the rule, as *justifying* the ascription of the general habit, the maxim? The alternative strategy would also assume an unwelcome form of particularism about understanding.

we are told (Rysiew, 2007; Lipton, 2007). First, recall that understanding is a *foundational* project: whatever undetected mistake would be made at the level of understanding a speaker will affect the extent to which his further intentions, like transmitting knowledge, will be realized. Peter Lipton applied Inference to the Best Explanation (IBE) to reasoning about communication: his core idea was that we believe what we are told when the *truth* of what we are told would figure in the best explanation of the fact that we were told it. We believe the fact uttered when its truth is part of the best *explanation* of the utterance. And here the concept of truth figures essentially in the explanation: if we are told that this sample of water is drinkable, we believe the speaker because its truth explains the fact that we are told it. Lipton's approach is quite explicit in embedding the principle in an explanatory project, and the principle once again involves the concept of truth: his proposed *explanation* of why we believe what we are told involves a detached mode of thinking: we assume what we are told as something we come to know because we think of it—the belief, the utterance—as being true, and not because we antecedently knew the proposition expressed was true. Shogenji (2006), as discussed in Rysiew (2007, p. 295) suggests that the presumption of truthfulness may be correct psychologically, but that there is no logical reason that the subject must make this presumption. In order to interpret utterances, the subject only needs the presumption that testimony is generally credible. But to this Rysiew replies, correctly in my view, with a key insight: the issue *is* a psychological one, for the search space of possible interpretations of a speaker must be reduced, and the mere *availability* of the hypothesis of truthfulness is not enough: the speaker must *adopt* (Rysiew, 2007, p. 295, slightly adapted). His argument explicitly appeals to truthfulness as a central explanatory assumption in the interpretation of assertoric speech acts, and one in which the concept of truth figures essentially.[12]

Learning new words—Arguing against the Locke-Quine picture of language acquisition and defending what he calls the 'Augustinian infant,' developmental psychologist Paul Bloom (2000) defends the importance of the child's capacity of mind reading and brings together evidence that shows that some capacity to understand the minds of others may be present in babies before they begin to speak. When this capacity is fully developed is of course an empirical question we need not go into here. Here's one interesting series of observations relevant for the perspective just sketched (Bloom, 2000, p. 63ff.): Baldwin (1993) tested babies in which they were

---

[12] Tyler Burge's much-discussed *Acceptance Principle* also employs the concept of truth: "A person is a priori entitled to accept a proposition that is presented to him as true and that is intelligible to him, unless there are stronger reasons not to do so" (Burge, 1993, p. 469).

given one object to play with while another object was put in a bucket that was in front of the experimenter. When the baby was looking at the object in front of her, the experimenter looked at the object in the bucket and said—in Bloom's words, "a new word," such as "It's a modi!" (a sentence, as it turns out). This, Bloom suggests, might give rise to a perfect Lockean correspondence between the new word and the object the baby was looking at. But 18-months-olds don't just *associate modi* with the object. Instead— and here comes the crucial observation—*they look at the experimenter and redirect their attention to what she (the experimenter) is looking at, i.e. the object in the bucket*. When later shown the two objects and asked to "find the modi" they assume that the word refers to the object the experimenter was looking at when she said the word—not the object that the child herself was looking at.

The experiment shows, first, that ascribing to the child the simple explanatory inference that since what is uttered—"that's a modi"—must be true, that it must look at the modi—cannot be correct: the child must first intend to *interpret* the utterance, and (s)he does that not just under the assumption that the sentence is true, but under the assumption that whatever it means will be explained by what the experimenter is looking at; that directs her attention from the experimenter to the object: she learns which state of affairs is supposed to make it true and the tacit, emerging assumption that holds these cognitive moves together is that whatever the experimenter believes, is *true*—a specimen of a blind truth ascription applied with a view to understanding. From our vantage point—the point of view from which we give a rational reconstruction of this process—the child could be described as *assuming* that a sentence not yet understood is true, and as trying to figure out what it means by reading off referential intentions of speakers, given that assumption.

Articulatory speech acts—The core intuition behind these experimental results can be related to further observation by Jamie Tappenden. Explaining what he calls the 'articulatory use' of language—a use quite reminiscent of the experiment just described—Tappenden argues that "we should turn our attention to the apparently uninteresting, but crucial activities like correcting the linguistic mistakes of others, teaching people the proper use, and the like. They are the oil changes and wheel realignments of a smoothly functioning language—the routine maintenance of the conventions of use" (Tappenden, 1993, p. 569). The 'articulatory' use of language assumes that what the speaker says is true (or at least not false), and then solves for the intended meaning of the term the speaker uses in the articulatory speech act. *Articulatory speech acts* explicitly aim at being accepted as true; the intended audience will benefit in the following obvious sense: it will learn what the newly introduced word means. Robert Koons, in a paper which criticizes certain versions of minimalism, points out that ". . . no one can understand a speech act as a stipulative definition without explicitly comprehending the concept of truth. Take a stipulation of the meaning of a term ϕ to be a speech act whose content is: so understand ϕ as to make the statement $S(\phi)$ true. An agent cannot participate in such a sophisticated linguistic practice without prior understanding of semantic notions like truth." He concludes: "Thus, no hypothetical origin for the concept of truth relies on such sophisticated practices as stipulative definition. This means that truth is essentially indefinable." (Koons, 2000, p. 183). But one might as well conclude that grasping the concept of truth seems to be involved in *coming to successfully understand stipulative introductions of new words*. Intriguingly, understanding a proposed *stipulative* definition of truth cannot introduce the *old* truth-predicate in the language; the concept of truth it expresses must be assumed to be present in the speaker's cognitive economy order to understand the point of a speaker's stipulation.

Conclusion

There are eminent reasons to assume that in rational reconstructions of how we come to correctly understand each other's utterances, truth-involving habits play an indispensable role. It remains an open question how (or whether) these habits are learned, to what extent grasp of the concept of truth must be present before the predicate 'is true' is learned, and how implicit or practical that grasp of the concept can be allowed to be to make the account plausible (see Hinzen 2006 for important suggestions). What I have tried to show is not that stating these habits requires using the predicate 'is true,' but that these habits correctly describe creatures who must grasp and apply truth-involving habits when they intend to understand other creatures with the similar capacities.

References

Azzouni, J. (2005). Tarski, Quine, and the transcendence of the vernacular 'true'. *Synthese*, 142, 273–288.
Bach, K. (1994). Conversational implicitures. *Mind and Language*, 9, 124–162.
Baldwin, D. A. (1993). Infants' ability to consult the speaker for clues to word reference. *Journal of Child Language*, 20, 395–418.
Bloom, P. (2000). *How Children Learn the Meaning of Words*. MIT Press, Cambridge.
Burge, T. (1993). Content preservation. *Philosophical Review*, 102, 457–488.
Burgess, J. (2002). Is there a problem about the deflationary theory of truth? In V. Halbach and L. Horsten (Eds.), *Principles of Truth* (37–57). Hänsel-Hohenhausen.

Carston, R. (2002). *Thoughts and Utterances: the Pragmatics of Explicit Communicatio.* Blackwell, Oxford.
Davidson, D. (1984). *Inquiries into Truth and Interpretatio.* Clarendon Press, Oxford.
Davidson, D. (1990). The structure and content of truth. *Journal of Philosophy.*
Davidson, D. (2004). *Truth and Predication.* Harvard University Press, Harvard.
Field, H. (1972). Tarski's theory of truth. *Journal of Philosophy,* 64 (13), 347–375.
Field, H. (1999). Deflationary views of meaning and content. In S. Blackburn and K. Simmons (Eds.), *Truth* (351–391), Oxford Readings in Philosophy. Oxford University Press, Oxford. (Original work published 1995).
Frege, G. (1999). The thought: A logical inquiry. In S. Blackburn and K. Simmons (Eds.), *Truth* (85–106), Oxford Readings in Philosophy. Oxford University Press, Oxford. (Original work published 1918).
Grice, P. (1989a). Logic and conversation. In Grice (1989b).
Grice, P. (1989b). *Studies in the Ways of Words.* Harvard University Press, Harvard.
Gupta, A. (1999). A critique of deflationism. In S. Blackburn and K. Simmons (Eds.), *Truth* (282–307), Oxford Readings in Philosophy. Oxford University Press, Oxford. (Original work published 1993).
Heal, J. (1988/1989). The disinterested search for truth. *Proceedings of the Aristotelian Society 1987–1988,* 97–108.
Hershfield, J. and Soles, D. (2003). Reinflating truth as an explanatory concept. *Pacific Philosophical Quarterly,* 84, 32–42.
Hinzen, W. (2006). Internalism about truth. *Mind & Society,* 5, 139–166.
Horwich, P. (1998a). *Truth* (2nd ed.). Oxford University Press, Oxford.
Horwich, P. (1998b). *Meaning.* Oxford University Press, Oxford.
Koons, R. (2000). 'Circularity and the hierarchy', in circularity and hierarchy. In *Circularity, Definition, and Truth* (177–198). Indian Council of Philosophical Research, New Delhi.
Kölbel, M. (2001). Two dogmas of Davidsonian semantics. *Journal of Philosophy,* 98(12), 613–635.
Lipton, P. (2004). What good is an explanation? In J. Cornwell (Ed.), *Understanding Explanation* (1–22). Oxford University Press, Oxford.
Lipton, P. (2007). Alien abduction: Inference to the best explanation and the management of testimony. *Episteme,* 4, 238–251.
Lynch, M. (2004a). Minimalism and the value of truth. *The Philosophical Quarterly,* 217, 497–517.
Lynch, M. (2004b). *True to Life.* MIT Press, Cambridge, MA.
Lynch, M. and Greenough, P. (Eds.). (2005). *Truth and Realism.* Oxford University Press, Oxford.
Quine, W. V. O. (1970). *Philosophy of Logic.* Harvard University Press, Cambridge, MA.
Ramsey, F. (1990). *Philosophical Papers* (H. Mellor, Ed.). Cambridge University Press, Cambridge.
Recanati, F. (2004). *Literal Meaning.* Oxford University Press, Oxford.
Rysiew, P. (2007). Beyond words: Communication, truthfulness, and understanding. *Episteme,* 4, 285–304.
Sher, G. (2004). In search of a substantive theory of truth. *Journal of Philosophy,* 101, 5–36.
Shogenji, T. (2006). A defense of reductionism about testimonial justification of beliefs. *Noûs,* 40(2), 331–346.
Soames, S. (2008). Drawing the line between meaning and implicature—and relating both to assertion. *Noûs,* 42(3), 529–554.
Tappenden, J. (1993). The liar and sorites paradoxes: Towards a unified treatment. *Journal of Philosophy,* 90, 551–776.

Tarski, A. (1999). The semantic conception of truth and the foundations of semantics. In S. Blackburn and K. Simmons (Eds.), *Truth* (115–144), Oxford Readings in Philosophy. Oxford University Press, Oxford. (Original work published 1944).

Williams, M. (1999). Meaning and deflationary truth. *Journal of Philosophy*, 96(11), 545–564.

Wright, C. (1999). Truth: A traditional debate reviewed. In S. Blackburn and K. Simmons (Eds.), *Truth*, Oxford Readings in Philosophy. Oxford University Press, Oxford. (Original work published 1992).

Deflationism and Reducibility

Martin Fischer

ABSTRACT

Deflationist theories of truth contain at least two claims: One is a minimalist claim. The other is the explanation of the function of the truth predicate as an expressive one. I will argue that it is not an easy task to fulfill both claims by showing that the most natural axiomatic theories, that are adequate in respect to a minimalist criterion, cannot adequately capture the expressive function of the truth predicate because they are reducible to the base theory.

I

The ideas underlying deflationism have roots that can be traced back to the beginning of the 20th century and philosophers like Frege, Ramsey or Tarski. Some of these ideas are as attractive as ever and there is a variety of different theories of truth that fall under the label of deflationism. Without trying to cover all versions of deflationism I will focus on deflationist theories that characterize the truth predicate as a special predicate that is simple, not 'substantial' and only functions as an expressive device. These informal criteria will be the basis for a more thorough characterization of the desiderata of deflationist theories. Furthermore it will be argued that these desiderata are not satisfiable in an adequate way.

The text is structured as follows: First I will introduce deflationist theories of truth as theories that are simple, minimal and ascribe an expressive function to the truth predicate. After that I will sketch two forms of arguments against deflationism which are well known: One is Tarski's critique that certain axiomatic theories of truth are deductively too weak. The other one is an argument given by Shapiro and Ketland, claiming that deflationism is committed to conservativity and being able to prove a reflection principle for the base theory, which cannot be fulfilled. In the third part I will present an alternative form of critique with a similar motivation. I will argue that the claim of expressive power commits a deflationist at least to a theory that is not reducible to the base theory in the sense that it is not interpretable in the base theory. I will then show that the most promising deflationist theories are reducible and therefore not adequate.

The claims connected with deflationism are often formulated informally and the vocabulary used to give a characterization how these claims are to be understood allow a variety of interpretations. One example is the claim that truth is not 'substantial.' I don't think that very much is gained by an explanation that says that truth is not a 'real' property. One expression that needs an explanation is replaced by another. So to be more concrete I will discuss deflationist theories as special formal theories of truth. There is also a more general reason for a formal investigation, which is nicely summarized by Leitgeb:

> But ultimately every successful philosophical theory of truth has to stand the test of formalization and every successful formal theory of truth must be supported by philosophical argumentation; . . . (Leitgeb, 2007, p. 276)

Since Tarski there has been an intensive investigation of formal theories of truth. One branch is concerned with axiomatic theories of truth, which is especially useful for the discussion of deflationist theories of truth, since deflationism is concerned with a simple notion of truth. Gupta for example describes deflationism in the following way:

> Deflationism, maintains that truth is a simple and clear concept and has no substantial role to play in philosophy. (Gupta, 2001, p. 58)

The simplicity claim is a good reason to consider axiomatic theories of truth as the best way of formalizing deflationist theories. I will only consider theories in classical first-order logic. An axiomatic theory of truth is based on a theory of syntax, so it is usually understood as an extension of Peano Arithmetic, in short PA. The theory of truth is formulated in an expansion of the language of arithmetic by a one-place truth predicate τ. An almost universally accepted requirement for a theory of truth is that it contains all T-sentences for the language of arithmetic, where a T-sentence is a sentence of the form: 'φ' is true iff φ. To be in accordance with the simplicity claim the axioms should be well motivated and simple, so optimally have the form of T-sentences.

The core of deflationism consists of a minimality claim. In the literature this is often expressed saying that the truth predicate is not 'substantial.' As already mentioned this is too informal and vague to be a satisfactory explanation of the minimalist claim and there is a dispute what it amounts to say that truth has no substance. In the recent literature it was suggested that it is best understood as a form of conservativity. One of the most influential participants in this discussion is Shapiro who notes:

> I submit that in one form or another, conservativeness is essential to deflationism. Suppose for example, that Karl correctly holds a theory [A] in a language that cannot express truth. He adds a truth predicate to the language and extends [A] to a theory [A'] using only axioms essential to truth. Assume that [A'] is not conservative over [A]. Then there is a sentence ψ in the original language (so that ψ does not contain the truth predicate) such that ψ is a consequence of [A'] but not a consequence of [A]. That is, it is logically possible for the axioms of [A] to be true and yet ψ false. This undermines the central deflationist theme that truth is insubstantial. Before Karl moved to [A'], $\neg\psi$ was possible. The move from [A] to [A'] added semantic content sufficient to rule out the falsity of ψ. But by hypotheses, all that was added in [A'] were principles essential to truth. Thus, those principles have substantial semantic content. (Shapiro, 1998, pp. 487ff.)

Before it is possible to judge if conservativity captures the deflationst intentions adequately a few specifications are to be made:

There are various forms of conservativity. We will consider the standard proof theoretic version of conservativity of a theory over another theory which is defined as follows: A theory S is **conservative over** a theory T in respect to Γ iff for all Γ-formulas φ: if $S \vdash \varphi$ then $T \vdash \varphi$.

Since conservativity is a relative notion the question arises conservativity over which theory. For Shapiro it seems to be important that conservativity holds for any theory. According to his criterion only theories of truth are acceptable for a deflationist, which are conservative over any substantial theory. To avoid trivializations one has to take care of the empty theory and consider extensions of a theory that implies infinite models. The base theory PA seems to be a good candidate for this.

My suggestion for the minimality criterion is:

Criterion 1. *A theory of truth T is minimal iff for any (substantial) theory A formulated in a language not containing the truth predicate: $T \cup A$ is conservative over $PA \cup A$ in respect to the language of A.*

If you follow this line of argumentation it is plausible to assume conservativity over the base theory as a necessary condition for deflationist theories. There are several philosophers who accept this criterion. For example Shapiro (1998), Ketland (1999), Tennant (2002) and Azzouni (1999).

Moderate deflationists usually do not only claim that truth is not substantial, but they also try to explain the function of the truth predicate as an expressive one. Quine describes the expressive function partly as disquotational: "The truth predicate is a device of disquotation." (Quine, 1970, p. 12). But disquotation is not the only expressive function of the truth predicate, it also serves as a device of generalizations.

> We may affirm the single sentence by just uttering it, unaided by quotation or by the truth predicate; but if we want to affirm some infinite lot of sentences that we can demarcate only by talking about the sentences, then the truth predicate has its use. We need it to restore the effect of objective reference when for the sake of some generalization we have resorted to semantic ascent. (Quine, 1970, p. 12)

So whereas in the case of disquotation the truth predicate is not essential, it is needed in the case of generalizations. To have the full expressive power the truth predicate should not be eliminable and the theory of truth should not be reducible to the base theory. The expressive function in the case of generalizations is often illustrated by examples like the following: All tautologies are true.

As far as I know there is no precise explication what this expressive function of the truth predicate exactly is. Later on I will try to be a little more specific, but for now I will just repeat the three important claims deflationists make for the expressive function of the truth predicate: First, it is a device of disquotation. Second, it allows generalizations. Third, it should not be eliminable.

II

The first natural candidate as a deflationist axiomatic theory of truth is the disquotational theory of truth, *TB*, which consists of all instances of the T-scheme for sentences of arithmetic and *PA* with induction for formulas of the language containing the truth predicate. But this theory faces serious difficulties. Already Tarski pointed out that *TB* is deductively too weak to be considered as an adequate theory of truth (see Tarski, 1935/1971). The reasons he gives are the following:

TB is a conservative extension of *PA* and therefore it cannot proof any new theorems in the language of arithmetic. Although this may have been a sufficient reason for Tarski to reject it as an interesting theory of truth, it does not have to be a reason to reject it as an adequate deflationist theory. On the contrary a conservative theory seems to be preferable for deflationism. Conservativity of a theory *A* over a theory *B* is also not sufficient to show that *A* has no more expressive power than *B*. Examples of interesting conservative extensions are *PA*, ACA_0 as well as *ZFC*, *NBG*. An extreme example would be Presburger arithmetic and *PA*.

Tarski's second reason is that *TB* does not prove generalizations one would expect to be provable by a theory of truth, like tertium non datur or that modus ponens preserves truth. Halbach gives an even stronger argument by showing that *TB* only proves trivial generalizations, that is finite generalizations (see Halbach, 1999). In contrast to the first reason this one is a real challenge for *TB* but it is not conclusive as an argument against deflationism. The deflationist has two options for handling sentences that are not provable in *TB*: Either argue that they do not have to be provable or choose a different axiomatic theory that proves it.

The first strategy could be used to answer a critique using special theorems, like Tarski's. But for Halbach's argument this strategy seems not helpful. At least one nontrivial generalization should be provable in a theory of truth in which the truth predicate is needed for generalizations.

For the second strategy a deflationist could choose a compositional theory of truth like TC^r or add the relevant T-sentences. A compositional theory of truth has axioms for atomic sentences and axioms for the commutation of truth with connectives and quantifiers. In these theories some generalizations are provable and it is possible to stay conservative by restricting the induction scheme to formulas of the language of arithmetic. If a deflationist prefers only axioms in the form of T-sentences he can extend his theory by adding the relevant T-sentences. That this is possible was shown by McGee: For a language containing a typefree truth predicate it can be proved that for any sentence in this language there is an equivalent T-sentence (see McGee, 1992).

There is another form of critique of deflationism that was given by Shapiro (1998) and Ketland (1999). In their argument they make use of a tension between two of the claims of deflationism. The basic idea is that the two criteria imply that a theory of truth should be a conservative extension of *PA* and prove a reflection principle for *PA*. But this is impossible since in a theory of truth *T* that proves a reflection principle for *PA* also a consistency statement for *PA* is provable. But this consistency statement is formulated in purely arithmetical terms and not provable in *PA* itself. Therefore *T* would be a nonconservative extension of *PA*. So no theory can possibly fulfill both criteria. If deflationism is really committed to both, it fails.

Sticking to first-order[1] there are at least three possible answers:

A first reply would be to argue that there is no real conflict. It is possible to stick to a non conservative extension of *B* and still argue that it is defla-

---

[1] There are more possibilities if one allows for higher-order logics, like Shapiro's own suggestion. I will ignore these possibilities here.

tionist if the non conservativity depends on an arithmetical principle and not on truth-theoretic principles. Field and others try to argue that this is the case with induction. Induction is an arithmetic principle which holds for all sets of numbers, it is indefinitely extensible. So an extension of the scheme of induction to an expanded language is part of the arithmetical concept of induction and not part of the concept of truth. This is a somewhat questionable move. For one thing the distinction between arithmetic axioms and truth theoretic axioms is not very clear. Another worry might be what concept of the base theory is assumed in this picture and if not a similar move can be made for other nonconservative theories, such that the criterion of conservativity becomes empty.

A second reply could be to challenge conservativity as an inadequate criterion. There are a lot of interesting axiomatic theories which are non conservative extensions of *PA*, so the criterion should be well motivated. In my opinion the universal criterion of conservativity is well motivated as a sufficient and necessary condition for a theory to be insubstantial. It gives a justification to claim that a theory of truth is insubstantial, which I take to be the core of deflationism. So conservativity over *PA* as a necessary condition is also well motivated. Even if a deflationist does not share these intuitions he has to answer the question what distinguishes deflationist theories from other theories. In this respect the conservativity criterion is a good criterion. If a deflationist rejects the conservativity criterion there remains at least the challenge to give an alternative criterion that captures the deflationist motivation adequately.

The last reply is the one I want to focus on. It challenges the claim that a deflationist theory has to prove reflection. The deflationist claim of the expressiveness is vague enough that it leaves open the question if a reflection principle for *PA* has to be provable or not. So a deflationist might simply deny the commitment.[2]

III

The argument of Shapiro and Ketland is correct in trying to criticize deflationism in a way that makes use of the tension between the minimalist claim and the claim of expressiveness. But these claims are not necessarily incom-

---

[2] It is also not the case that a deflationist if he cannot prove a reflection principle for *PA* he cannot account for what Tennant (2002) calls the Gödel phenomena. Tennant shows that it is not obligatory to use a substantial concept of truth to make sense of the claim that I somehow grasp that the Gödel sentence is true even if I cannot prove it in *PA*.

patible. The argument is not as strong as it could be because it focuses only on one particular generalization which can always be refuted as a necessary requirement for deflationism. More convincing would be an argument that makes use of a more general implication of the claim of expressiveness, namely that at least some interesting statements about truth should be provable even if I cannot give a specific example. In the following I will sketch such an argument.

The aim of the argument is to show that there are no theories of truth available for a deflationist that fulfill the three desiderata: simplicity, minimality and expressiveness. The argument proceeds in four steps:

(A) According to deflationism the truth predicate has an irreducible expressive function.

(B) A theory of truth that is interpretable in the base theory cannot account for an irreducible expressive function of the truth predicate.

(C) According to deflationism truth is simple.

(D) The most simple theories of truth are *TB* and its variants. More complicated but still reasonably simple are *TC* and its variants.

(E) According to deflationism truth is minimal.

(F) A theory of truth is only minimal if it is a conservative extension of *PA*.

(G) It is provable that the variants of *TB* and *TC* which are conservative extensions of *PA* are also interpretable in *PA*.

(H) There are no adequate deflationist theories.

There are some simplifications in the presentation which will be clarified in the following.

The first thing will be a justification of (B). According to deflationism the truth predicate has an irreducible expressive function. This means that a theory of truth should prove at least some interesting facts. Interesting facts about the truth predicate are facts that do not trivialize the role of the truth predicate. So at least one theorem should be provable that has no counterpart in the base theory. In other words the truth theory should not be reducible to the base theory. Our claim is that in this context the correct concept of reducibility is interpretability. A theory of truth that is interpretable trivializes the function of a truth predicate. A deflationist is therefore committed to a theory that is not interpretable in the base theory.

One step in this argumentation that might appear problematic is the assumption that a theory of truth which is interpretable in the base theory trivializes the function of the truth predicate. Definability in the base theory is one way in which a truth predicate is trivialized. In this case the theory of syntax alone would have the same expressive power as the theory of truth. But definability is only sufficient and not necessary. Halbach's proof that a disquotational theory of truth only proves trivial generalizations is an example of a trivialization without explicit definability.[3]

An interpretation of a theory S in a theory T is a translation I of the language of S in the language of T, such that: Identity will be kept fixed. Every n-place predicate of $\mathscr{L}_S$ is translated into a formula of $\mathscr{L}_T$ with n free variables. The translation preserves logical structure, so the translation of a negation is equal to the negation of the translation, the same for connectives and quantifiers without a relativization. Furthermore every theorem of S will be translated into a theorem of T. This form of interpretation without relativizing the quantifiers is also called strict or direct interpretation in contrast to relative interpretations.

My argument depends on the correctness of the statement that a theory of truth that is interpretable in PA does not prove any interesting facts. But why should interpretability be a relevant criterion?

One reason is that interpretability is sometimes understood as a form of reduction. Examples in this respect are: PA is relatively interpretable in ZF. This is understood as a reduction of numbers to sets and thereby as a step in the foundational program to show that classical mathematics is reducible to set theory. Another example is the mutual interpretability of arithmetic and syntax. In one direction Gödel used it to show that arithmetic can handle syntax. In the other direction Quine showed that syntax can be used to do arithmetic. A third example is the mutual interpretability of PA and ZF^-, set theory without the axiom of infinity.

There are examples of relative interpretations which are not as good examples of reduction like the interpretability of $ZF + V = L$ in ZF. But in this case the relativization seems necessary and I do not know any example of a strict interpretation which is not a reduction. Furthermore I don't want to argue that interpretation is 'the' adequate explication of reduction, only that in this special case the role of the truth predicate will be trivialized.

In the context of truth theories T and PA, interpretability results can be strengthened. If T is interpretable in PA and conservative over PA, then T and PA are mutually faithfully interpretable. So they are in an equivalence class of theories with a stricter equivalence relation than just mutual interpretability.

---

[3] That TB is not a subtheory of a definitorical extension of PA is an easy consequence of Tarski's undefinability theorem.

In relevant cases non interpretability is connected with an increase of expressive power. Take as an example Presburger arithmetic and Q. Presburger arithmetic is formulated in a language of arithmetic that contains only one two place function symbol. In this case the additional function symbol increases the expressive power in a significant way. Presburger arithmetic is maximally consistent and therefore decidable. Typically Q is not interpretable in Presburger arithmetic. Another example is PA and ACA_0. This subsystem of second order arithmetic can be used to formalize parts of model-theory. Compared to PA, ACA_0 is expressively strong. Although ACA_0 is a conservative extension of PA it is not relatively interpretable in PA.

I claim that a theory of truth T which is interpretable in PA does not have more expressive power than PA itself. The reason is simply that for every sentence φ provable in T, there is a corresponding sentence ψ provable in PA. In a sense the sentences of the theory of truth and their translations in the language of arithmetic would have the same inferential role.

But here it could be argued that an interpretation is just a syntactical translation and there is no guarantee that this translation preserves meaning. Since for the translation not only a formula is given which corresponds to the truth predicate, but also translations of the arithmetical part of the language will be given, the meaning of those sentences will be changed. So for example where in φ occurred $<$ in the translation ψ there will be a formula $\psi_<$ in place of $<$. But in order to make this form of criticism effective one has to explain wherein the difference of meaning lies.

The difference cannot be in the inferential role. $<$ and $\psi_<$ play the same role in all formulas and derivations in which they occur. The interpretation guarantees that for every derivation in T there is a corresponding derivation in PA. The faithfulness of the interpretation guarantees that there are not more derivations in PA and the strictness preserves logical form. This is not the place to start a discussion of what the correct theory of meaning is, but I just want to point out that there is a burden of proof on the opponent's side to show how the difference in meaning is to be understood.

One way the opponent could try to explain in what the difference lies is via the standard model $\mathcal{N}$. The meaning of the arithmetical vocabulary is given by its standard interpretation. But the restriction of a model of the theory T to the language of arithmetic is not the standard model in the general case and it is not possible to expand the standard model in a straightforward way to a model of T. Because of the liar it is not possible to preserve all the extensions of the arithmetical vocabulary in $\mathcal{N}$. There must be a difference in the extensions of the translations in $\mathcal{N}$. So at least the extension of one predicate P has to be different from its translation ψ_P. So $P^{\mathcal{N}} \neq \{\bar{a} | \mathcal{N} \models \psi_P[\bar{a}]\}$.

But the interpretation I of T in PA gives us a method to construct a model $\mathfrak{M}^I$ of T for any model $\mathfrak{M}$ of PA. In the case of the standard model $\mathcal{N}$ there is a model $\mathcal{N}^I$ with special properties. Take $\mathcal{N}^I$ to be the structure $(N, \mathbf{0}^{\mathcal{N}^I}, \mathbf{S}^{\mathcal{N}^I}, +^{\mathcal{N}^I}, \cdot^{\mathcal{N}^I}, <^{\mathcal{N}^I}, \tau^{\mathcal{N}^I})$ with

$N := \{a \in \mathbb{N} \mid \text{there is a } b \in \mathbb{N}, a = \ulcorner I(\bar{b}) \urcorner\},$

$\mathbf{0}^{\mathcal{N}^I} := \{a \in N \mid \text{there is a } b \in \mathbb{N}, \mathcal{N} \models \mathbf{0}(x)[b] \& a = \ulcorner I(\bar{b}) \urcorner\},$

$\mathbf{S}^{\mathcal{N}^I} := \{\langle a, b \rangle \in N \mid \text{there are } c, d \in \mathbb{N},$
$\qquad \mathcal{N} \models \mathbf{S}(x,y)[c,d] \& a = \ulcorner I(\bar{c}) \urcorner \& b = \ulcorner I(\bar{d}) \urcorner\},$

$+^{\mathcal{N}^I} := \{\langle a, b, c \rangle \in N \mid \text{there are } d, e, f \in \mathbb{N},$
$\qquad \mathcal{N} \models +(xyz)[d,e,f] \& a = \ulcorner I(\bar{d}) \urcorner \& b = \ulcorner I(\bar{e}) \urcorner \& c = \ulcorner I(\bar{f}) \urcorner\},$

$\cdot^{\mathcal{N}^I} := \{\langle a, b, c \rangle \in N \mid \text{there are } d, e, f \in \mathbb{N},$
$\qquad \mathcal{N} \models \cdot(xyz)[d,e,f] \& a = \ulcorner I(\bar{d}) \urcorner \& b = \ulcorner I(\bar{e}) \urcorner \& c = \ulcorner I(\bar{f}) \urcorner\},$

$<^{\mathcal{N}^I} := \{\langle a, b \rangle \in N \mid \text{there are } c, d \in \mathbb{N},$
$\qquad \mathcal{N} \models (x<y)[c,d] \& a = \ulcorner I(\bar{c}) \urcorner \& b = \ulcorner I(\bar{d}) \urcorner\},$

$\tau^{\mathcal{N}^I} := \{a \in N \mid \varphi \in \mathcal{L}_A \& \mathcal{N} \models \psi_\tau(x)[a] \& a = \ulcorner \varphi \urcorner\}.$

But $\mathcal{N}^I \upharpoonright \mathcal{L}_A$ is isomorphic to $\mathcal{N}$.

Theorem 1. *Let $I : T \preccurlyeq PA$ and $\mathcal{N}^I$ the T structure given by $\mathcal{N}$ and I. Then $\mathcal{N}^I \upharpoonright \mathcal{L}_A$ is isomorphic to $\mathcal{N}$.*

Proof. Define $h : \mathbb{N} \to N$ by $h(n) := \ulcorner I(\bar{n}) \urcorner$. I will show that h is an isomorphism between $\mathcal{N}$ and $\mathcal{N}^I \upharpoonright \mathcal{L}_A$.

h is surjective by the definition of N.

h is injective since $\ulcorner \urcorner$ is and PA proofs that *num* is.

If $a \in \mathbf{0}^{\mathcal{N}}$, then since $h(a) = \ulcorner I(\bar{a}) \urcorner$ it follows that $h(a) \in \mathbf{0}^{\mathcal{N}^I}$.

If $h(a) \in \mathbf{0}^{\mathcal{N}^I}$, then there is $b \in \mathbb{N}$ with $b \in \mathbf{0}^{\mathcal{N}}$ and $h(a) = \ulcorner I(\bar{b}) \urcorner$. But since $h(a) = \ulcorner I(\bar{a}) \urcorner$ and h injective $a = b$ and therefore $a \in \mathbf{0}^{\mathcal{N}}$.

For the other predicates the proof is analog. □

With this result it seems hard to explain a difference in meaning between the arithmetical predicates P and their translations ψ_P. The expressions of the arithmetical part of the expanded language still refer to a structure which is isomorphic to the standard model. In what sense do they have a different meaning even in a realist semantics for which I grant an access to the standard model.

This point is even more serious in this situation in which we do not pretend to talk directly of a language but use a theory of arithmetic as an acceptable theory for the syntax of a language. One justification for this move is that theories of syntax can be interpreted in theories of arithmetic. But from a realist point of view a better justification would be to say that the syntactical structure of a language and the structure of the natural numbers are basically the same backing this claim up by saying that there is an isomorphism between the two structures. But if an isomorphism is sufficient for a realist to accept that a theory of arithmetic can be used to talk about the syntax of a language, she should also accept that an arithmetical theory can be used to talk about the syntax and 'truth' if the truth theory is interpretable in it.

So to sum up the reasons why a theory of truth T that is interpretable in PA cannot explain the expressive function of the truth predicate: it is because T is reducible to PA. For every sentence provable in T there is a corresponding sentence provable in PA. If T is interpretable in PA and T is conservative over PA, then T and PA are mutually faithful interpretable. So for every sentence φ, φ and its translation $I(\varphi)$ have the same inferential role and there is an interpretation that restricted to arithmetical vocabulary is isomorphic to the standard model.

Another step in the argument makes use of the simplicity claim to restrict the class of relevant theories. I already argued for axiomatic theories of truth. Axiomatic theories of truth have a great advantage over semantic theories like Tarski's or Kripke's. Whereas the latter force an 'essentially stronger' metatheory, the former are compatible with a metatheory that is not essentially stronger especially a metatheory which is conservative over the base theory. This is very much along the lines of a deflationist minimality claim.

For disquotationalism we would like to have a pretty simple set of T-sentences. In a conception of deflationism that is not as strict as disquotationalism in the choice of axioms other axioms could be allowed. The most famous one is an axiomatization of Tarski's theory TC. Although TC is not conservative over PA it's variant with restricted induction TC^r is.

So if one is willing to accept the formal counterparts of the claims, this gives three criteria of adequacy for deflationism. A deflationist theory of truth should be an axiomatic theory of truth, a conservative extension of PA and not interpretable in PA.

But it can be shown that most of the suggested axiomatic theories of truth do not fulfill these criteria. So for example the disquotational theory of truth TB is interpretable in PA. The interpretability result does not only hold for this theory but for any extension of TB that is still conservative over PA.

So the first strategy to overcome Tarski's critique of the deductive weakness of *TB* by extending it with further T-sentences does not work for the deflationist. But also the second strategy to use a compositional theory of truth is problematic because the most natural candidate, Tarski's compositional theory with restricted induction, is interpretable in *PA*.

The argument makes use of a theorem:[4]

Theorem 2.

(I) *Every subtheory and every pure extension of TB is interpretable in PA.*

(II) *TC^r is interpretable in PA.*

The first part of the theorem covers a wide range of disquotational theories of truth. It is not only an argument against *TB* and its variants *TB*↾, *UTB* and *UTB*↾ against which Halbach's argument is already quite convincing. It is also an argument against all pure extensions. The theorem shows that these extensions are either nonconservative over *PA* or interpretable in *PA*. In both cases they are not adequate as deflationist theories of truth. So disquotational theories are not satisfactory. In so far the range of theories is broader than in the case of the usual arguments against disquotationalism along the lines of deductive weakness.

The second part of the theorem covers the most simple versions of a compositional theory of truth especially TC^r and all it's subtheories. But in this case there is no strengthened version for all pure extensions.[5]

IV

Compared to the arguments presented in the second section the argument of the last section is advantageous. On the one hand it has a broader scope than arguments against the deductive weakness of special disquotational theories of truth. On the other hand it does not rely on a claim that a certain sentence, a reflection principle, has to be derivable.

In the end the argument leaves the deflationist with options that are not very promising. The first option would be to choose an alternative form of

---

[4] For a proof see (Fischer, 2008).

[5] It can be shown that there are theories of truth which are conservative extensions of *PA* and not interpretable in it. But these axioms are not as simple as the former, especially since they have a version of induction which is somewhere in between the restricted version and full induction. See (Fischer, 2008).

explication to axiomatization or to retreat from a formal explication. Since this would either leave the concept of truth unexplained or would be difficult to combine with a deflationist conception of a simple form of truth.

The second option would be to give one of the three desiderata of deflationism a different interpretation. In the case of minimality this would at least amount to the formulation of a criterion that explains the non substantiality of truth as good as conservativity and does not imply it. Since there is no suggestion in the literature the argument can be understood as a challenge for deflationism. In the case of the explanation of the expressive function of the truth predicate an acceptance of the interpretability of the truth theory in the base theory amounts to an acceptance of reducibility which would be a retreat to a redundancy conception that cannot explain the function of the truth predicate. The case of simplicity seems to be the most promising, although there is no 'simple' set of T-sentences or T-axioms known. Most of the known versions either are subject to the argument or they are somewhat artificial. But if there would be an axiomatization which is conservative and not interpretable and has independent justification this could be a solution for deflationism. But as long as there are no suggestions this can only be seen as a challenge.

References

Azzouni, J. (1999). Comments on Shapiro. *The Journal of Philosophy*, 96, 541–544.
Fischer, M. (2008). *Davidsons semantisches Programm und deflationäre Wahrheitskonzeptionen*. Ontos Verlag, Heusenstamm.
Gupta, A. (2001). A critique of deflationism. In M. P. Lynch (Ed.), *The Nature of Truth: Classical and Contemporary Perspectives* (527–557). MIT Press, Cambridge MA.
Halbach, V. (1999). Disquotationalism and infinite conjunctions. *Mind*, 108, 1–22.
Ketland, J. (1999). Deflationism and Tarski's paradise. *Mind*, 108, 69–94.
Leitgeb, H. (2007). What theories of truth should be like (but cannot be). *Philosophy Compass*, 2(2), 276–290.
McGee, V. (1992). Maximal consistent sets of instances of Tarski's schema (T). *The Journal of Philosophical Logic*, 21, 235–241.
Quine, W. V. O. (1970). *Philosophy of Logic*. Englewood Cliffs.
Shapiro, S. (1998). Proof and truth: Through thick and thin. *The Journal of Philosophy*, 95, 493–521.
Tarski, A. (1971). Der Wahrheitsbegriff in den formalisierten Sprachen. In K. Berka and L. Kreiser (Eds.), *Logik-Texte*, (445–546). Berlin. (Original work published 1935).
Tennant, N. (2002). Deflationism and the Gödel-phenomena. *Mind*, 111, 551–582.

On a Necessary Use of Truth in Epistemology*

Leon Horsten

Abstract

It is argued that theories of truth that are stronger than the disquotational theory are needed in order to validate Fitch's argument.

Deflationism about truth claims that the notion of truth does not play a central role in the resolution of metaphysical, epistemological, and semantic problems. Horwich expresses this sentiment as follows (Horwich, 1998, p. 52):

> A deflationist attitude toward truth is inconsistent with the usual view of it as a deep and vital element of philosophical theory. Consequently the many philosophers who are inclined to give the notion of truth a central role in their reflections in metaphysical, epistemological, and semantic problems must reject the minimalist account of its function. Conversely, those who sympathize with deflationary ideas about truth will not wish to place much theoretical weight on it. They will maintain that philosophy may employ the notion only in its minimalist capacity . . . and that theoretical problems must be resolved without it.

Let us take a closer look at the question whether truth is conservative over one particular philosophical discipline: epistemology. We shall adopt the attitude of Zermelo when he sought to defend the Axiom of Choice. He scrutinized the standard textbooks of mathematical analysis with an eye on

---

* I am grateful for the comments on my presentation that were given at the ECAP conference.

essential uses of the Axiom of Choice. Likewise, we shall look at 'textbook epistemology' with an eye on essential uses of strong principles of truth.

Take, for instance, the "traditional analysis of knowledge." Its central commitment is that knowledge is true justified belief, which is most naturally expressed along the following lines:

$$\forall x \in \mathscr{L} : K(x) \leftrightarrow T(x) \wedge J(x) \wedge B(x),$$

where $\mathscr{L}$ is some language that we need not specify in detail, K is a knowledge predicate, J a justification predicate, and B a belief predicate. In what follows, I shall be somewhat sloppy in notation. In particular, in the interest of readability I shall omit some of the notational details of Gödel coding. If the reader so wishes, he or she can supply these details and verify that this abuse of notation does not affect any of the points that are made in this paper.

Clearly, if we want to use this principle to derive that some particular proposition is known, we need truth axioms. So, in this sense, truth is not conservative over epistemology.

Of course in a way this is cheating. For it is clear that for many applications, we do not need to express the central commitment of the traditional analysis of knowledge in one single sentence, using a truth predicate. For the ordinary applications of the traditional analysis of knowledge, the *schematic* version of the central commitment will do just as well. And this schematic version can be expressed without the truth predicate:

$$K(\phi) \leftrightarrow \phi \wedge J(\phi) \wedge B(\phi),$$

where ϕ ranges over all sentences of $\mathscr{L}$. To give a trite example, suppose our epistemological theory entails $0 = 0$, $J(0 = 0)$, and $B(0 = 0)$. Then the schematic version of our central commitment entails $K(0 = 0)$.

But the use of the truth predicate is not *always* so easily eliminated in epistemology. We shall demonstrate this on the basis of a variation on an epistemological argument that has been widely discussed recently. Fitch has constructed an argument to show that a certain version of verificationism is untenable (Fitch, 1963). Williamson has convincingly argued that Fitch's argument is sound—even if it should be left open whether the conclusion of Fitch's argument is a faithful rendering of a main tenet of verificationism (Williamson, 2000, Chap. 12).

Fitch's argument is usually not formulated in modal-epistemic first order logic, but in modal-epistemic propositional logic where quantification over propositions is allowed. We work in an intensional language $\mathscr{L}_P$ that contains a possibility operator $\Diamond$, a knowledge operator K ('it is known that'). The argument then runs roughly as follows.

In this language $\mathcal{L}_P$ we formulate two verificationist principles:

WV $\forall p[p \to \Diamond Kp]$

SV $\forall p[p \to Kp]$

The principle *WV* ('weak verificationism') has been taken by many philosophers to have some plausibility. The principle *SV* ('strong verificationism'), by contrast, has been taken by most philosophers to be false: it seems that we know that there are unknown truths. Fitch now shows how using plausible principles, *SV* can be derived from *WV*. This argument can then be taken as a refutation of weak verificationism.

Aside from the principles of the minimal modal logic **K**, the principles that are used in Fitch's argument are:

FACT $Kp \to p$

DIST $K(p \wedge q) \to [Kp \wedge Kq]$

Fitch's derivation of *SV* from *VW* goes as follows:

Proposition 1. $WV \vdash SV$

Proof.
1. $\forall p[p \to \Diamond Kp]$ WV
2. $\forall p[(p \wedge \neg Kp) \to \Diamond K(p \wedge \neg Kp)]$ Logic, 1
3. $\forall p[(p \wedge \neg Kp) \to \Diamond(Kp \wedge K \neg Kp)]$ DIST, 2
4. $\forall p[(p \wedge \neg Kp) \to \Diamond(Kp \wedge \neg Kp)]$ FACT, 3
5. $\forall p[\neg(p \wedge \neg Kp)]$ Logic, 4
6. $\forall p[p \to Kp]$ Logic, 5 □

Our common understanding of quantification is in terms of objectual quantification. A formula of the form $\exists p : p$ simply appears to be ill-formed, because an object is not a candidate for having a truth value. The received view is that from the conventional objectual quantification point of view sense can be made of propositional quantification, using a truth predicate (Kripke, 1976). A sentence of the form $\exists p : p$ is then taken to be short for a sentence of the form $\exists x : x \in \mathcal{L} \wedge Tx$. If this line is adopted, then Fitch's argument is really an argument that involves a truth predicate. It is worth spelling out this argument in detail, for it will tell us something about the role of the concept of truth in epistemology.

We work in an intensional *first-order* language $\mathcal{L}$ that contains a possibility operator $\Diamond$, a knowledge operator K ('it is known that'), and a Tarskian truth predicate T for $\mathcal{L}^- = \mathcal{L} \backslash \{T\}$. It is assumed that the language $\mathcal{L}^-$ contains the required coding machinery.

Let **F** be the theory which consists of:

1. The axioms of first-order logic and of the minimal normal logic **K**
2. $\forall x[Sent_{\mathscr{L}^-}(x) \to \Box(T(Kx) \to Tx)]$
3. $\forall x \forall y[Sent_{\mathscr{L}^-}(x) \land Sent_{\mathscr{L}^-}(y) \to \Box(T(K(x \land y)) \to T(Kx) \land T(Ky))]$
4. $\forall x[Sent_{\mathscr{L}^-}(x) \to (\neg T(x) \leftrightarrow T(\neg x))]$
5. $\forall x \forall y[Sent_{\mathscr{L}^-}(x) \land Sent_{\mathscr{L}^-}(y) \to (T(x \land y) \leftrightarrow T(x) \land T(y))]$

F has the axioms of Peano Arithmetic as its theory of syntax. **F** is the theory in which Fitch's argument can be formalized.

The first three of the principles of **F** (logic, FACT, DIST) are used in the derivation of the orthodox version of Fitch's argument. The next two principles are versions of the Tarskian compositional truth clauses for propositional logical connectives. Since T is intended to be a truth predicate for the language $\mathscr{L}^-$, they are unproblematic.

Weak and strong verificationism can be expressed as follows:

WV^* $\forall x[Sent_{\mathscr{L}^-}(x) \to (T(x) \to \Diamond T(Kx))]$
SV^* $\forall x[Sent_{\mathscr{L}^-}(x) \to (T(x) \to T(Kx))]$.

Now we can reformulate Fitch's argument without quantification over propositions:

Proposition 2. $WV^* \vdash_{\mathbf{F}} SV^*$

Proof.
1. $\forall x[Sent_{\mathscr{L}^-}(x) \to \Box[T(K(x \land \neg Kx)) \to (T(Kx) \land T(K \neg Kx))]]$
   DIST
2. $\forall x[Sent_{\mathscr{L}^-}(x) \to \Box[T(K(x \land \neg Kx)) \to (T(Kx) \land T(\neg Kx))]]$
   FACT, 1
3. $\forall x[Sent_{\mathscr{L}^-}(x) \to \Box[T(K(x \land \neg Kx)) \to (T(Kx) \land \neg T(Kx))]]$
   Comp Ax for $\neg$, 2
4. $\forall x[Sent_{\mathscr{L}^-}(x) \to \Box \neg T(K(x \land \neg Kx))]$
   K, 3
5. $\forall x[Sent_{\mathscr{L}^-}(x) \to (\neg \Diamond T(K(x \land \neg Kx)) \to \neg T(x \land \neg Kx))]$
   WV^*
6. $\forall x[Sent_{\mathscr{L}}(x) \to (\neg \Diamond T(K(x \land \neg Kx)) \to \neg(T(x) \land T(\neg Kx)))]$
   Comp Ax for $\land$, 5
7. $\forall x[Sent_{\mathscr{L}^-}(x) \to (\neg \Diamond T(K(x \land \neg Kx)) \to \neg(T(x) \land \neg T(Kx)))]$
   Comp Ax for $\neg$, 6
8. $\forall x[Sent_{\mathscr{L}^-}(x) \to (\neg \Diamond T(K(x \land \neg Kx)) \to (T(x) \to T(Kx)))]$
   Logic, 7
9. $\forall x[Sent_{\mathscr{L}^-}(x) \to (T(x) \to T(Kx))]$
   Logic, 4,8 □

The principles concerning K and $\Diamond$ that play a role in this argument are those that are used in Fitch's original argument. The principles concerning truth (roughly) state that truth commutes with the propositional logical connectives.

There is absolutely no threat of paradox here: in the truth axioms objectlanguage and metalanguage are scrupulously kept apart. Indeed, a simple consistency proof goes as follows. Consider first the translation τ that erases all occurrences of $\Box$ in a given proof of **F**. τ translates proofs of **F** into proofs of a system **F*** which has as its axioms all the sentences $\tau(\phi)$ such that ϕ is an axiom of **F**. Thus for a consistency proof for **F**, it suffices to show that **F*** has a model. We construct a model $\mathfrak{M}$ for **F*** as follows. The domain of $\mathfrak{M}$ consists of the natural numbers, and the arithmetical vocabulary is given its standard interpretation by $\mathfrak{M}$. A sentence ϕ is in the extension of the truth predicate according to $\mathfrak{M}$ if and only if the result of erasing all occurrences of $\Box$ and of K from ϕ results in an arithmetical truth. Then it is routine to verify that $\mathfrak{M}$ makes all axioms of **F*** true.

In sum, the version of Fitch's argument where propositional quantification is dispensed with by using a Tarskian truth predicate, seems unobjectionable. This shows that Fitch's argument cannot be faulted on account of its use of supposedly ungrammatical quantification over propositions.

It is important to highlight that in this reconstruction of Fitch's argument we had to use more than restricted Tarski-biconditionals. If one believes (as Horwich does) that *DT* is truth-theoretically complete, and if one also believes that propositional quantification has to be interpreted using a truth predicate (as received opinion has it), then one simply cannot accept Fitch's argument as valid. On the other hand, it also deserves remark that for the reconstruction of Fitch's argument the full compositional truth theory *TC* is not needed. The principle stating that truth commutes with the quantifiers plays no role in the argument. (Also, it is immaterial for the argument whether the truth predicate is allowed in the induction scheme.)

Fitch's argument crucially involves the notion of knowledge. And it relies on basic epistemological principles. So it seems fair to characterize it as an epistemological argument. Weak and Strong Verificationism involve the notion of truth as well. So our argument for $\neg WV$ does not show that truth is in the technical sense of the word nonconservative over epistemology. But it does appear to show that the theory of truth plays a substantial role in epistemology.

An objection to this line of reasoning runs as follows.[1] Weak Verificationism is, so the objection goes, a verificationist and hence substantial

---

[1] Thanks to Igor Douven for formulating it.

theory of *truth*. The fact that the compositional theory of truth that is part of the theory **F** can be used to refute a substantial theory of truth shows that the former is itself substantial and thus non-deflationist. Indeed, Horwich might see the fact that Fitch's argument does not go through if only the truth principles of *DT* are used as an argument in favor of *DT*. If deflationism is correct, then our theory of truth should be neutral in substantial philosophical disputes. The compositional truth theory that is part of **F** is not neutral in the dispute about Weak Verificationism, so it cannot possibly be an acceptable truth theory. *DT* does remain neutral in this dispute, so it is a more likely candidate for being a satisfactory theory of truth.

But this line of reasoning is unacceptable. We have no independent reasons for thinking that the compositional theory of truth is unsound. To reiterate, it seems hard to imagine any consequence of *TC* that is untoward. The fact that it can be used to refute Weak Verificationism may be surprising. But it is not a sufficient reason for taking *TC* to be unsound.

The discussion whether Weak Verificationism is a chapter in epistemology or in the theory of truth strikes me as unprofitable. The thesis *WV* involves both the concept of knowledge and the concept of truth. And Fitch's argument against *WV* uses both laws of epistemology (such as *FACT*) and compositional truth laws (laws of *TC*). In any case, Weak Verificationism is a substantial philosophical thesis. And if one wants to appeal to Fitch's argument to argue that Weak Verificationism is false, then one had better accept more laws of truth than just the restricted Tarski-biconditionals.

References

Fitch, F. B. (1963). A logical analysis of some value concepts. *Journal of Symbolic Logic*, 28, 135–142.
Horwich, P. (1998). *Truth* (2nd ed.). Clarendon Press, Oxford.
Kripke, S. (1976). Is there a problem with substitutional quantification? In G. Evans and J. McDowell (Eds.), *Truth and Meaning. Essays in Semantics* (325–419). Oxford University Press, Oxford.
Williamson, T. (2000). *Knowledge and Its Limits*. Oxford University Press, Oxford.

Metaphysics and Its Methods

Generals and Particulars

Andrea Borghini

ABSTRACT

Is it true that some entities are general, while others are particular? Ramsey famously challenged this distinction, and more recently Fraser McBride has revived the challenge. In this paper I argue that there are at least five substantial distinctions among entities, and that the distinction between general and particular entities should be made to correspond to one of those substantial distinctions.

In this paper I wish to wonder about what, for simplicity, I will label THE QUESTION: What tells apart a general from a particular entity? That *prima facie* there is such a distinction is hard to deny. We appear to use it everyday. Consider, for example, a war: this is one, but it is also many. As I am writing, the war is in Congo, it is in Sudan, it is in Somalia. To be there, in each of those countries, is the same entity, the war; and the being of the war in one country is independent from the being of the war in another country. An end to the war in Congo will not affect the war in Somalia. Different wars are fought, although it is the same entity to be involved in every case: the war.

Saddam Hussein, on the other hand, was a particular entity. There was at most one entity corresponding to the name 'Saddam Hussein.' There are several entities closely resembling Saddam Hussein, and in fact the secret services had some work to do to find the right one. Yet, once he was found, Saddam Hussein was captured. The whole Saddam, not only a part of it (at least, the whole Saddam *at that moment*, if Saddam has temporal parts and whether those have been wholly or partially captured is a question we should keep out of the picture here). More generally, each person is particular; many of our values depend upon such conviction and our actions are guided by it too.

To adopt a more mundane example, consider two standard cans of Coca-Cola (as packaged for the U.S. market), that we will name respectively MIO and MIA. Each can is unique. MIO is only here on the right, and MIA only here on the left. On the other hand, their colors are the same: white lettering over a bright red background. And they have the same typical can-shape. So, for example, *Redness* and *Being a can* are both *here* and *here* (pointing at the cans' backgrounds); and *Redness* and *Being a can* are here in MIO in its entirety. We could destroy MIA and all the general entities related to it without affecting *Redness* and *Being a can* in MIO.

One more example. Here are two copies of the same book, the *Commedia* by Dante Alighieri. They are two copies, two particular entities. There is only one of each. On the other hand, both are books, i.e. *Being a book* is in both of them. The *Commedia* is also in both of them. *Being paper* is in both of them too, as is *Being a cover* or the letters N-E-L-M-E-Z-Z-O-D-E-L-C-A-M-M-I-N-D-I-N-O-S-T-R-A-V-I-T-A, and so on. So here we have two particular entities and a multiplicity of general ones. This, *prima facie*, seems a plausible thing to say.

Like the war, *Redness*, and the *Commedia*, there are many other *prima facie* general entities: *Generosity*, *Mass*, *Being a person*, *Motherhood*, and so on. And like Saddam Hussein, MIO, MIA, and these books, there are many other *prima facie* particular entities: each of us sitting in this room, Sherlock Holmes, the dog Rubino, my basil plant, this chair, and so on. In what follows, I shall not be concerned with pinning down which ones of those *prima facie* entities are 'ultimate' or 'real'; what I wish to do is to throw some light on what sets apart these two kinds of entities.

* * *

That there is a distinction between general and particular entities is, however, a *prima facie* impression, which clashes with other *prima facie* thoughts that we might have (on this point, cfr. Ramsey (1990)). Here are two of them:

1. Although many entities might seem easy to classify as general or particular, there are several controversial cases: are numbers general or particular? Is a law a general or a particular? Are ideas or concepts, such as the idea of beauty, general or particular?

2. Besides, even the examples I mentioned before might turn out to be controversial. Is the war really general? After all, every war is different from any other, and sometimes it's hard to find compelling elements of resemblance. The first Punic War was quite different from the Gulf War, so different that we might want to classify them even

as two distinct categories of events. Or, above I mentioned *Being a person*; but, is there really such a general entity? People are all so different; to what extent can we claim that there is this universal, *Being a person*?

On the other hand, particulars are no less immune from difficulties. MIA may not be, after all, so unrepeatable: we seem to have many exact replicas; by placing them in spatio-temporal regions that are perfectly identical to the one that MIA now occupies, we could have MIA two, three, or infinite times. The same would go for each of us, for the dog Rubino, for the basil plant, and so on. Ultimately, what is it about each of them that is unrepeatable?

The distinction between what is general and what is particular tends thus to blur at a closer look. Maybe no case is uncontroversial. And, certainly, no case is uncontroversial till we have a definition of what a general and a particular entity are.

Now, what the best definition might be (if one is there) is a vexed old question. And it is a question recurring across most philosophical fields and styles. That is, one may reasonably expect any philosopher to have an opinion on what makes a particular a particular and what makes a universal a universal in the same way than one may reasonably expect any philosopher to have an opinion on what makes good good and what makes evil evil. This fact has a positive and a challenging aspect. In other words, the point at stake here is of broad interest. On the other hand, precisely because most of those who are versed in Philosophy will already harbor an answer to THE QUESTION, it is a delicate endeavor to ask for reconsideration.

Needless to say, I won't be able to do justice to every relevant answer that has been given to THE QUESTION. My goal will be to compare and contrast my own proposal to those that are closest. While doing this, I hope also to accomplish two further tasks: (i) to uncover some piece of methodology in theoretical philosophy; (ii) to pitch a relevant piece of metaphysical theory that I embrace. I will cover those tasks in order. I will first address the methodological issue: whether THE QUESTION concerns a logical, cognitive, or metaphysical distinction. I will argue for the latter. I shall, then, concentrate on four metaphysical distinctions that have been proposed, and argue in favor of the latter I will present.

Before entering into the details, I should add a disclaimer. The views I will discuss have been defended in various forms throughout the centuries and the various continents. Although I draw inspiration from a number of model authors and I value the close study of their texts, it is not here my pretense to accurately present the view of any specific author. Rather, my goal is to present, to the best of my capacities, each view in the way I believe one might defend it today.

1. A Matter of Methodology

Let us thus start from the methodological aspect of THE QUESTION. Sometimes the distinction between general and particular entities has been identified with a certain distinction between either logical roles (dating back to Aristotle's *Categories*—cfr. Mann (2000), Strawson (1974), Westerhoff (2005)) or cognitive roles (cfr. Frege (1952a), Frege (1952b), Dummett (1981)). Here are two examples:

> *Example of a Distinction in Terms of Logical Roles*
> Particular entities are those that cannot appear in the role of predicates; general entities, on the contrary, can appear both as predicates and as logical subjects.

> *Example of a Distinction in Terms of Cognitive Roles*
> Particular entities are the most immediate subjects of knowledge (in other words: our experiential knowledge is knowledge of particulars); general entities are known only through our knowledge of particular entities.

I tend to be sympathetic with both these attempts; however, I don't think that, by themselves, they can provide a satisfying answer to THE QUESTION. For two reasons, one general and the other specific to each role.

General reason. To try to tell apart general from particular entities in terms of some features of our language or our cognitive attitudes is like trying to understand what the score of a game is by looking at the face of a spectator. There is a connection between the two, but they are two separate phenomena. My language does not (and cannot) make what I am looking at what it is; more specifically, a certain word cannot make something be particular or general. In the same way, my attitude does not (and cannot) make something what it is; more specifically, a certain attitude cannot make something be a particular or a general entity. There has to be something about the entities in our domain of discourse, which makes them particular or general.

Specific reason—Logical Role. Looking at the specifics of the Logical Role, we should note that logical roles can be switched or eliminated. It is futile to try and find whether an entity is particular or general by looking at the expression(s) denoting it. Expressions that function as predicates can be rendered logical subjects, and *vice versa* as in:

(1) Socrates runs

which can be transformed, arguably without changing its meaning, in

(2) Running socratizes.

A language can be regimented in several different ways. For example, as Ayer showed in "Individuals," you can construe a language void of reference to particulars.

Specific reason—Cognitive Role. As for the cognitive role, there simply is no agreement about which are the most immediate subjects of knowledge. Each party will claim that the entities envisaged by its theory are the immediate subjects of knowledge. Here are some foremost examples.

a) *Nominalists* claim that the most immediate subjects of knowledge are particulars. For example, in "Speaking of Objects," Quine claims:

> In one's earliest phase of word-learning, terms like 'mama' and 'water' were learned which may be viewed retrospectively as names each of an observed spatio-temporal object. (Quine, 1957, p. 12)

b) Some, like Wittgenstein in the *Tractatus*, claim we perceive facts: "The world is the totality of facts." (Wittgenstein, 1986, 1.1)

c) *Universalists* claim we perceive general entities. For example, Marcuse, in *One-Dimensional Man*, writes:

> Talking of a beautiful girl, a beautiful landscape, a beautiful picture, I certainly have very different things in mind. What is common to all of them—"beauty"—is neither a mysterious entity, nor a mysterious word. On the contrary, nothing is perhaps more directly and clearly experienced than the appearance of "beauty" in various beautiful objects. (Marcuse, 1991, p. 210)

Russell defended a similar view in his later writings, for example in *Enquiry into Meaning and Truth* and *Human Knowledge*.

d) Finally, others claim that both types of entities may play a role. Here is a passage from Strawson's "Particular and General":

> There is, for example, the suggestion that general, unlike particular, things cannot be perceived by means of the senses; and this seems most plausible if one is thinking of the things designated by certain abstract nouns. It is not with the eyes that one is said to see hope. But one can quite literally smell blood or bacon, watch cricket, hear music or thunder; and there are, on the other hand, certain particulars which it makes dubious sense to say one perceives. (Strawson, 1992, p. 235)

In the end, there certainly is no common agreement as to the role that one or the other type of entity plays in cognition. We may take side with one party or another, but by doing so we would fail to draw a distinction that attempts to appeal to a vast majority of the disputants. For this reason, it does not seem viable to try and answer THE QUESTION on the basis of which entity are the most immediate subjects of our knowledge.

The general and specific considerations offered suggest—I believe— that we refrain from drawing a distinction between general and particular entities based on any of these lines: a certain logical role that expressions may cover in our language, or, a certain role that an entity may play in our cognition. There is a link between our language and the entities we talk about; there is a link between our cognitive attitudes and the entities we entertain; yet these links are not going to tell us what those entities are. We need to do metaphysics in order to answer THE QUESTION. It is thus to metaphysical distinctions that I shall now turn.

2. Metaphysical Distinctions

There are four metaphysical answers to THE QUESTION that seem most deserving of attention. I think only one of them is tenable. I will present it for last and offer my interpretation of it. Before doing that, however, I will discuss the other three answers and explain what I find missing in them.

2.1. Role in Instantiation

The first answer draws from the role that entities play in the so-called instantiation relation, a binary relation occurring between an arbitrary n-tuple of entities, some of which are said to instantiate and the other(s) to be instantiated. For example, if MIO is red, MIO instantiates *Redness* (and *Redness* is instantiated by MIO). In a recent paper, Jonathan Lowe claims:

> As for the distinction between universal [here: general entity] and particular, I simply define it in terms of the instantiation relation. A particular is that which has (or, in a stronger version, that which can have) no instances, whereas a universal is that which has (or, in a weaker version, that which can have) instances. (Lowe, 2004, p. 303)

A general entity, say, *Wisdom*, can instantiate other general entities (e.g. *Virtuosity*) but can also be instantiated by other entities, such as Socrates. Socrates, on the other hand, cannot be instantiated, but only instantiate.

General and particular entities, thus, behave differently with respect to the relation of instantiation, which is, according to most philosophers, a necessary relation for the existence of an entity. General entities can have instances, while particular entities cannot. Thus we have these two classes of entities, *T1* and *T2*, and a relation *R*, such that the one in *T1* can be on both ends of *R*, while the others can be only on one end.

This much for the answer. Now let's see what is unpalatable about it. The major problem with this way of telling apart general from particular entities is that the relation of instantiation is a rather obscure entity.

If we intend it as a normal relation, we run into a famous regress: what relates the relation to each of the relata? If another relation, what will relate such relation to each of the relata and the instantiation relation?

On the other hand, if we intend the instantiation relation as a special relation, a primitive tie between entities, how are we supposed to intend such a tie? Most authors reply that, since it is a primitive, the instantiation relation does not need to be explained.

Now, it is not unusual to have primitives in Philosophy (probably it is even unavoidable); but, it seems plausible to expect that our primitives be in some way self-explanatory and compelling. For example, what makes Descartes's *Cogito* and proof of the existence of God particularly compelling is the fact that the first-person intuition, as well as the intuition of infinity and perfection, can be regarded both as self-explanatory and compelling. They are not just evident, but evident by "natural light," as Descartes puts it. Is the instantiation something that can be maintained to be evident by natural light? My impression is that we have a quite approximate and confused idea of it. One way to define the instantiation relation proceeds as follows:

IR: a binary, non-symmetric and non-transitive relation occurring between an arbitrary *n*-tuple of entities, some of which are said to instantiate and the other(s) to be instantiated.

As *IR* shows, the relation is unspecified in most of its aspects: all that we know about it is that it is non-symmetric and non-transitive. But this is hardly enough for identifying *a specific* relation. Moreover, to try and make the relation more specific would be circular, as Strawson already pointed out (Strawson, 1992): to try and explain the instantiation relation in terms of logical roles (i.e., subject and predicate) is just another way to reaffirm that it is the distinction between entities that can and those that cannot have instances.

2.2. Mode of Instantiation

Other authors concentrated still on the instantiation relation, but on a different aspect of it. According to these, some entities (i.e. the general ones) enter in a relation of instantiation with the same number of entities; on the other hand, other entities (i.e. the particular ones) enter in a relation of instantiation with a variable number of entities. Our MIA, here, is instantiating—say—seven general entities; but they might change: for example, they might be eight if I would paint part of its surface with blue paint. Maybe there is a minimum number of general entities that MIA has to instantiate, yet the overall number is variable. On the other hand, *Being a cap*, for example, always enters into an instantiation relation with one entity, viz. the one which has the cap; the same goes for *Redness*, or *Being a can*.

Recently, this view has been at the center of a debate between David Armstrong—who defended it—and Fraser McBride—who criticized it (cfr. Armstrong (1997), McBride (1998a), McBride (1998b), McBride (1999), McBride (2004), McBride (2005)). The latter argued that the number of entities with which a general entity enters into a relation of instantiation is not an 'intrinsic characteristic' of the entity; therefore, it can change without changing the general entity, thus dissolving the distinction between modes of instantiation.

The problem with this objection is that it does not specify what an intrinsic characteristic of an entity is, nor why the number of entities with which an entity enters into a relation of instantiation should be variable.

I think there is a simpler way of addressing the problem with the proposed distinction between general and particular entities, a way which is in agreement with McBride's point, if I understand it.

Consider, for example, the entity *Being a can-maker*, and suppose that MIO and MIA were made by the same entity (probably a machine), call it Frank, which made only those two cans. There are two ways of portraying *Being a can-maker* when belonging to Frank. You could think that Frank has *Being a can-maker* twice, thus each time *Being a can-maker* relates two entities, viz. the maker with a can. On the other hand, you could think that *Being a can-maker* belongs only once to Frank: it belongs to him all along, and it relates Frank with MIO and MIA, thus being a three-place relation.

Now, if you are sympathetic to the latter case, then there will be a time at which *Being a can-maker* relates Frank and (say) MIO only, and a subsequent time at which it relates Frank and both MIO and MIA. Thus *Being a can-maker* enters into relations of instantiation which vary in the number of their *relata*.

A similar consideration holds for other general entities. Consider for example *Being a father*. If you have two children at two different times, do you instantiate *Being a father* twice (so that each time *Being a father* is a binary relation) or once (first as a binary and then as a triadic relation)? Is *Being a professor* binary or does it relate a variable number of entities?

At this point, to answer such questions, one might want to apply a test similar to the one I applied at the beginning to explain why *War* or *Redness* seemed to be general entities. Suppose Frank would have made only one of the two cans, say MIO. It would still instantiate *Being a can-maker* (supposing that to make one can is enough for instantiating *Being a can-maker*). Thus:

(a) Frank would instantiate *Being a can-maker* even if it would have never made MIA.

But, an analogous counterfactual situation in which MIO and MIA are switched roles can also be imagined. Thus:

(b) Frank would instantiate *Being a can-maker* even if it would have never made MIO.

Now, since the making of a can cannot change Frank's relation to the making of the other can, from (a) and (b) we can conclude that:

(c) Frank instantiated *Being a gnome-maker* twice.

Since a similar argument could be run for any other general entity (claim the defender of the distinction we are considering), the instantiation relation is always *minimal*: it is binary for binary predicates, triadic for triadic predicates, n-ary for n-ary predicates. Thus, although particulars can instantiate more than one general entity at a time, general entities always enter into instantiation relation with an identical number of entities. (Of course the same general entity can be instantiated more than once. But, the point here is to see how an entity behaves with respect to a specific case of instantiation.)

To this argument I have a reply. The making of one extra can does indeed change the relation between Frank and the other can. Relations among entities change even after one of them ceases to exist, as Aristotle's paradox of a good life in *Nichomachean Ethics* teaches us. A rockstar can have tons of fans as she is alive and lose them all after a few years her death, upon the discovery of some dishonoring deed of her.

Upon making another can, Frank's relationship with the first can will be changed. It won't be any longer its only can, perhaps it won't be any longer its best can, perhaps it will change our mind on the quality of the first can made by Frank.

But, suppose that not all instantiation relations are like *Being a canmaker*. Consider *Being two feet apart*. I am—say—two feet apart from MIO; and now I am two feet apart also from MIA. The latter certainly did not change my *Being two feet apart* from MIO. Thus, *Being two feet apart* is here instantiated twice. Once with respect to me and MIO, the other with respect to me and MIA.

The example of *Being two feet apart* just shows that *some* general entities enter into relation with the same number of entities all the times. What the defender of the distinction should be able to show is that all relations are of this sort. This seems quite implausible, however. Consider, for example, *Generosity*. It can be ascribed to one person, but also to two people who made a joint donation. *Courage* can be of one person, but also of all the players in a team. If you become a father of a second child your relationship as a father to the first child is going to change.

Maybe, I am confused as to what the claim exactly is when one says that general entities always enter into relations of instantiation with the same number of entities, while particular entities don't. What do we mean when we say that MIA instantiates seven general entities? Here are two options:

A) *MIA has one instantiation relation with seven entities*. If so, why each of such general entities does not have, in turn, a relation with seven entities, and why isn't then such number variable as it is for MIA?

B) *MIA has seven instantiation relations with general entities, each of which is minimal (as proved above)*. If so, isn't also MIA always in an instantiation relation with an identical number of entities, else, if instantiation is not always *minimal*, doesn't the number of entities which entertain the instantiation relation vary in the same way both for MIA and for the general entities it instantiates?

Either way, I do not see why this answer to THE QUESTION should stand. Thus, I conclude, this way to tell apart general from particular entities is not viable.

2.3. Completeness

The third distinction I would like to consider is the one between complete and incomplete entities. Here is how Peter Strawson, in *Individuals*, introduces it:

> A subject-expression is one which, in a sense, presents a fact in its own right and is to that extent complete. A predicate-expression is one which in no sense presents a fact in its own right and is to that extent incomplete. (Strawson, 1959, p. 187)

This much for what concerns the semantic distinction. At the ontological level, we have a parallel one:

> So the fundamental picture, or metaphor, I offer is that of the particular resting on, or unfolding into, a fact. It is in this sense that the thought of a definite particular is a complete thought. (Strawson, 1959, p. 211)

Hence, for Strawson it seems that:

A) A particular is an entity, which necessarily exists in a specific place or context, thus it has to rest on a fact.

B) A general entity, on the other hand, can rest on no specific place or context.

Although such interpretation probably loads Strawson's thought with more metaphysical speculation than he would have been willing to admit, I take this to be his main metaphysical lesson. It's a lesson that can be derived also from the distinction he endorses in "Particular and General":

> It is a necessary condition for a thing's being a general thing that it can be referred to by a singular substantival expression, a unique reference for which is determined solely by the meaning of the words making up that expression; and it is a necessary condition of a thing's being a particular thing that it cannot be referred to by a singular substantival expression, a unique reference for which is determined solely by the meaning of the words making up that expression. (Strawson, 1992, p. 257)

The reference of expressions referring to particulars cannot be grasped solely by grasping their meaning since they contain contextual elements; this is because particulars are necessarily rooted in facts, they have to exist in a specific place or context, which is reflected by the expressions referring to them. The reference of the name "Barack Obama" cannot be determined solely by knowing Barack under some description. We need to look into this room and point at him.

On the other hand, the reference of expressions picking out general entities can be grasped by grasping their meaning. And you may know what *Redness* means without the need of looking into any specific place. General entities are not necessarily rooted in any fact, thus we can think of them also outside of any specific context.

But, if this is the case—I surmise—then a general entity may exist in no specific place, while it is necessary for a particular entity to exist in a specific place. Thus, general entities might have an incomplete existence, an existence that is not determinate, while particular entities always have a complete (i.e. determinate) existence.

I find myself in disagreement with both Strawson's semantic and metaphysical theses. But, I shall not talk about the semantic view here. As for the metaphysical thesis, I believe that it may be interpreted in two ways, both of which might be philologically inaccurate. But, my pretense is of course not to push Strawson to defend theses he always refrained from defending. My goal is instead to delve into two ways of maintaining, in a metaphysical sense, that general entities are incomplete while particular ones are complete. With this spirit, let us explore the two interpretations.

1. The first interpretation forces upon us a Platonist conception of general entities. *Being a can* could exist in no specific place because general entities do exist in no specific place. For many, however, it is simply false that *Being a can* could exist in no specific place. The fact that we can refer to *Being a can* without calling into question any specific can does not suffice to entail this conclusion. This suggests that the way we form concepts of general entities differs from the way we form concepts of particular ones. In the case of an expression referring to a general entity, we seem to abstract from the specific instances of the general entity we encounter; on the contrary, we do not do this when it comes to particular entities. Yet, to claim that this is so because general entities can exist without being in any specific place means to reify our abstraction. And this is not what most of us want to do.

This conception also brings with itself a host of problems. If general entities can be in no specific place, do they exist in a different realm? Do they exist everywhere? And what relationship is there between general and particular entities? If general entities exist also in space, are the instances of a general entity that are in space and those that are not the same type of instances?

I am not claiming that those questions cannot receive an answer. The point is that the distinction between general and particular entities should not cut across a distinction so fundamental as the one between Platonists and Aristotelians on general entities. Probably, as I said at the outset, it is impossible not to distinguish between general and particular without partially relying on one's own philosophical inclinations. But, to imbue the distinction with another fundamental metaphysical distinction regarding general entities seems to cut off the intelligibility of the account too deeply.

GENERALS AND PARTICULARS

2. According to the second interpretation, the fact that general entities can exist in no specific place just means that they are more imperfect than particular ones. Particulars always have a completely determinate existence, while general entities are by nature incomplete. They become complete only upon being instantiated by a particular.

This interpretation strikes me as more plausible than the first one, at least in its form. In fact, it does not presuppose additional metaphysical distinctions or doctrines. On the other hand, it seems the least tenable interpretation from a theoretical standpoint. Consider MIO, and the general entity *Being a can*. It seems to me that, as we are able to form the concept of the general entity *Being a can* in a way that abstracts from the present context, so we are able to form an idea of MIO which abstracts from the present context. That's precisely what we do when we consider MIO in some counterfactual situation. And note (this is key) that most of the times such counterfactual situations do not pick out a fully specified context. When considered in this respect, MIO does not seem to me less complete or incomplete than *Being a can* or *Redness*. Certainly I can think of the latter without thinking at a specific context, yet I can similarly think of MIO without thinking of it in a specific context. The concept of a particular abstracts from a specific context as much as the one of a general entity.

Hence, this reading of Strawson's claim strikes me as simply false. General and particular entities are complete and incomplete in an analogous way.

2.4. Repeatability

We now come to the fourth and last distinction I will present, the one I prefer. It is a classical distinction, which is formulated with different terminologies, such as:

(A) General entities are *repeatable*; particular entities cannot repeat.

(B) General entities can be *wholly present* at more than one place/time; particular entities cannot be wholly present at more than one place/time.

(C) General entities *can be in their entirety* at more than one place/time; particular entities cannot be in their entirety at more than one place/time.

For example, *Redness* is both here (MIO) and here (MIA); while each can cannot be in its entirety in more than one place at a time.

Most authors introduce expressions such as "wholly present" or "present in its entirety" without any further clarification, assuming that it is evident what they mean. Although I believe (and this will turn out to be crucial later) that these expressions do appeal to an intuition, I also believe that by themselves they are not explanatory enough for the distinction. In fact, most authors introduce them without pretending that to include entities of this sort into their ontology is in some way paradoxical or, at least, controversial. On the contrary, I believe that the distinction between general and particular entities should bring to light the theoretical burdens of accepting general entities into one's ontology. In other words: we owe an explanation of such expressions. So, the main task is to explain what repeatability/whole presence/entire existence are. Out of the three possible expressions, I will consider repeatability as the main term for tracing the distinction. Our question, hence, is: what is repeatability?

We should start our answer by availing ourselves of an example. Consider the following two sentences, representing two propositions:

P: "*Being a can* exists"
Q: "*Socrates* exists"

P and *Q* represent propositions, which are fully specified by the expressions contained in *P* and *Q*. There is no hidden intended context or any other contextual element which should be supplied in order for us to understand the propositions that *P* or *Q* represent.

I believe that there is a key difference in the way in which the entities that *P* and *Q* are about exist. It is this difference which gives us the meaning of 'repeatable' or 'unrepeatable'; thus, it is this difference that tells us what the distinction between a particular and a general entity is. The main difference lies—I argue—in the possible number of the truth-makers of the propositions expressed by *P* and *Q*. Here is why:

Q-FEATURES

- *Q* ("Socrates exists") has at most one truth-maker.

- For every possible scenario that we consider, either Socrates exists or he does not exist in the scenario.

- *Q* is true in all the scenarios in which Socrates exists.

P-FEATURES

- *P* ("*Being a can* exists") can have multiple, possibly infinite, truth-makers.

Generals and Particulars

- For every possible scenario that we consider, *Being a can* could 'exist one, two, . . ., infinite times' if you pass me this expression.

- *P* is true in all the scenarios in which *Being a can* exists.

- But—and this is the key difference from *Q*—the truth of *P* can be *over-determined*. *P* might be true twice over, three times over, or infinite times over.

This is what it means to say that there are multiple, possibly infinite regions in each scenario where *Being a can* might be wholly/in its entirety.

Here you might suspect that the distinction I am drawing is in some way 'linguistic,' in that I am introducing it by appealing to sentences. However, this is not the case. I am appealing to propositions, and their truth-makers. I believe that a proposition is a metaphysical entity, that it is 'a piece' of the world (probably not a primitive piece, but still a piece).

More in details, *P* and *Q* express simple propositions, single 'pieces' of the world. That some of those 'pieces' exist at most once, while others exist two, three, or infinite times is an intuitive fact about the world. It is the intuitive fact with which I started off this talk and that the distinction in terms of possible number of truth-makers purports to spell out.

Thus, it is not that particular entities are complete while general ones are not. In a sense, general entities, are even more complete: they can wholly exist more than once! We need to look at the way the world is to establish the truth-value sentences regarding both sorts of entities.

The distinction is that a sentence such as *P*, expressing a proposition constituted by a general entity, can be over-determinedly true, while this is not the case for a sentence expressing a proposition constituted by a particular entity.

Moral

The distinction between general and particular entities belongs to metaphysics. Among the definitions that have been proposed, the ones appealing to instantiation are circular or inconclusive. The one appealing to completeness strikes the wrong metaphysical cord. I believe my distinction in terms of the number of truth-makers that a sentence can have has two main virtues:

(i) It preserves the chief intuitive distinction, i.e. that particular entities are unrepeatable while general ones can repeat;

(ii) It points the finger to the distinguishing feature, the possible number of truth-makers.

It is the fact that the truth of a proposition like *P* can be over-determined to render general entities unpalatable to many. You might not want to accept that there are entities that exist more than once. My stance is that we constantly use to think that there are. If you deny this, then you are a Particularist. But, the definition of particularity and generality here offered stands. And this is what I aimed to achieve.

References

Armstrong, D. M. (1997). *A World of States of Affairs*. Cambridge University Press, Cambridge.
Ayer, A. J. (1952). Individuals. *Mind*, 61, 441–457.
Dummett, M. (1981). *Frege: Philosophy of Language*. Harvard University Press, Cambridge, MA.
Frege, G. (1952a). On concept and object. In P. T. Geach and M. Black (Eds.), *Translations from the Philosophical Writings of Gottlob Frege* (42–55). Basil Blackwell, Oxford.
Frege, G. (1952b). On sense and reference. In P. T. Geach and M. Black (Eds.), *Translations from the Philosophical Writings of Gottlob Frege* (56–78). Basil Blackwell, Oxford.
Lowe, E. J. (2004). Some formal ontological relations. *Dialectica*, 58, 297–316.
Mann, W. R. (2000). *The Discovery of Things: Aristotle's Categories and Their Context*. Princeton University Press, Princeton, NJ.
Marcuse, H. (1991). *One-Dimensional Man*. Beacon Press, Boston. (Original work published 1964).
McBride, F. (1998a). On how we know what there is. *Analysis*, 58, 27–37.
McBride, F. (1998b). Where are particulars and universals? *Dialectica*, 52, 203–237.
McBride, F. (1999). Could Armstrong have been a universal? *Mind*, 108, 471–501.
McBride, F. (2004). Whence the particular-universal distinction? *Grazer Philosophische Studien*, 67, 181–194.
McBride, F. (2005). The particular-universal distinction: a dogma of metaphysics? *Mind*, 114, 565–614.
Quine, W. V. O. (1957). Speaking of objects. *Proceedings and Addresses of the American Philosophical Association*, 31, 5–22.
Ramsey, F. P. (1990). Universals. In D. H. Mellor (Ed.), *F. P. Ramsey: Philosophical Papers* (8–30). Cambridge University Press, Cambridge. (Original work published 1925 in *Mind*, 34, 401–417).
Russell, B. (1940). *Inquiry into Meaning and Truth*. Allen and Unwin, London.
Russell, B. (1948). *Human Knowledge: Its Scope and Limits*. George Allen and Unwin, London.
Strawson, P. F. (1959). *Individuals*. Methuen & Co. LTD, London.
Strawson, P. F. (1974). *Subject and Predicate in Logic and Grammar*. Methuen & Co. LTD, London.
Strawson, P. F. (1992). Particular and general. In A. B. Schoedinger (Ed.), *The Problem of Universals* (212–231). Humanities Press, London. (Original work published 1954, *Proceedings of the Aristotelian Society*).
Westerhoff, J. (2005). *Ontological Categories*. Clarendon Press, Oxford.
Wittgenstein, L. (1986). *Tractatus Logico-Philosophicus*. Routledge, New York.

Index of Authors

Anderson, C. A., 291, 311–313, 317, 319
Aristotle, 56, 92, 169, 177, 203, 289, 382, 387
Armstrong, D. M., 386
Arndt, H., 176
Arnheim, R., 64
Aspect, A., 106, 137
Auden, W. H., 342
Augustine, 56
Axtell, R., 248, 249
Ayer, A. J., 383
Azzouni, J., 343, 359

Bach, K., 324–326, 334, 350
Baldwin, D. A., 351
Baltag, A., 208
Barbour, J., 60
Barwise, K. J., 70, 71
Batterman, R., 207
Baumol, W. J., 175
Bealer, G., 269, 307
Bell, J., 106, 115
Belnap, N., 105–107, 119, 122, 125, 211, 212
Benacerraf, P., 38
van Benthem, J., 206
Bergson, H., 56
Berkeley, G., 22–24, 35–38
Birkhoff, G., 208
Blass, A., 79
Block, N., 162
Bloom, P., 351, 352
Bogen, J., 213
Bolzano, B., 289, 291
Bovens, L., 207
Bradley, F. H., 49, 53, 54, 56

Bradley, R., 228
Braithwaite, J., 171
Brandom, R., 262, 270
Brentano, F., 157
Broad, C. D., 48, 49, 56–62
Bromberger, S., 37
Buchanan, A., 169, 175
Buchanan, J. M., 183
Buchanan, R., 324, 332
Burge, T., 263, 351
Burgess, J., 32, 346
Byrne, A., 155–158

Calmette, J., 87
Cantor, G., 57
Cappelen, H., 328
Carnap, R., 204, 262, 265–270, 280, 286, 288, 291, 299, 307, 322
Carston, R., 340, 350
Batterman, R., 207
Chalmers, D., 34
Chapman, J. W., 168
Church, A., 265, 269, 274, 289, 291, 292, 295, 299, 300, 305, 307, 308, 310, 311, 318, 319
Clauser, J. F., 115, 137
Clay, E. R., 55
Cleugh, M. F., 50
Coleman, J. L., 169, 171
Cordato, R. E., 176
Cortens, A., 97
Crane, T., 155, 156, 161
Cresswell, M. J., 262, 269, 270, 273, 300, 304, 305, 307
Currie, G., 52

D'Andria, L. J., 248–250
Dalibard, J., 106

Darwin, C., 99
Davidson, D., 339, 342, 345, 347–349
Descartes, R., 24, 38–41, 385
Dietrich, F., 207
Douven, I., 193, 206, 246, 248, 252, 254, 255, 257
Duží, M., 280, 282, 303, 304, 311, 314, 316–318
Dummett, M., 101, 316, 382
Duncker, K., 64
Dworkin, R., 168

Earman, J., 194, 208
Eco, U., 94
Einstein, A., 60, 61
Elgin, C. Z., 93
Epstein, J., 248, 249
Ewers, H.-J., 173

Fagin, R., 303
Farkas, K., 52
Febvre, L., 90
Feinberg, J., 170
Feldman, R., 197
Ferraris, M., 98
Field, H., 31, 362
Fine, K., 17, 18, 210, 211
Fischer, M., 368
Fitch, F. B., 371–376
Fitelson, B., 206
Fodor, J. A., 315
Frege, G., 67, 68, 88, 262–268, 272, 289–291, 294, 296, 299, 307–309, 311, 319, 322, 325, 343, 357, 382
Fritsch, M., 173, 175

Gale, R., 52
Gärdenfors, B. P., 194, 196
Galileo, G., 204
Galton, A., 92
Ganter, B., 289
Gaskin, R., 307
Gauthier, D., 175, 182
Gaylord, R. J., 248–250
Geach, P., 60
de Gijsel, P., 180

Goodman, N., 90, 91, 97, 99
Gödel, K., 60, 61, 194, 364
Graf, G., 173
Grampp, W. D., 174
Grasshoff, G., 140
Greenberger, D., 106
Grice, P., 340, 345, 346
Gunn, A., 52, 58
Gupta, A., 348, 358
Gurevich, Y., 79

Halbach, V., 361, 364, 368
Halpern, J. Y., 303
Hansson, S. O., 193, 196, 201
Harman, G., 153
Hartmann, S., 193, 207, 248
Haslanger, S., 6, 7
Haslinger, F., 174
Hawley, K., 4, 7, 8
Hawthorne, J., 206
Hayek, F. A., 175, 179, 180, 183
Heal, J., 346
Hegel, G. W. F., 56
Hegselmann, R., 245–248, 250
van Heijenoort, J., 308
Heller, M., 14, 99
Hellie, B., 149, 152
Henkin, L., 70, 72, 74
Hintikka, J., 71, 72, 74, 303
Hinzen, W., 353
Hodes, H. T., 308
Hodges, W., 76
Hofer-Szabó, G., 106, 140
Hoffmann, D., 204
Hofweber, T., 3, 4
Holt, R. A., 115, 137
Horák, A., 294
Horne, M. A., 106, 115, 137
Horsten, L., 193, 206
Horwich, P., 339, 344, 345, 349, 371, 375, 376
Höffe, O., 171
Höffner, J., 178, 179
Huber, F., 206
Humphreys, P., 248
Hunt, E. K., 176

Husserl, E., 96, 157, 299, 308

Jackson, B., 23
Jacobson, P., 26
James, W., 55, 56, 63
Jespersen, B., 269, 280–282, 303, 304, 321
Johnson, M., 63
Johnson-Laird, P. N., 315
Joyce, R., 31, 32

Kant, I., 56, 168, 204
Kaplan, D., 269
Kaulla, R., 178
Keeling, S. V., 50
Keil, G., 214
Kemeny, J. G., 230
Kerry, B., 290
Ketland, J., 357, 359, 361, 362
King, J. C., 268, 269, 280, 324
Kirkham, R. L., 266
Kirzner, I. M., 175
Kishida, K., 107, 212
Koller, P., 168, 170, 172, 176, 182
Koons, R., 353
Koslowski, P., 178
Kowalski, T., 106
Kölbel, M., 349
Krause, U., 245–248, 250
Kripke, S. A., 305, 306, 367, 373
Kuhn, T. S., 205
Kvart, I., 200

Lahav, R., 323, 328
Lakoff, G., 63
Langholm, O., 178, 179
Leibniz, G. W., 35, 36, 38, 56
Leitgeb, H., 206, 358
Lepore, E., 328
Lesson, R., 95
Levesque, H. J., 303
Levi, I., 196
Lewis, D., 3, 4, 6, 7, 11, 35, 36, 38, 96, 99, 194–196, 228, 269, 307
Linnaeus, C., 98
Lipton, P., 348, 351
Locke, J., 94, 351

Lowe, E. J., 6, 8, 9, 384
Lucas, J. R., 168, 171
Lynch, M., 346, 350

Mach, E., 203
Mackie, J. L., 122
Maher, P., 206
Mann, W. R., 382
Marcuse, H., 383
Martin, M. G. F., 153
Marx, K., 178
Materna, P., 270, 280, 282, 289, 294, 304, 316–318
May, R., 69
McBride, F., 379, 386
McCraw, T. K., 175
McGee, V., 361
McTaggart, J. M. E., 47–62, 64
Mellor, H., 60
Mermin, D. N., 106
Merricks, T., 7, 8
Mill, J. S., 179
Miller, D., 168, 170
Minkowski, H., 61
Minowitz, P., 174
de Molina, L., 179
Montague, R., 208, 262, 265, 266, 271
Moore, G. E., 23, 24, 56, 59, 147–152, 162
Morgenstern, O., 76
Moschovakis, Y. N., 262, 271, 307
Moses, Y., 303
Muskens, R., 312
Müller, J., 303
Müller, T., 106, 107, 122, 123, 135, 141, 212
Münsterberg, H., 63
Myhill, J., 305

Nagel, E., 204, 207
Neale, S., 332
Neurath, O., 204
Newton, I., 56
Nordhaus, W. D., 174
Nowak, L., 246
Nozick, R., 168, 181, 183

Nutzinger, H. G., 179
Nyíri, K., 62

Olsson, E., 196
Oppy, G. R., 194
Ostertag, G., 324, 332

Parmenides, 56
Peacocke, C., 160
Pears, D., 59, 60
Penco, C., 308
Pennock, J. R., 168
Peregrin, J., 262, 270
Perloff, M., 211
Perry, J., 58, 62
Petrželka, J., 289
Pettit, P., 171
Pigozzi, G., 207
Placek, T., 106, 107, 119, 122, 212
Plantinga, A. P., 194
Portmann, S., 140
Prawitz, D., 300
Priest, G., 304
Prior, A., 15, 211, 307
Putnam, H., 97

Quine, W. V. O., 24–30, 36, 100, 262, 270, 277, 344, 351, 360, 364, 383

Ramsey, F. P., 210, 340, 341, 357, 379
Rawls, J., 168, 170
Recanati, F., 325, 340, 350
Reisch, G. A., 204
Rédei, M., 106
Richard, M., 310
Richardson, A., 205
Riegler, A., 246, 248, 252, 254, 255, 257
Rochelle, G., 49, 61
Roger, G., 106
de Roover, R., 178, 179
Rorty, R., 97
Rosen, G., 31
Rothschild, K. W., 176, 182
Rott, H., 195, 196
Runggaldier, E., 52

Russell, B., 56, 267, 268, 270, 280, 307, 308, 319, 322, 383
Rysiew, P., 351

Salmon, N., 311
Salmon, W. C., 194
Samuelson, P. A., 174
Sandu, G., 74, 81
Sattig, T., 61
Scheffler, I., 91
Schneider, J., 174
Schopenhauer, A., 56
Schroeder, M., 228
Schumpeter, J. A., 175
Searle, J., 89
Sellars, W., 47, 48, 59, 62
Sevenster, M., 79, 81
Shapiro, S., 357, 359, 361, 362
Sher, G., 68, 70, 72, 342
Sherman, H., 176
Shimony, A., 106, 115, 137
Shogenji, T., 351
Sidelle, A., 100
Sider, T., 3, 4, 7, 8, 60, 61
Simons, P. M., 91
Skyrms, B., 78, 194
Smets, S., 208
Smith, A., 174, 179
Smith, B., 88, 89
Snell, J. L., 230
Soames, S., 269, 340
Socrates, 94
Sokal, R. R., 94
Spinoza, B., 56
Stadler, F., 204
Stamos, D. N., 93
Sterba, J. P., 171
Stiglitz, J. E., 174, 176
Strawson, P. F., 280, 284, 382, 383, 385, 388–391
Streissler, E., 174, 176
Sturn, R., 175, 178
Sullivan, A., 305
Sundholm, G., 308, 309
Sunstein, C. R., 168
Szabó L., 106, 107

Szabó, Z., 32, 38

Tappenden, J., 352
Tarski, A., 194, 275, 304, 344, 347, 349, 357, 358, 360, 361, 364, 367, 368, 375, 376
Tennant, N., 359, 362
Thales, 28
Thomas Aquinas, 178
Thomason, R. H., 211, 306, 313
Tichý, P., 262, 263, 269, 271, 290, 292, 295, 307, 310–313, 316, 321, 328
Trusen, W., 178, 179
Tye, M., 152, 153, 156, 161

Uebel, T., 204
Ullmann-Margalit, E., 221

van Cleve, J., 96
van Fraassen, B., 31
van Inwagen, P., 8, 16
Vardi, M. Y., 303
Velleman, D., 3, 4
de Vitoria, F., 179
von Neumann, J., 76, 80, 208
Vranas, P. B. M., 206

Walras, L., 179, 180
Wansing, H., 196–199
Wasserman, R., 7
Wein, T., 173
Weiner, M., 107, 125, 212
Westerhoff, J., 382
Whitrow G. J., 57
Wille, R., 289
Williams, B., 198
Williams, M., 348
Williamson, T., 342, 372
Wilson, J., 92
Winsberg, E., 248
Wittgenstein, L., 60, 159, 160, 210, 262, 270, 322, 383
Wood, D., 178
Woods, W. A., 315
Woodward, J., 213
Wright, C., 345, 347

Wroński, L., 107, 122
Wüthrich, A., 140

Xu, M., 211

Yablo, S., 32

Zahavi, D., 157
Zalta, E., 307
Zeilinger, A., 106
Zeno of Elea, 56, 57
Zermelo, E., 371
Zimmerman, D., 8

www.ingramcontent.com/pod-product-compliance
Lightning Source LLC
Chambersburg PA
CBHW051623230426
43669CB00013B/2165